AF838441

Handbook of Analytic Operator Theory

CRC Press/Chapman and Hall Handbooks in Mathematics Series

Series Editor: Steven G. Krantz

Handbook of Analytic Operator Theory
Kehe Zhu, Editor

https://www.crcpress.com/CRC-PressChapman-and-Hall-Handbooks-in-Mathematics-Series/book-series/CRCCHPHBKMTH

Handbook of Analytic Operator Theory

Edited by

Kehe Zhu

CRC Press
Taylor & Francis Group
Boca Raton London New York

CRC Press is an imprint of the
Taylor & Francis Group, an **informa** business
A CHAPMAN & HALL BOOK

CRC Press
Taylor & Francis Group
6000 Broken Sound Parkway NW, Suite 300
Boca Raton, FL 33487-2742

© 2019 by Taylor & Francis Group, LLC
CRC Press is an imprint of Taylor & Francis Group, an Informa business

No claim to original U.S. Government works

Printed on acid-free paper
Version Date: 20190415

International Standard Book Number-13: 978-1-138-48641-6 (Hardback)

This book contains information obtained from authentic and highly regarded sources. Reasonable efforts have been made to publish reliable data and information, but the author and publisher cannot assume responsibility for the validity of all materials or the consequences of their use. The authors and publishers have attempted to trace the copyright holders of all material reproduced in this publication and apologize to copyright holders if permission to publish in this form has not been obtained. If any copyright material has not been acknowledged please write and let us know so we may rectify in any future reprint.

Except as permitted under U.S. Copyright Law, no part of this book may be reprinted, reproduced, transmitted, or utilized in any form by any electronic, mechanical, or other means, now known or hereafter invented, including photocopying, microfilming, and recording, or in any information storage or retrieval system, without written permission from the publishers.

For permission to photocopy or use material electronically from this work, please access www.copyright.com (http://www.copyright.com/) or contact the Copyright Clearance Center, Inc. (CCC), 222 Rosewood Drive, Danvers, MA 01923, 978-750-8400. CCC is a not-for-profit organization that provides licenses and registration for a variety of users. For organizations that have been granted a photocopy license by the CCC, a separate system of payment has been arranged.

Trademark Notice: Product or corporate names may be trademarks or registered trademarks, and are used only for identification and explanation without intent to infringe.

Library of Congress Cataloging-in-Publication Data

Names: Zhu, Kehe, 1961- author.
Title: Handbook of analytic operator theory / Kehe Zhu.
Description: Boca Raton : CRC Press, Taylor & Francis Group, 2019. | Includes bibliographical references and index.
Identifiers: LCCN 2019004988 | ISBN 9781138486416
Subjects: LCSH: Operator theory. | Holomorphic functions. | Function spaces.
Classification: LCC QA329 .Z475 2019 | DDC 515/.724--dc23
LC record available at https://lccn.loc.gov/2019004988

Visit the Taylor & Francis Web site at
http://www.taylorandfrancis.com

and the CRC Press Web site at
http://www.crcpress.com

Contents

Preface

There has been broad interest in analytic function spaces and related operator theory over the past few decades. In particular, several monographs appeared in this period, including "Composition Operators on Spaces of Analytic Functions" by Cowen and MacCluer, "Bergman Spaces" by Duren and Schuster, "Interpolation and Sampling in Spaces of Analytic Functions" by Seip, and "Analysis on Fock Spaces", "Spaces of Holomorphic Functions in the Unit Ball", "Operator Theory in Function Spaces" all by Zhu.

The present volume consists of eleven articles in the general area of analytic function spaces and operators on them, which nicely complement the monographs mentioned above. The overlap between this volume and existing books in the area is minimal. The material on two-variable weighted shifts by Curto, the Drury-Arveson space by Fang and Xia, the Cowen-Douglas class by Misra, and operator theory on the bi-disk by Yang has never appeared in book form before.

All the papers here are surveys in nature, although some of them do contain results that have not been published before. Toeplitz operators on various function spaces have been studied extensively in the literature and discussions about them can be found in many books. However, the exposition in the papers by Coburn, Upmeier, and Zhao and Zheng purposely avoided material that has already been widely disseminated.

The topics covered in the volume are diverse, so I will be happy if every reader finds one article (or more) interesting and useful. There are also several important topics that should have been included here but are missing, for example, function theory and operator theory on Hilbert spaces of Dirichlet series on the complex plane. I invited experts in these areas to make contributions to the volume, but for one reason or another, I did not eventually receive these articles.

I want to thank Steven Krantz for inviting me to edit this volume. I also want to thank all the contributors for your hard work and great papers. I enjoyed working with you all!

Kehe Zhu
Albany, NY
March, 2019

Chapter 1

Fock Space, the Heisenberg Group, Heat Flow, and Toeplitz Operators

Lewis A. Coburn
Department of Mathematics, SUNY at Buffalo, Buffalo, NY 14260, USA
lcoburn@buffalo.edu

CONTENTS

1.1 Introduction

I discuss some work which has given significant insight into the analytic structure of a space first studied in the late 1920's by the distinguished Soviet physicist Vladimir A. Fock [16]. Here, $\mathbf{C}^n$ is complex n-space and, for z in $\mathbf{C}^n$, we consider the standard family of Gaussian measures

$$d\mu_t(z) = (4\pi t)^{-n} \exp(-|z|^2/4t) dv(z), \qquad t > 0,$$

where dv is Lebesgue measure and t is any positive number. We consider the Hilbert space L_t^2 of all μ_t-square integrable complex-valued measurable functions on $\mathbf{C}^n$ and the closed subspace of all square-integrable entire functions, H_t^2. The space H_t^2 is often called the ("bosonic") Fock space (or, sometimes, the Segal-Bargmann space).

For f measurable and h in H_t^2 with fh in L_t^2, we consider the Toeplitz and Hankel operators

$$T_f^{(t)}h = P^{(t)}(fh), \qquad H_f^{(t)}h = (I - P^{(t)})(fh),$$

where $P^{(t)}$ is the orthogonal projection from L_t^2 onto H_t^2. For essentially bounded f ($f \in L^\infty$), these are bounded operators. For "nice" unbounded f, these operators are densely defined and may be unbounded.

In what follows, we write $z \cdot w = z_1\overline{w}_1 + \cdots + z_n\overline{w}_n$ for complex numbers z_j, w_j : $j = 1, 2, \ldots\ldots, n$, where $\overline{w}_j$ is the complex conjugate of w_j. Then, we have $|z|^2 = z \cdot z$. As usual, the inner product on L_t^2 is given by

$$\langle f, g \rangle_t = \int f\bar{g}\, d\mu_t,$$

with the corresponding norm $\sqrt{\langle f, f \rangle_t}$, and we may suppress the subscript t. For bounded operators from H_t^2 into L_t^2, we have the usual norm $\|\cdot\|_t$. We note that $\|T_f^{(t)}\|_t \leq \|f\|_\infty$ and $\|H_f^{(t)}\|_t \leq \|f\|_\infty$.

I note that many results about Fock space and its operators–published up until 2012–have been usefully incorporated into [21]. I will often refer to [21] in what follows, but my aim is to deal mostly with results that were not handled in [21].

As a brief historical note, the seminal paper of Fock [16] presents the Hilbert space H_t^2 in a close to recognizable form and deals primarily with the "creation" and "annihilation" operators T_{zj} and $T_{\bar{z}j}$: the translated version [17, pp. 51-67] is readable and the preceeding article [17, pp. 33-50] is also recommendable. There is a "follow-on" paper by P. Dirac [14]. Apart from some lecture notes by Irving Segal, "modern" analysis on H_t^2 begins in the paper of Valentin Bargmann [1]. Bargmann recognized that the H_t^2 are natural representation spaces for irreducible infinite-dimensional unitary representations of the Heisenberg group. It turns out that Toeplitz operators play a role in this enterprise. I next turn to a brief description of this early appearance of Toeplitz operators.

1.2 Toeplitz operators and the Heisenberg group

To begin this discussion we need the fact that H_t^2 is a Bergman (reproducing kernel) space. In particular, for any a in $\mathbf{C}^n$ and f in H_t^2, we have

$$f(a) = \int f(z) \exp\left(\overline{z \cdot a/4t}\right) d\mu_t(z) = \left\langle f, \exp\left(\frac{z \cdot a}{4t}\right)\right\rangle_t.$$

In fact, for any g in L_t^2 we have

$$(P^{(t)}g)(a) = \int g(z) \exp(\overline{z \cdot a/4t})\, d\mu_t(z).$$

The reproducing kernel functions $K^{(t)}(z,a) = \exp(z \cdot a/4t)$ will be used heavily in what follows. We will also need the (normalized) kernel functions

$$k_a^{(t)} = K^{(t)}(\cdot,a)/\sqrt{K^{(t)}(a,a)}.$$

We consider the Weyl operators $(W_a^{(t)} f)(z) = k_a^{(t)}(z) f(z-a)$. Direct calculation shows that these operators are isometries on L_t^2 and satisfy the equation

$$W_a^{(t)} W_b^{(t)} = \exp\left(\frac{b \cdot a - a \cdot b}{8t}\right) W_{a+b}^{(t)}. \tag{1.1}$$

It follows that the $W_a^{(t)}$ satisfy $W_a^{(t)} W_{-a}^{(t)} = I$ and so $W_a^{(t)*} = W_{-a}^{(t)}$ and the $W_a^{(t)}$ are unitary operators both on L_t^2 and on H_t^2.

Next, we consider the Heisenberg group $\mathbf{H}_n$, with underlying space $\mathbf{C}^n \times \mathbf{R}$ for $\mathbf{R}$ the real numbers. Multiplication on $\mathbf{H}_n$ is given by

$$(a,r)(b,s) = \left(a+b,\ r+s+\frac{b \cdot a - a \cdot b}{2i}\right).$$

For each $t > 0$, there is a strongly continuous unitary representation of $\mathbf{H}_n$ on L_t^2 or H_t^2 given by

$$\rho_t(a,r) = e^{ir/4t} W_a^{(t)}. \tag{1.2}$$

This representation is irreducible on H_t^2.

For the connection to Toeplitz operators–which goes back to [9, 10] for $t = \frac{1}{2}$– we first introduce the Berezin transform [7], mapping operators on H_t^2 to smooth functions on $\mathbf{C}^n$. For operators $X^{(t)}$, this transform is given by

$$\widetilde{X}^{(t)}(b) = \langle X^{(t)} k_b^{(t)}, k_b^{(t)} \rangle_t.$$

It is well-known that the map $X^{(t)} \to \widetilde{X}^{(t)}$ is injective for all bounded operators $X^{(t)}$ and for the large class of unbounded operators for which $domain(X^{(t)})$ contains $\{k_b^{(t)} : b \in \mathbf{C}^n\}$. Using the Cauchy-Schwarz inequality, we have $\|\widetilde{X}^{(t)}\|_\infty \leq \|X^{(t)}\|_t$. For Toeplitz operators, we have

$$\widetilde{T}_f^{(t)}(b) = \langle f k_b^{(t)}, k_b^{(t)} \rangle_t = (4\pi t)^{-n} \int_{\mathbf{C}^n} f(z) \exp(-|z-b|^2/4t)\, dv(z) \equiv \tilde{f}^{(t)}(b),$$

just the "heat transform" of f at time t. For any f in L_t^2, $\tilde{f}^{(s)}$ is well-defined for all s with $0 < s < 2t$.

We can now check that, for the character

$$\chi_a(z) = \exp\left(\frac{z \cdot a - a \cdot z}{2}\right),$$

we have

$$\widetilde{T}_{\chi_a}^{(t)}(b) = \exp(-t|a|^2)\chi_a(b),$$

while for $W_a^{(t)}$ on H_t^2, we have

$$\widetilde{W}_{2at}^{(t)}(b) = \exp\left(-\frac{t}{2}|a|^2\right)\chi_a(b). \tag{1.3}$$

It follows that

$$W_{2at}^{(t)} = \exp\left(\frac{t}{2}|a|^2\right)T_{\chi_a}^{(t)}. \tag{1.4}$$

The C^*-algebra generated by the Weyl unitaries $W_a^{(t)}$ is, thus, exactly the C^*-algebra $\tau(AP)$ of Toeplitz operators on H_t^2 with almost-periodic symbols. Note that combining equations (1.2) and (1.4) gives

$$\rho_t(a,r) = \exp\left(\frac{ir}{4t} + \frac{1}{8t}|a|^2\right)T^{(t)}_{\chi_{\{a/(2t)\}}}. \tag{1.5}$$

1.3 Toeplitz operators and the Bargmann transform

One of Bargmann's seminal contributions was to examine the relationship between Toeplitz operators on H_t^2 and pseudodifferential operators on $L^2(\mathbf{R}^n, dv)$. Let $k = (k_1, k_2, \cdot, \cdot, \cdot, k_n)$ with the k_j non-negative integers. We write $|k| = k_1 + \cdots + k_n$ and $k! = k_1!k_2!\cdots k_n!$. For z in $\mathbf{C}^n$, we write $z^k = z_1^{k_1}z_2^{k_2}z_3^{k_3}\cdots z_n^{k_n}$. For H_t^2 we have the standard orthonormal basis

$$e_k^{(t)} = ((4t)^{|k|}k!)^{-1/2}z^k.$$

The Bargmann isometry, β_t, maps $L^2(\mathbf{R}^n, dv)$ onto H_t^2. Following [21], we consider an orthonormal basis for $L^2(\mathbf{R}^n, dv)$ consisting of Hermite functions which have been "adapted" to the parameter t

$$w_k^{(t)} = (2\pi t)^{-\frac{n}{4}}(2^{|k|}k!)^{-\frac{1}{2}}H_k(2t^{-1/2}x)e^{-\frac{1}{4t}x^2}, \tag{1.6}$$

where $az = a_1z_1 + \cdots + a_nz_n$, $z^2 = zz$, and $a\cdot z = a\bar{z}$ with $|z|^2 = z\cdot z$ for $\bar{z}_j$ the complex conjugate of z_j and

$$H_k(x) = (-1)^{|k|}e^{x^2}D^k e^{-x^2}$$

for

$$D^k = \left(\frac{\partial}{\partial x_1}\right)^{k_1}\cdots\left(\frac{\partial}{\partial x_n}\right)^{k_n}.$$

In what follows, I will also write

$$\partial^k = \left(\frac{\partial}{\partial z_1}\right)^{k_1}\cdots\left(\frac{\partial}{\partial z_n}\right)^{k_n}, \qquad \bar{\partial}^k = \left(\frac{\partial}{\partial \bar{z}_1}\right)^{k_1}\cdots\left(\frac{\partial}{\partial \bar{z}_n}\right)^{k_n}.$$

The isometry β_t is given by

$$\beta_t w_k^{(t)} = e_k^{(t)}.$$

The Heisenberg group, discussed above, is a nilpotent Lie group which plays a key role in quantum mechanics and in the representation theory of nilpotent Lie groups. It is standard that Lebesgue measure on $\mathbf{C}^n \times \mathbf{R}$ is biinvariant Haar measure on $\mathbf{H}_n$. The representation theory for $\mathbf{H}_n$ is well-known. The infinite-dimensional strongly-continuous irreducible unitary representations are traditionally parametrized by non-zero values of "Planck's constant" and act on $L^2(\mathbf{R}^n, dv)$. In particular, again following [21], the Schrödinger representation is given, for a in $\mathbf{C}^n$ and for p,q,x in $\mathbf{R}^n$, with r in $\mathbf{R}$, $f(x)$ in $L^2(\mathbf{R}^n, dv)$ and $a = p + iq$, by

$$[\rho_t^S(p+iq,r)f](x) = e^{ir\alpha - 2i\alpha qx + i\alpha pq} f(x-p).$$

where $\alpha = 1/4t$. Computations in [21, pp.225,226] together with equation (1.3) show that, for $a = p + iq$,

$$\rho_t^S(p+iq,r) = \beta_t^{-1}\rho_t(a,r)\beta_t.$$

In fact, the Bargmann isometry has the property [21] that $\beta_t^{-1} T_f^{(t)} \beta_t = W_{\sigma_t}$, where W_{σ_t} is the Weyl pseudodifferential operator with Weyl symbol $\sigma_t(x,\eta) = \tilde{f}^{(t/2)}(x+i\eta)$ for $z = x + i\eta$ with x and η in $\mathbf{R}^n$.

Related to these facts, there is a long-standing [8]

Conjecture 3.1 *For f such that $fk_a^{(t)}$ is in L_t^2 for all a in $\mathbf{C}^n$, $T_f^{(t)}$ is bounded if and only if $\tilde{f}^{(t/2)}$ is bounded.*

Some interesting examples of Fock space Toeplitz operators were provided in [8, pp. 581, 582] for the case $t = 1/2$ and $n = 1$. These examples extend to general $t > 0$ and $n \geq 1$. We consider the functions $g(\lambda, z) = \exp(\lambda |z|^2)$, for λ in $\mathbf{C}$, and the Toeplitz operators $T_{g(\lambda,\cdot)}^{(t)}$ on H_t^2 (with $\mathrm{Re}\lambda < 1/8t$). We see that these operators are diagonal with respect to the standard orthonormal basis $\{e_k^{(t)}\}$ and we have

$$T_{g(\lambda,\cdot)}^{(t)} e_k^{(t)} = (1-4t\lambda)^{-(|k|+n)} e_k^{(t)}.$$

Moreover, we obtain the heat transform formula

$$g(\lambda,\cdot)^{\tilde{}(s)}(w) = (1-4s\lambda)^{-n} \exp\{\lambda(1-4s\lambda)^{-1}|w|^2\},\ s < 2t.$$

The operators $T_{g(\lambda,\cdot)}^{(t)}$ are trace class for $|\lambda - 1/4t| > 1/4t$ and unitary for $|\lambda - 1/4t| = 1/4t$. Note that $\mathrm{Re}\lambda > 0$ occurs in these cases, so that $g(\lambda,\cdot)$ can have quadratic exponential growth. The heat transform $g(\lambda,\cdot)^{\tilde{}(s)}$ has the "semigroup property" [8, p. 566] for positive s, s' with $s + s' < 2t$. Also, note that

$$\lim_{t\to 0} g(\lambda,\cdot)^{\tilde{}(s)}(w) = g(w)$$

for all w.

There is a considerable literature about Fock space Toeplitz operators which, to some extent, parallels the literature on pseudo-differential operators. One unresolved issue is the problem of composing Toeplitz operators. There has been some progress [2, 12] toward determining the largest class of functions φ, ψ for which there is a "twisted multiplication" $\varphi \diamond_t \psi$ so that $T_\varphi^{t)} T_\psi^{(t)} = T_{\varphi\diamond_t\psi}^{(t)}$. In particular, for φ, ψ in $B_a(\mathbf{C}^n)$, the algebra of Fourier-Stieltjes transforms of compactly supported, complex, bounded regular Borel measures on $\mathbf{C}^n$, $T_\varphi^{(t)} T_\psi^{(t)}$ is a Toeplitz operator $T_{\varphi\diamond_t\psi}^{(t)}$ and there is a "Moyal-type" formula

$$\varphi \diamond_t \psi = \sum_k \frac{(-4t)^{|k|}}{k!} (\partial^k \varphi)(\bar{\partial}^k \psi), \tag{1.7}$$

where the series on the right converges uniformly and absolutely to a function in $B_a(\mathbf{C}^n)$. It is also known that for φ, ψ arbitrary polynomials in z and $\bar{z}$, $T_\varphi^{(t)} T_\psi^{(t)}$ is a (generally unbounded) Toeplitz operator $T_{\varphi\diamond_t\psi}^{(t)}$, with $\varphi \diamond_t \psi$ given again by equation (1.7). For a different perspective, with related results, see Theorems 13.11 and 13.18 in [22].

For other interesting results of this type, see [2]. In particular, for $f, g \in L^\infty$, [2] shows that

$$T_{\tilde{f}^{(t)}}^{(t)} T_{\tilde{g}^{(t)}}^{(t)} = T_{\tilde{f}^{(t)}\diamond_t \tilde{g}^{(t)}}^{(t)}. \tag{1.8}$$

For an example of an unbounded f for which $T_f^{(1/2)}$ is bounded but $(T_f^{(1/2)})^2$ is not a Toeplitz operator, see [12].

The next application of the Bargmann transform concerns the relationship between the *Gabor-Daubechies* windowed Fourier localization operators on $L^2(\mathbf{R}^n, dv)$ and Toeplitz operators on H_t^2. Underpinning this analysis is the well-known "square-integrability modulo the center" of irreducible representations of $\mathbf{H}_n$. Precisely, the *Gabor-Daubechies* localization operator, $L_{\varphi,t}^w$, with "symbol" (or "weight function") φ and "window" w, is defined on $L^2(\mathbf{R}^n, dv)$ by

$$\begin{aligned}\langle L_{\varphi,t}^w f, g\rangle &= (4\pi t)^{-n} \int_{\mathbf{C}^n} \varphi(a) \langle f, \rho_t^S(a,0)w\rangle \langle \rho_t^S(a,0)w, g\rangle \, dv(a) \\ &= (4\pi t)^{-n} \int_{\mathbf{C}^n} \varphi(a) \langle \beta_t f, W_a^{(t)} \beta_t w\rangle_t \langle W_a^{(t)} \beta_t w, \beta_t g\rangle_t \, dv(a).\end{aligned}$$

Typically, the window w is a unit vector in $L^2(\mathbf{R}^n, dv)$. For $w = w_0^{(t)}$, we have $\beta_t w = 1$ and it is not hard to check that $\beta_t L_{\varphi,t}^w \beta_t^{-1} = T_\varphi^{(t)}$. Thus, the Bargmann isometry allows the amalgamation of the substantial work already done in analyzing $T_\varphi^{(t)}$ and $L_{\varphi,t}^w$.

There has been a significant extension of this result by M. Lo and M. Englis [15, 18] (see also [13]). For c_k in $\mathbf{C}$ and $w_k^{(t)}$ the Hermite functions from equation (1.6), let

$$w = \sum_{|k| \le r} c_k w_k^{(t)},$$

and let φ have moderate growth (satisfying condition M_r of [15, p.1046]). Then

Theorem 3.2 *There is a constant coefficient linear partial differential operator $D(w,t)$ such that*

$$\beta_t L^w_{\varphi,t}\beta_t^{-1} g = T^{(t)}_{D(w,t)\varphi} g$$

for all g in the dense linear span of $\{k^{(t)}_a : a \in \mathbf{C}^n\}$.

For $t = 1/2$, explicit examples were given in [13] for the Hermite functions $w^{(t)}_1, w^{(t)}_2$ with $1 = (1,0,0,....,0)$ and $2 = (2,0,0,....,0)$. Here, we have

$$D(w^{(t)}_1, t) = I + 4t(\bar{\partial}^1\partial^1), \qquad D(w^{(t)}_2, t) = I + 16t^2(\bar{\partial}^1\partial^1) + 4t(\bar{\partial}^1\partial^1)^2.$$

1.4 Limit behavior of $\|T^{(t)}_f\|_t$ and $\|H^{(t)}_f\|_t$ as $t \to 0$

A classical problem for the heat equation is the recovery from $\tilde{f}^{(t)}$ of the initial value f in the limit as $t \to 0$. The limiting behaviours of $\|T^{(t)}_f\|_t$ and $\|H^{(t)}_f\|_t$ are also of interest. I now turn to the "Toeplitz quantization" problem in which we study the asymptotic behavior of Toeplitz operators and their commutators for small values of the parameter t.

For f, g bounded, with "sufficiently many" bounded derivatives, it is known (see [11]) that we have the *deformation quantization* conditions:

(i)

$$\lim_{t\to 0} \|T^{(t)}_f\|_t = \|f\|_\infty,$$

(ii)

$$\lim_{t\to 0} \|T^{(t)}_f T^{(t)}_g - T^{(t)}_{fg}\|_t = 0,$$

(iii)

$$\lim_{t\to 0} \left\| \frac{1}{it}[T^{(t)}_f, T^{(t)}_g] - T^{(t)}_{\{f,g\}} \right\|_t = 0,$$

where $[A,B] = AB - BA$ and $\{\cdot,\cdot\}$ is the Poisson bracket. Recently, in [5, 6] Bauer, Hagger and I have considerably sharpened these results in the case of equations (i) and (ii).

I note, first, that a key identity in our work is the standard fact that

$$T^{(t)}_f T^{(t)}_g - T^{(t)}_{fg} = -(H^{(t)}_{\bar{f}})^* H^{(t)}_g. \tag{1.9}$$

It is, thus, interesting to give sharp conditions on g so that $\lim_{t\to 0} \|H^{(t)}_g\|_t = 0$ and on f so that (i) holds.

To start, we consider the scaling transformations $(A_\alpha f)(z) = f(\alpha z)$ for real $\alpha > 0$ and note that

$$A_{\sqrt{s/t}} : L^2_s \to L^2_t$$

is a surjective isometry which restricts as surjective isometry from H^2_s to H^2_t. It is easy to check that, for both Toeplitz and Hankel operators,

$$A_{\sqrt{t/s}} T_f^{(t)} A_{\sqrt{s/t}} = T^{(s)}_{f(\cdot\sqrt{t/s})}, \qquad A_{\sqrt{t/s}} H_f^{(t)} A_{\sqrt{s/t}} = H^{(s)}_{f(\cdot\sqrt{t/s})}. \tag{1.10}$$

We recall the definitions [6]

$$MO^t(f)(w) = (\widetilde{|f|^2}^{(t)}(w) - |\tilde{f}^{(t)}(w)|^2,$$

with seminorm

$$\|f\|_{BMO^t_*} = \sup_{w\in\mathbf{C}^n} \{MO^t(f)(w)\}^{1/2},$$

and note that

$$\|f\|_{BMO^t_*} = \|\bar{f}\|_{BMO^t_*}.$$

For $t = 1/4$ there is a non-trivial norm estimate [3]

$$\|H_f^{(1/4)}\|_{1/4} \leq C\|f\|_{BMO^{1/4}_*}.$$

An easy calculation shows

$$\left\{ f\left(\cdot\sqrt{\frac{t}{s}}\right)\right\}^{\sim(s)}(z) = \tilde{f}^{(t)}\left(z\sqrt{\frac{t}{s}}\right)$$

and, taking $s = 1/4$, gives

$$A_{2\sqrt{t}} H_f^{(t)} A^{-1}_{2\sqrt{t}} = H^{(1/4)}_{f(\cdot 2\sqrt{t})},$$

while

$$\|f(\cdot 2\sqrt{t})\|_{BMO^{1/4}_*} = \|f\|_{BMO^t_*}.$$

It follows that

$$\begin{aligned}\|H_f^{(t)}\|_t = \|A_{2\sqrt{t}} H_f^{(t)} A^{-1}_{2\sqrt{t}}\|_{1/4} &= \|H^{(1/4)}_{f(\cdot 2\sqrt{t})}\|_{1/4} \\ \leq C\|f(\cdot 2\sqrt{t})\|_{BMO^{1/4}_*} &= C\|f\|_{BMO^t_*}\end{aligned} \tag{1.11}$$

where the constant C is **independent of** t. The argument above, for t-independence, was pointed out by Jingbo Xia.

A partial reverse estimate is standard [21, p. 291]:

$$\|f\|_{BMO^t_*} \leq \|H_f^{(t)}\|_t + \|H_{\bar{f}}^{(t)}\|_t. \tag{1.12}$$

I will need estimates (1.11) and (1.12) in the following discussion.

As in [6], we consider the set

$$A = \left\{ f \in L^\infty : \lim_{t\to 0} \|T_f^{(t)} T_g^{(t)} - T_{fg}^{(t)}\|_t = \lim_{t\to 0} \|T_g^{(t)} T_f^{(t)} - T_{fg}^{(t)}\|_t = 0, g \in L^\infty \right\}.$$

Proposition 4.1 *A is a norm-closed, conjugate-closed subalgebra of L^∞ and we have*

$$A = \left\{ f \in L^\infty : \lim_{t\to 0} \|H_f^{(t)}\| = \lim_{t\to 0} \|H_{\bar{f}}^{(t)}\| = 0 \right\}. \tag{1.13}$$

Proof For $\{f_n\}$ in A, with $f_n \to f$ (in L^∞), it is easy to check that f is in A. For f, h in A and g in L^∞, we have

$$T_{fh}^{(t)}T_g^{(t)} - T_{fhg}^{(t)} = (T_{fh}^{(t)} - T_f^{(t)}T_h^{(t)})T_g^{(t)} + T_f^{(t)}(T_h^{(t)}T_g^{(t)} - T_{hg}^{(t)}) + (T_f^{(t)}T_{hg}^{(t)} - T_{f(hg)}^{(t)}).$$

Hence, $fh \in A$. It is now clear that A is an algebra which is invariant under complex conjugation. The equality of sets in equation (1.13) follows from equation (1.9).

Remark 4.2 . *Using equation (1.9), we see that A is maximal among all conjugate-closed subalgebras of L^∞ whose functions satisfy (ii).*

We also have

Proposition 4.3 *The algebra $A = B$ where*

$$B = \left\{ f \in L^\infty : \lim_{t\to 0} \|f\|_{BMO_*^t} = 0 \right\}. \tag{1.14}$$

Proof This follows immediately from inequalities (1.11) and (1.12).

To complete this analysis, we need a more standard characterization of the elements of B. To do this, we recall the space of bounded functions on $\mathbf{C}^n$ having "vanishing mean oscillation" $VMO(\mathbf{C}^n) \cap L^\infty$. The primary reference is the paper of Sarason [19]. Since that paper is mostly concerned with the underlying space $\mathbf{R}$ instead of $\mathbf{R}^n$ or $\mathbf{C}^n$, I give here a brief exposition taken from [6]. We consider locally integrable functions $f : \mathbf{R}^n \to C$ with average value

$$f_E = \frac{1}{|E|}\int_E f$$

on a bounded measurable subset E with finite measure $|E|$. We consider the variance of f on E

$$\mathrm{Var}_E(f) = \frac{1}{|E|}\int_E |f - f_E|^2$$

as well as the corresponding quantity

$$Osc_E(f) = \frac{1}{|E|}\int_E |f - f_E|.$$

We say f is in $BMO(\mathbf{R}^n)$ if the set

$$\{Osc_E(f) : E \text{ each n-cube in } \mathbf{R}^n\}$$

is bounded. We say f is in $VMO(\mathbf{R}^n)$ if f is in $BMO(\mathbf{R}^n)$ and, for all n-cubes E,

$$\limsup_{a\to 0}\{Osc_E(f) : |E| \leq a\} = 0.$$

If we replace $Osc_E(f)$ by $Var_E(f)$, we get new sets BMO_2 and VMO_2. The Cauchy-Schwarz inequality shows that

$$BMO_2 \subset BMO, \qquad VMO_2 \subset VMO.$$

Direct calculation shows

Lemma 4.4 *We have, for arbitrary bounded measurable subsets E,*

$$Var_E(f) = \frac{1}{2|E|^2}\int_E\int_E |f(z)-f(w)|^2 dv(z)dv(w)$$

so that, for $F \subset E$,

$$Var_E(f) \geq \frac{|F|^2}{|E|^2} Var_F(f).$$

Remark 4.5 *Because of Lemma 4.4, BMO_2 and VMO_2 have the additional property that n-cubes can be replaced by n-balls in their definitions. We need only consider the inscribed and circumscribed balls for a given cube.*

In the analysis that follows, we identify $\mathbf{C}^n$ with $\mathbf{R}^{2n}$ in the standard way and consider the set VMO(C^n)$\cap L^\infty$, the essentially bounded functions in VMO($\mathbf{C}^n$). It is easy to see that the bounded uniformly continuous functions $BUC(\mathbf{C}^n)$ are a subset of $VMO_2(\mathbf{C}^n)$. There are also discontinuous functions in $VMO(\mathrm{C}^n)\cap L^\infty$ [20, p. 290]. We have

Proposition 4.6 *$VMO(\mathbf{C}^n)\cap L^\infty$ is a sup-norm closed, conjugate closed subalgebra of L^∞ with $VMO(\mathbf{C}^n)\cap L^\infty = VMO_2(\mathbf{C}^n)\cap L^\infty$.*

Proof An easy estimate using $|f - f_E| \leq 2\|f\|_\infty$ shows that

$$VMO(\mathrm{C}^n)\cap L^\infty = VMO_2(\mathrm{C}^n)\cap L^\infty.$$

Using Lemma 4.4 and integrating the inequality

$$|f(z)g(z)-f(w)g(w)|^2 \leq 2\|f\|_\infty^2|g(z)-g(w)|^2 + 2\|g\|_\infty^2|f(z)-f(w)|^2$$

over $E\times E$ shows that for f,g in $VMO_2(\mathbf{C}^n)\cap L^\infty$, we also have fg in $VMO_2(\mathbf{C}^n)\cap L^\infty$. The remainder of the proof is standard.

Remark 4.7 *Proposition 4.6 establishes, for C^n (or R^n), a result proved by Sarason [19] for the circle, by quite different methods .*

I can now state

Proposition 4.8 *The algebra* $VMO(\mathbf{C}^n) \cap L^\infty \equiv B \equiv A$.

Proof The proof requires direct but somewhat laborious estimates [6, Theorems 4.8 and 5.7] and is omitted.

1.5 Examples

I first consider a smooth counter example from [6], namely, I provide a pair of bounded, real analytic functions F, G such that (ii) fails. Precisely, for $n = 1$ and for $F(z) = \exp(i|z|^2)$ and $G(z) = \exp(-i|z|^2)$, we have

$$\|T_F^{(t)} T_G^{(t)} - T_{FG}^{(t)}\|_t \equiv 1$$

for all $t > 0$. We see that $T_F^{(t)}, T_G^{(t)}$ are diagonal in the orthonormal basis $\{e_k^{(t)} : k = 0, 1, 2, \cdots\}$ with eigenvalues

$$s_k(F,t) = (1 - 4ti)^{-(k+1)}, \qquad s_k(G,t) = (1 + 4ti)^{-(k+1)}.$$

Since $F(z)G(z) \equiv 1$, we have

$$\|T_F^{(t)} T_G^{(t)} - T_{FG}^{(t)}\|_t = \sup\{1 - (1 + 16t^2)^{-(k+1)} : k = 0, 1, 2, \cdots\} = 1$$

for all $t > 0$. Note that $[T_F^{(t)}, T_G^{(t)}] = 0$ while $\{F, G\} = 0$ so (iii) holds trivially. By direct computation, we have

$$T_F^{(t)} T_G^{(t)} = T^{(t)}_{\exp\{-4t|z|^2\}}.$$

I next consider an example and theorem from [6] for unbounded f, g. Standard calculation shows that, for $n = 1$,

$$\begin{aligned} T_z^{(t)} e_k^{(t)} &= \{(4t)(k+1)\}^{1/2} e_{k+1}^{(t)}, \qquad k = 0, 1, 2, \cdots, \\ T_{\bar{z}}^{(t)} e_{k+1}^{(t)} &= \{(4t)(k+1)\}^{1/2} e_k^{(t)}, \qquad k = 0, 1, 2, \cdots, \\ T_{\bar{z}}^{(t)} e_0^{(t)} &= 0. \end{aligned}$$

It follows that

$$[T_{\bar{z}}^{(t)}, T_z^{(t)}] e_k^{(t)} = 4t e_k^{(t)}, \qquad k = 0, 1, 2, \cdots.$$

Note that

$$T_{\bar{z}}^{(t)} T_z^{(t)} = T_{\bar{z}z}^{(t)},$$

which yields

$$\|T_z^{(t)}T_{\bar{z}}^{(t)} - T_{z\bar{z}}^{(t)}\|_t = 4t.$$

I note that, while $T_{\bar{z}}^{(t)}, T_z^{(t)}$ are unbounded,

$$\left(H_{\bar{z}}^{(t)}\right)^* H_{\bar{z}}^{(t)} = 4tI,$$

so that $\|H_{\bar{z}}^{(t)}\|_t = 2\sqrt{t}$. Since $f(z) = \bar{z}$ is uniformly continuous (UC), this suggests [6, Theorem 3.4]:

Proposition 5.1 *For f in $UC(C^n)$ and g in L^∞ or in UC, we have*

$$\lim_{t\to 0} \|T_f^{(t)}T_g^{(t)} - T_{fg}^{(t)}\|_t = 0.$$

Proof We show that $\lim_{t\to 0}\|H_f^{(t)}\|_t = 0$ for f in UC. The argument is technical and is, therefore, omitted.

Remark 5.2 *Direct computational checks of Proposition 5.1 in the diagonal case bring us to–or, even, over–the edge of what is possible using Stirling's approximation. For an example, consider the estimation of $\|T_{|z|}^{(t)}T_{|z|}^{(t)} - T_{|z|^2}^{(t)}\|_t$.*

I next consider a maximal strengthening of (i). Recalling that

$$\tilde{f}^{(t)}(w) = \int_{\mathbf{C}^n} f(w-z)d\mu_t(z)$$

for the Berezin transform of f (the heat transform), I can now state the best version of deformation quantization condition (i) from [6]:

Proposition 5.3 *For f any L^∞ function on C^n, we have*

$$\lim_{t\to 0}\|\tilde{f}^{(t)}\|_\infty = \lim_{t\to 0}\|T_f^{(t)}\|_t = \|f\|_\infty.$$

Proof The proof uses the Hardy-Littlewood maximal function f^* to estimate $\tilde{f}^{(t)}$. A technical argument shows that $\lim_{t\to 0}\tilde{f}^{(t)}(w) = f(w)$ except on a set of measure 0. The rest follows since it is standard that

$$\|\tilde{f}^{(t)}\|_\infty \leq \|T_f^{(t)}\|_t \leq \|f\|_\infty.$$

Remark 5.4 *In [4], Bauer and I gave a related application of the heat flow. For f in UC (bounded or unbounded), we have for $t > 0$,*

$$\|\tilde{f}^{(t)} - f\|_\infty < \infty,$$

and

$$\lim_{t\to 0} \|\tilde{f}^{(t)} - f\|_\infty = 0,$$

with $\tilde{f}^{(t)}$ Lipschitz and real-analytic. An analogous result in [4] holds for all bounded symmetric domains in $\mathbf{C}^n$ with their Bergman metrics. Here, the heat flow $\tilde{f}^{(t)}$ is replaced by the one-parameter "Berezin, Harish-Chandra flow of weights." In particular, this result holds for the open unit ball with hyperbolic metric.

Finally I consider an example of Proposition 5.3. For $B(a,r)$ the closed ball in $\mathbf{C}^n$ centered at a with radius r, we let $f(z)$ be the characteristic function of $B(0,1)$. The heat flow of f is given by

$$\tilde{f}^{(t)}(w) = \int_{\mathbf{C}^n} f(w-z)d\mu_t(z) = \int_{B(w,1)} d\mu_t(z).$$

Note first that

$$\frac{|B(0,1)|}{(4\pi t)^n} \geq \tilde{f}^{(t)}(w) \geq 0,$$

so that

$$\lim_{t\to+\infty} ||\tilde{f}^{(t)}||_\infty = 0.$$

It should be pointed out that, despite the above result, it is standard that for all g in $L^1(\mathbf{C}^n, dv)$, we have, for all $t > 0$,

$$\int_{\mathbf{C}^n} \tilde{g}^{(t)}(w)dv(w) = \int_{\mathbf{C}^n} g(w)dv(w).$$

By Theorem 6.2 of [6], we know that

$$\lim_{t\to 0} \tilde{f}^{(t)}(w) = f(w)$$

for almost all w. Here we can get a more precise result.

For $|w| > 1$, we have for any z in $B(w,1)$, $|w| - 1 \leq |z|$ so that

$$\int_{B(w,1)} d\mu_t(z) \leq \frac{\exp\{-(|w|-1)^2/4t\}}{(4\pi t)^n} |B(0,1)|$$

and it follows that $\lim_{t\to 0} \tilde{f}^{(t)}(w) = 0 = f(w)$.

For $|w| < 1$, we have $B(0, 1-|w|) \subset B(w,1)$ and writing $R = 1-|w|$ gives

$$1 \geq \tilde{f}^{(t)}(w) \geq \int_{B(0,R)} d\mu_t(z).$$

For $\rho = R/\sqrt{n}$ and $B(0,\rho)$ in $\mathbf{C}$, we have

$$B(0,\rho) \times B(0,\rho) \times \cdots \times B(0,\rho) \subset B(0,R).$$

Using

$$\tilde{f}^{(t)}(w) \geq \left(\int_{B(0,\rho)} d\mu_t(z) \right)^n,$$

it is now easy to check that

$$\lim_{t\to 0} \tilde{f}^{(t)}(w) = 1 = f(w).$$

Thus, we have $\lim_{t\to 0} \tilde{f}^{(t)}(w) = f(w)$ on $\{w \in \mathbf{C}^n : |w| \neq 1\}$.

References

[1] Bargmann, V., On a Hilbert space of analytic functions and an associated integral transform, *Comm. Pure Appl. Math.* **14** (1961), 187-214.

[2] Bauer, W., Berezin-Toeplitz quantization and composition formulas, *J. Funct. Anal.* **256** (2009), 3107-3142.

[3] Bauer, W., Mean oscillation and Hankel operators on the Segal-Bargmann space, *Integral Eq. Oper. Thy.* **52** (2005), 1- 15.

[4] Bauer, W. and Coburn, L. A., Heat flow, weighted Bergman spaces and real analytic Lipschitz approximation, *J. reine angew. Math.* **703** (2015), 225-246.

[5] Bauer, W. and Coburn, L. A., Uniformly continuous functions and quantization on the Fock space, *Bol. Soc. Mat. Mex.* **22** (2016), 669-677.

[6] Bauer, W., Coburn, L. A. and Hagger, R., Toeplitz quantization on Fock space, *J. Funct. Anal.* **274** (2018), 3531-3551.

[7] Berezin, F. A., Covariant and contravariant symbols of operators, *Math. USSR-Izv.* **6** (1972), 1117-1151.

[8] Berger, C. A. and Coburn, L. A., Heat flow and Berezin-Toeplitz estimates, *Amer. J. Math.* **116** (1994), 563-590.

[9] Berger, C. A. and Coburn, L. A., Toeplitz operators and quantum mechanics, *J. Funct. Anal.* **68** (1986), 273-299.

[10] Berger, C. A. and Coburn, L. A., Toeplitz operators on the Segal- Bargmann space, *Trans. Amer. Math. Soc.* **301** (1987), 813-829.

[11] Borthwick, D., Microlocal techniques for semiclassical problems in geometric quantization, In: Perspectives on quantization *Contemporary Mathematics* **214** (1998), 23-37, AMS, Providence.

[12] Coburn, L. A., On the Berezin-Toeplitz calculus, *Proc. Amer. Math. Soc.* **129** (2001), 3331-3338 .

[13] Coburn, L. A., The Bargmann isometry and Gabor-Daubechies wavelet localization operators, in: *Operator Thy. Adv. Appl.* **129**, Birkhauser (2001) Basel, 169-178.

[14] Dirac, P., La seconde quantification, *Ann. Inst. H. Poincare* **11** (1949), 15-47.

[15] Englis, M., Toeplitz operators and localization operators, *Trans. Amer. Math. Soc.* **361** (2009), 1039-1052.

[16] Fock, V. A., Verallgemeinerung und Losing der Diracschen statistischen Gleichung, *Zeitschrift für Physik* **49** (1928), 339-357.

[17] Fock, V. A., *Selected works: quantum mechanics and quantum field theory*, Chapman and Hall/CRC (2004), Boca Raton.

[18] Lo, M., The Bargmann transform and windowed Fourier localization, *Integral Eq. Operator Thy.* **57** (2007), 397-412.

[19] Sarason, D., Functions of vanishing mean oscillation, *Trans. Amer. Math. Soc.* **207** (1975), 391-405.

[20] Shih, X. and Torchinsky, A., Functions of vanishing mean oscillation, *Math. Nachr.* **133** (1987), 289-296.

[21] Zhu, K., *Analysis on Fock Spaces*, Springer, New York, 2012.

[22] Zworski, M., *Semiclassical Analysis*, American Mathematical Society Graduate Studies in Mathematics **138**, Providence, 2012.

Chapter 2

Two-Variable Weighted Shifts in Multivariable Operator Theory

Raúl E. Curto
Department of Mathematics, The University of Iowa, Iowa City, Iowa 52242, USA
raul-curto@uiowa.edu

CONTENTS

2.1 Introduction

Over the last fifty years, several generations of operator theorists have been acquainted with the expository writings of Paul R. Halmos, John B. Conway, Ronald G. Douglas and Allen P. Shields. A common theme in these works is the ubiquity of unilateral weighted shifts acting on the Hilbert space $\ell^2(\mathbb{Z})$ and $\ell^2(\mathbb{Z}_+)$, or its function-theoretic counterparts, the multiplication operators on the Hardy space, Bergman space, Dirichlet space, weighted Bergman spaces and more generally reproducing kernel Hilbert spaces. While weighted shifts represent the core of the celebrated survey article by Allen P. Shields [80], they are also explicitly mentioned in at least 29 of the 250 problems listed by P.R. Halmos in [65] (see also [64]. Weighted shifts have been, and continue to be, the source of countless examples and counterexamples in spectral theory, invariant subspace theory, C^*-dynamical system theory, and subnormal and hyponormal operator theory and its generalizations. Here's a sample elementary result.

Lemma 1.1 *([80]) All norm-one hyponormal unilateral weighted shifts have a spectral picture identical to the spectral picture of the (unweighted) unilateral shift U_+; that is, the spectrum is the closed unit disc $\bar{\mathbb{D}}$, the essential spectrum is the unit circle $\mathbb{T}$, and the Fredholm index of $U_+ - \lambda$ is -1 in the open unit disc $\mathbb{D}$ and 0 in the complement of $\bar{\mathbb{D}}$.*

The purpose of this article is to bring attention to the role and significance of the natural analogues of unilateral weighted shifts in bivariate operator theory. These are the so-called 2-variable weighted shifts, which are commuting pairs of operator weighted shifts, acting on the space $\ell^2(\mathbb{Z}_+^2)$, and shifting in orthogonal directions. The first component shifts to the right (or eastward direction) and the second component shifts up (or northward direction). Of special interest to us are the notions of subnormality and joint k-hyponormality, the lifting problem for commuting subnormals, operator transforms such as the spherical Aluthge transform, bivariate truncated moment problems, and subnormal completion problems. It is in these areas that we will focus our attention.

Before we proceed, we need some basic terminology and elementary facts. Let $\mathcal{H}$ be a complex Hilbert space and let $\mathcal{B}(\mathcal{H})$ denote the algebra of bounded linear operators on $\mathcal{H}$. For $S,T \in \mathcal{B}(\mathcal{H})$ let $[S,T] := ST - TS$. We say that an n-tuple $\mathbf{T} = (T_1,\dots,T_n)$ of operators on $\mathcal{H}$ is (jointly) *hyponormal* ([5], [6], [42]) if the operator matrix

$$[\mathbf{T}^*,\mathbf{T}] := \begin{pmatrix} [T_1^*,T_1] & [T_2^*,T_1] & \cdots & [T_n^*,T_1] \\ [T_1^*,T_2] & [T_2^*,T_2] & \cdots & [T_n^*,T_2] \\ \vdots & \vdots & \cdots & \vdots \\ [T_1^*,T_n] & [T_2^*,T_n] & \cdots & [T_n^*,T_n] \end{pmatrix}$$

is positive on the direct sum of n copies of $\mathcal{H}$ (cf. [5] , [42]); $\mathbf{T}$ is weakly hyponormal if $\lambda_1 T_1 + \cdots + \lambda_n T_n$ is hyponormal for every $\lambda_1, \cdots, \lambda_n \in \mathbb{C}$ [16]. The n-tuple $\mathbf{T}$ is said to be *normal* if $\mathbf{T}$ is commuting and each T_i is normal, and *subnormal* if $\mathbf{T}$ is the restriction of a normal n-tuple to a common invariant subspace. Clearly, normal $\Rightarrow$ subnormal $\Rightarrow$ hyponormal. It is also straightforward to prove that each component $(T_i$ of a subnormal (resp. hyponormal) n-tuple $\mathbf{T}$ is subnormal (resp. hyponormal). Moreover, the Bram-Halmos criterion states that an operator $T \in \mathcal{B}(\mathcal{H})$ is subnormal if and only if it is k-hyponormal for all $k \geq 1$, that is, if the k-tuple $(T, T^2, ..., T^k)$ is hyponormal for all $k \geq 1$. The notion of k-hyponormality admits a simple extension to two or more variables (cf. [33]). For instance, a commuting pair (T_1,T_2) is 2-hyponormal if the 5-tuple $(T_1, T_2, T_1^2, T_1T_2, T_2^2)$ is hyponormal; moreover, a suitable version of the Bram-Halmos Theorem for n-tuples was proved in [33].

2.1.1 Unilateral weighted shifts

For $\alpha \equiv \{\alpha_n\}_{n=0}^{\infty}$ a bounded sequence of positive real numbers (called *weights*), let $W_\alpha : \ell^2(\mathbb{Z}_+) \to \ell^2(\mathbb{Z}_+)$ be the associated unilateral weighted shift, defined by

$$W_\alpha e_n := \alpha_n e_{n+1} \text{ (all } n \geq 0),$$

where $\{e_n\}_{n=0}^{\infty}$ is the canonical orthonormal basis in $\ell^2(\mathbb{Z}_+)$. When $\alpha_k = 1$ (all $k \geq 0$), $W_\alpha = U_+$, the (unweighted) unilateral shift. In general, (i) $W_\alpha = U_+ D_\alpha$ (polar decomposition); (ii) $\|W_\alpha\| = \sup_k \alpha_k$; (iii) $W_\alpha^n e_k = \alpha_k \alpha_{k+1} \cdots \alpha_{k+n-1} e_{k+n}$, so

$$W_\alpha^n \cong \bigoplus_{i=0}^{n-1} W_{\beta(i)}.$$

The moments of α are given as

$$\gamma_k \equiv \gamma_k(\alpha) := \left\{ \begin{array}{cc} 1 & \text{if } k = 0 \\ \alpha_0^2 \cdot \ldots \cdot \alpha_{k-1}^2 & \text{if } k > 0 \end{array} \right\}.$$

We will usually denote W_α by $\text{shift}(\alpha_0, \alpha_1, \alpha_2, \cdots)$. It is easy to see that W_α is never normal, and that it is hyponormal if and only if $\alpha_0 \leq \alpha_1 \leq \alpha_2 \leq ....$

We now recall a well known characterization of subnormality for single variable weighted shifts, due to C. Berger (cf. [15, III.8.16]) and independently established by R. Gellar and L.J. Wallen [62]: W_α is subnormal if and only if there exists a probability measure ξ supported in $[0, \|W_\alpha\|^2]$ such that

$$\gamma_k(\alpha) := \alpha_0^2 \cdot \ldots \cdot \alpha_{k-1}^2 = \int t^k \, d\xi(t) \ \ (k \geq 1).$$

If W_α is subnormal, and if for $h \geq 1$ we let $\mathcal{M}_h := \bigvee \{e_n : n \geq h\}$ denote the invariant subspace obtained by removing the first h vectors in the canonical orthonormal basis of $\ell^2(\mathbb{Z}_+)$, then the Berger measure of $W_\alpha|_{\mathcal{M}_h}$ is $\frac{1}{\gamma_h} t^h d\xi(t)$.

It is easy to see that the Berger measure of U_+ is δ_1, and for $0 < a < 1$, the Berger measure of $S_a :=$ shift $(a, 1, 1, \ldots)$ is $(1-a^2)\delta_0 + a^2\delta_1$. Moreover, the Berger measure of the Bergman shift $B_+ :=$ shift $\left(\sqrt{\frac{1}{2}}, \sqrt{\frac{2}{3}}, \sqrt{\frac{3}{4}}, \cdots\right)$ is Lebesgue measure on the interval $[0, 1]$.

2.2 2-variable weighted shifts

Consider now double-indexed positive bounded sequences $\alpha_{\mathbf{k}}, \beta_{\mathbf{k}} \in \ell^\infty(\mathbb{Z}_+^2)$, $\mathbf{k} \equiv (k_1, k_2) \in \mathbb{Z}_+^2 := \mathbb{Z}_+ \times \mathbb{Z}_+$ and let $\ell^2(\mathbb{Z}_+^2)$ be the Hilbert space of square-summable complex sequences indexed by $\mathbb{Z}_+^2$. (Recall that $\ell^2(\mathbb{Z}_+^2)$ is canonically isometrically isomorphic to $\ell^2(\mathbb{Z}_+) \bigotimes \ell^2(\mathbb{Z}_+)$.) We define the 2-variable weighted shift $\mathbf{T} \equiv (T_1, T_2)$ by

$$T_1 e_{\mathbf{k}} := \alpha_{\mathbf{k}} e_{\mathbf{k}+\varepsilon_1}$$

$$T_2 e_{\mathbf{k}} := \beta_{\mathbf{k}} e_{\mathbf{k}+\varepsilon_2},$$

where $\varepsilon_1 := (1, 0)$ and $\varepsilon_2 := (0, 1)$. Clearly,

$$T_1 T_2 = T_2 T_1 \Longleftrightarrow \beta_{\mathbf{k}+\varepsilon_1} \alpha_{\mathbf{k}} = \alpha_{\mathbf{k}+\varepsilon_2} \beta_{\mathbf{k}} \ \ (\text{all } \mathbf{k}). \tag{2.1}$$

Associated to a 2-variable weighted shift is a 2-dimensional weight diagram (see Figure 2.1). In an entirely similar way one can define multivariable weighted shifts.

Trivially, a pair of unilateral weighted shifts W_a and W_β gives rise to a 2-variable weighted shift $\mathbf{T} \equiv (T_1, T_2)$, if we let $\alpha_{(k_1,k_2)} := \alpha_{k_1}$ and $\beta_{(k_1,k_2)} := \beta_{k_2}$ (all $k_1, k_2 \in \mathbb{Z}_+^2$). In this case, $\mathbf{T}$ is subnormal (resp. hyponormal) if and only if so are T_1 and T_2; in fact, under the canonical identification of $\ell^2(\mathbb{Z}_+^2)$ and $\ell^2(\mathbb{Z}_+) \bigotimes \ell^2(\mathbb{Z}_+)$, $T_1 \cong I \bigotimes W_a$ and $T_2 \cong W_\beta \bigotimes I$, and $\mathbf{T}$ is also doubly commuting. For this reason, we do not focus attention on shifts of this type, and use them only when the above mentioned triviality is desirable or needed.

We also recall the notion of moment of order $\mathbf{k}$ for a pair (α, β) satisfying (2.1). Given $\mathbf{k} \in \mathbb{Z}_+^2$, the moment of (α, β) of order $\mathbf{k}$ is

$$\gamma_{\mathbf{k}} \equiv \gamma_{\mathbf{k}}(\alpha, \beta) := \left\{ \begin{array}{cc} 1 & \text{if } \mathbf{k} = 0 \\ \alpha_{(0,0)}^2 \cdot \ldots \cdot \alpha_{(k_1-1,0)}^2 \cdot \beta_{(k_1,0)}^2 \cdot \ldots \cdot \beta_{(k_1,k_2-1)}^2 & \text{if } \mathbf{k} \in \mathbb{Z}_+^2, \mathbf{k} \neq 0 \end{array} \right\}. \tag{2.2}$$

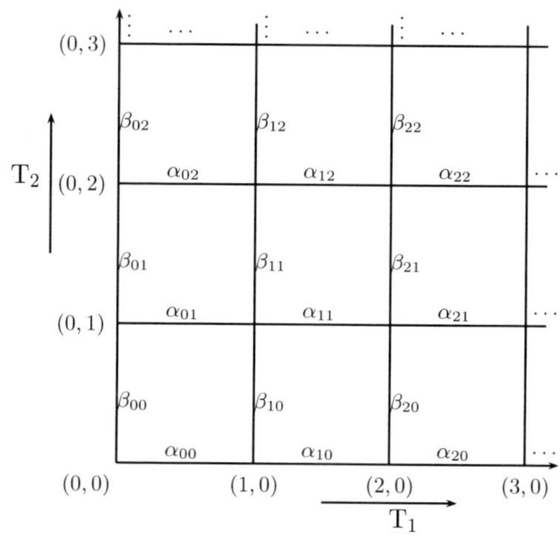

Figure 2.1
Weight diagram of a generic 2-variable weighted shift.

We remark that, due to (2.1), $\gamma_{\mathbf{k}}$ can be computed using any nondecreasing path from $(0,0)$ to (k_1,k_2).

The natural analog of Berger's Theorem is true, as we now state.

Theorem 2.1 *(Berger's Theorem in two variables) ([67]) A 2-variable weighted shift* $\mathbf{T} \equiv (T_1,T_2)$ *admits a commuting normal extension if and only if there is a probability measure* μ *defined on the 2-dimensional rectangle* $R = [0,a_1] \times [0,a_2]$ *(*$a_i := \|T_i\|^2$*) such that* $\gamma_{\mathbf{k}} = \int_R \mathbf{t}^{\mathbf{k}} d\mu(\mathbf{t}) := \int_R t_1^{k_1} t_2^{k_2}\, d\mu(t_1,t_2)$ *(all* $\mathbf{k} \in \mathbb{Z}_+^2$*).*

2.3 The lifting problem for commuting subnormals

As we observed in the previous section, each component T_i of a subnormal 2-variable weighted shift $\mathbf{T} \equiv (T_1,T_2)$ must be subnormal. On the other hand, if we only know that each of T_1, T_2 is subnormal, and that they commute, can we assert that the pair (T_1,T_2) is subnormal?

Problem 3.1 *(Lifting Problem for Commuting Subnormals (LPCS)) Find necessary and sufficient conditions on* T_1 *and* T_2 *to guarantee the subnormality of* $\mathbf{T} \equiv (T_1,T_2)$.

Example 3.2 *Given two* 1*-variable unilateral weighted shifts* W_α *and* W_β*, with Berger measures* σ *and* τ*, resp., the* 2*-variable weighted shift* $(I \bigotimes W_\alpha, W_\beta \bigotimes I)$ *is always subnormal, with Berger measure* $\mu := \sigma \times \tau$*.*

It is well known that the above mentioned necessary conditions (commutativity together with subnormality of each component) do not suffice (cf.[19]). It is also known that if a pair $\mathbf{T} \equiv (T_1, T_2)$ of commuting subnormals is such that $p(T_1, T_2)$ is subnormal for every bivariate polynomial p of degree at most 5, then $\mathbf{T}$ is subnormal [61]. While this result holds great promise as a tool to verify subnormality, the calculations needed involve fifth powers of the components, and at present they are intractable, even in the case of 2-variable weighted shifts. The result has some points of contact with the result proved by M. Putinar and the author in the early 1990's, on the existence of polynomially hyponormal weighted shifts which are not subnormal ([21], [44], [45]).

In what follows, we will first consider (joint) hyponormality as an additional condition. We will also reformulate LPCS as a reconstruction-of-measure problem. Concretely, we can visualize T_1 as

$$T_1 \cong \bigoplus_{j=0}^{\infty} W_{\alpha^{(j)}},$$

with $\alpha_i^{(j)} := \alpha_{(i,j)}$; then $W_{\alpha^{(j)}}$ has associated Berger measure $d\xi_j(t_1) := \frac{1}{\gamma_{(0,j)}} \int_{[0,a_2]} t_2^j d\Phi_{t_1}(t_2)$, where $d\mu(t_1,t_2) \equiv d\Phi_{t_1}(t_2) d\sigma(t_1)$ is the canonical disintegration of μ by horizontal slices (cf. Subsection 2.7 below). Similarly,

$$T_2 \cong \bigoplus_{i=0}^{\infty} W_{\beta^{(i)}},$$

with $\beta_j^{(i)} := \beta_{(i,j)}$, so that $W_{\beta^{(i)}}$ has associated Berger measure $d\eta_i(t_2) := \frac{1}{\gamma_{(0,i)}} \int_{[0,a_1]} t_1^i d\Psi_{t_2}(t_1)$, where $d\mu(t_1,t_2) \equiv d\Psi_{t_2}(t_1) d\tau(t_2)$ is the canonical disintegration of μ by vertical slices.

Problem 3.3 *(Reconstruction of the Measure Problem (ROMP)) Consider the marginal, one-variable measures* $\{\xi_j\}_{j=0}^{\infty}$ *and* $\{\eta_i\}_{i=0}^{\infty}$ *associated with* T_1 *and* T_2*, respectively. Under what conditions does there exist a* 2*-variable measure* μ *correctly interpolating all the powers* $t_1^{k_1} t_2^{k_2}$ $(k_1, k_2 \geq 0)$*?*

We will discuss Problem 3.3 in some detail in Section 2.8.

2.4 Hyponormality, 2-hyponormality and subnormality for 2-variable weighted shifts

To detect hyponormality for 2-variable weighted shifts, there is a simple criterion involving a base point $\mathbf{k}$ in $\mathbb{Z}_+^2$ and its five neighboring points in $\mathbf{k}+\mathbb{Z}_+^2$ at path distance at most 2 (cf. Figure 2.2).

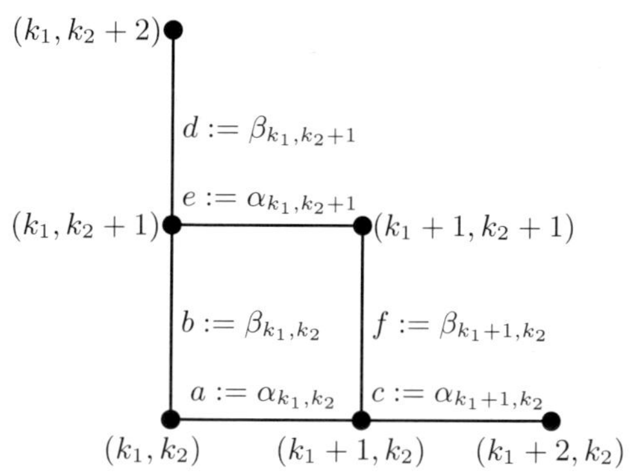

Figure 2.2
Weight diagram used in the Six-point Test.

Theorem 4.1 *([19]) (Six-point Test) Let* $\mathbf{T} \equiv (T_1, T_2)$ *be a 2-variable weighted shift, with weight sequences* α *and* β*. Then*

$$[\mathbf{T}^*, \mathbf{T}] \geq 0 \Leftrightarrow (([T_j^*, T_i]e_{\mathbf{k}+\varepsilon_j}, e_{\mathbf{k}+\varepsilon_i}))_{i,j=1}^2 \geq 0 \; (\mathit{all}\ \mathbf{k} \in \mathbf{Z}_+^2)$$

$$\Leftrightarrow \begin{pmatrix} \alpha_{\mathbf{k}+\varepsilon_1}^2 - \alpha_{\mathbf{k}}^2 & \alpha_{\mathbf{k}+\varepsilon_2}\beta_{\mathbf{k}+\varepsilon_1} - \alpha_{\mathbf{k}}\beta_{\mathbf{k}} \\ \alpha_{\mathbf{k}+\varepsilon_2}\beta_{\mathbf{k}+\varepsilon_1} - \alpha_{\mathbf{k}}\beta_{\mathbf{k}} & \beta_{\mathbf{k}+\varepsilon_2}^2 - \beta_{\mathbf{k}}^2 \end{pmatrix} \geq 0 \; (\mathit{all}\ \mathbf{k} \in \mathbf{Z}_+^2).$$

There is a similar test for 2-hyponormality [33], based on the 15 points whose path distance to $\mathbf{k}$ is at most 4 (see Figure 2.3).

Remark 4.2 *In terms of* a, b, c, d, e, f*, Theorem 4.1 implies that*

$$(ef - ab)^2 \leq (c^2 - a^2)(d^2 - b^2).$$

It follows that when $a = c$ *one must have* $ef = ab$*. Since we also know that* $be = af$*, we immediately get* $e = a$*, and a fortiori we have* $e = c = a$ *and* $f = b$*. That is, in the presence of joint hyponormality, the equality of two horizontally consecutive*

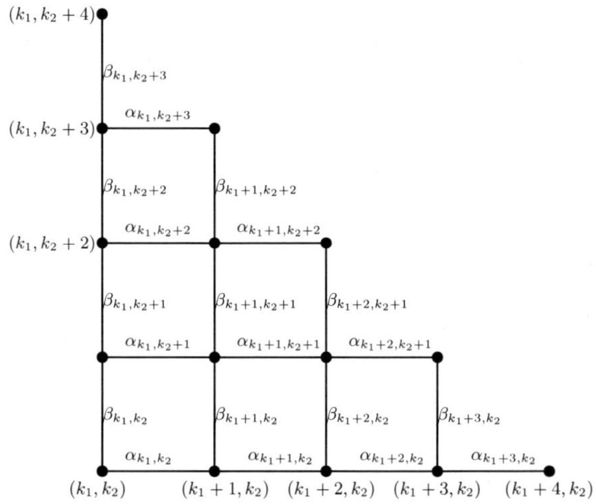

Figure 2.3
Weight diagram used in the 15-point Test.

α-weights ($\alpha_{k_1,k_2} = \alpha_{k_1+1,k_2}$) readily implies the equality of the associated two vertically consecutive α-weights ($\alpha_{k_1,k_2} = \alpha_{k_1,k_2+1}$). In addition, the associated consecutively horizontal β-weights are also equal ($\beta_{k_1,k_2} = \beta_{k_1+1,k_2}$). We call this phenomenon L-shaped propagation, and it is an important measure of weight rigidity that joint hyponormality does impose.

2.5 Existence of nonsubnormal hyponormal 2-variable weighted shifts

Unlike the single variable case, in which there is a clear separation between hyponormality and subnormality (cf. [26], [23], [40]), much less is known about the multivariable case. We will first briefly describe the three conceptually different families of examples originally obtained by J. Yoon and the author in [48]. These examples settled the following conjecture in the negative.

Conjecture 5.1 *([42]) Let $\mathbf{T} \equiv (T_1, T_2)$ be a pair of commuting subnormal operators on $\mathscr{H}$. Then $\mathbf{T}$ is subnormal if and only if $\mathbf{T}$ is hyponormal.*

We mention in passing that M. Dritschel and S. McCullough, working independently, were able to produce a separate example ([56]). It turns out that the Dritschel/McCullough example is a special case of a general construction that produces nonsubnormal hyponormal pairs with $T_1 \cong T_2$ (Subsection 2.5.3). These examples require an improved version of a result proved in [20, Proposition 8].

Proposition 5.2 *(Subnormal backward extension of a 1-variable weighted shift) (cf [48]) Let W_α be a weighted shift whose restriction to $\mathcal{M} := \vee\{e_1, e_2, \cdots\}$ is subnormal, with associated Berger measure $\mu_{\mathcal{M}}$. Then W_α is subnormal (with associated measure μ) if and only if*
(i) $\frac{1}{t} \in L^1(\mu_{\mathcal{M}})$
(ii) $\alpha_0^2 \leq (\|\frac{1}{t}\|_{L^1(\mu_{\mathcal{M}})})^{-1}$
In this case, $d\mu(t) = \frac{\alpha_0^2}{t} d\mu_{\mathcal{M}}(t) + (1 - \alpha_0^2 \|\frac{1}{t}\|_{L^1(\mu_{\mathcal{M}})}) d\delta_0(t)$, where δ_0 denotes Dirac measure at 0. In particular, T is never subnormal when $\mu_{\mathcal{M}}(\{0\}) > 0$.

Corollary 5.3 *Let W_α be a subnormal unilateral weighted shift, and let $\mathcal{M}_n$ be the subspace of $\ell^2(\mathbb{Z}_+)$ generated by the orthonormal basis vectors $e_n, e_{n+1}, \cdots$. Then α_k is completely determined by the restriction of W_α to $\mathcal{M}_{k+1}$, for each $k \geq 1$. As a consequence, if two subnormal unilateral weighted shifts W_α and $W_{\widetilde{\alpha}}$ have identical restrictions to $\mathcal{M}_{k+1}$ for some $k \geq 1$, then they also have identical restrictions to $\mathcal{M}_1$.*

Proof From Proposition 5.2, we know that for $k \geq 1$, $\alpha_k^2 = (\|\frac{1}{t}\|_{L^1(\mu_{\mathcal{M}_{\|+\infty}})})^{-1}$.

The following result is a very special case of the Reconstruction-of-the-Measure Problem. First, we need to define the 2-variable analogs of the subspace $\mathcal{M}$ in Proposition 5.2. For $\mathbf{k} \equiv (k_1, k_2) \in \mathbb{Z}_+^2$ let $\mathcal{M}_{\mathbf{k}}$ denote the subspace of $\ell^2(\mathbb{Z}_+^2)$ generated by the orthonormal basis vectors e_{k_1+m, k_2+n} $(m, n \geq 0)$. When $\mathbf{k} = \varepsilon_2$ (resp. $\mathbf{k} = \varepsilon_1$, we denote $\mathcal{M}_{\mathbf{k}}$ simply by $\mathcal{M}$ (resp. $\mathcal{N}$).

Theorem 5.4 *(Subnormal backward extension of a 2-variable weighted shift [48]) Consider the 2-variable weighted shift whose weight diagram is given in Figure 2.1. Assume that $\mathbf{T}|_{\mathcal{M}}$ is subnormal, with associated measure $\mu_{\mathcal{M}}$, and that $H_0 := \mathrm{shift}(\alpha_{00}, \alpha_{10}, \cdots)$ is subnormal with associated measure ν. Then $\mathbf{T}$ is subnormal if and only if*
(i) $\frac{1}{t} \in L^1(\mu_{\mathcal{M}})$;
(ii) $\beta_{00}^2 \leq (\|\frac{1}{t}\|_{L^1(\mu_{\mathcal{M}})})^{-1}$;
(iii) $\beta_{00}^2 \|\frac{1}{t}\|_{L^1(\mu_{\mathcal{M}})} (\mu_{\mathcal{M}})_{ext}^X \leq \nu$.

Moreover, if $\beta_{00}^2 \|\frac{1}{t}\|_{L^1(\mu_{\mathcal{M}})} = 1$ then $(\mu_{\mathcal{M}})_{ext}^X = \nu$. In the case when $\mathbf{T}$ is subnormal, the Berger measure μ of $\mathbf{T}$ is given by

$$\begin{aligned} d\mu(s,t) &= \beta_{00}^2 \left\| \frac{1}{t} \right\|_{L^1(\mu_{\mathcal{M}})} d(\mu_{\mathcal{M}})_{ext}(s,t) \\ &+ (d\nu(s) - \beta_{00}^2 \left\| \frac{1}{t} \right\|_{L^1(\mu_{\mathcal{M}})} d(\mu_{\mathcal{M}})_{ext}^X(s)) d\delta_0(t). \end{aligned}$$

On occasion, we will write shift$(\alpha_0, \alpha_1, ...)$ to denote the weighted shift with weight sequence $\{\alpha_k\}_{k=0}^{\infty}$. We also denote by $U_+ :=$ shift$(1,1,...)$ the (unweighted) unilateral shift, and for $0 < a < 1$ we let $S_a :=$ shift$\{a,1,1,...)$. Observe that the Berger measures of U_+ and S_a are δ_1 and $(1-a^2)\delta_0 + a^2\delta_1$, respectively, where δ_p denotes the point-mass probability measure with support the singleton $\{p\}$. Finally, we let B_+ denote the Bergman shift, whose Berger measure is Lebesgue measure on the interval $[0,1]$; the weights of B_+ are given by the formula $\alpha_n := \sqrt{\frac{n+1}{n+2}}$ $(n \geq 0)$.

An important class of subnormal unilateral weighted shifts is obtained by considering measures μ with exactly two atoms t_0 and t_1. These shifts arise naturally in the Subnormal Completion Problem [26] and in the theory of truncated moment problems (cf. [24], [27]). For $t_0, t_1 \in \mathbb{R}_+$, $t_0 < t_1$, and for $\rho_0, \rho_1 > 0$ with $\rho_0 + \rho_1 = 1$, consider the 2-atomic measure $\mu := \rho_0\delta_{t_0} + \rho_1\delta_{t_1}$, where δ_p denotes the point-mass probability measure with support the singleton $\{p\}$. The moments of μ, $\gamma_k = \rho_0 t_0^k + \rho_1 t_1^k$ $(k \geq 0)$ satisfy a 2-step recursive relation $\gamma_{n+2} = \varphi_0\gamma_n + \varphi_1\gamma_{n+1}$ $(n \geq 0)$, where $\varphi_0, \varphi_1 \in \mathbb{R}$; at the weight level, this can be written as $\alpha_{n+1}^2 = \frac{\varphi_0}{\alpha_n^2} + \varphi_1$ $(n \geq 0)$. More generally, any finitely atomic Berger measure corresponds to a recursively generated weighted shift (i.e., one whose moments satisfy an r-step recursive relation); in fact, $r = \text{card supp}\, \mu$ (see Theorem 5.5). In the special case of $r = 2$, the theory of recursively generated weighted shifts makes contact with the work of J. Stampfli in [82], in which he proved that given three positive numbers $\alpha_0 < \alpha_1 < \alpha_2$, it is always possible to find a subnormal weighted shift, denoted $W_{(\alpha_0,\alpha_1,\alpha_2)^\wedge}$, whose first three weights are α_0, α_1 and α_2. In this case, the coefficients of recursion (cf. [25, Example 3.12], [26, Section 3], [23, Section 1, p. 81]) are given by

$$\varphi_0 = -\frac{\alpha_0^2\alpha_1^2(\alpha_2^2 - \alpha_1^2)}{\alpha_1^2 - \alpha_0^2} \text{ and } \varphi_1 = \frac{\alpha_1^2(\alpha_2^2 - \alpha_0^2)}{\alpha_1^2 - \alpha_0^2}, \tag{2.3}$$

the atoms t_0 and t_1 are the roots of the equation

$$t^2 - (\varphi_0 + \varphi_1 t) = 0, \tag{2.4}$$

and the densities ρ_0 and ρ_1 uniquely solve the 2×2 system of equations

$$\begin{cases} \rho_0 + \rho_1 & = & 1 \\ \rho_0 t_0 + \rho_1 t_1 & = & \alpha_0^2 \end{cases}. \tag{2.5}$$

A more general result is actually true.

Theorem 5.5 *[27] Let W_α be a subnormal unilateral weighted shifts, with Berger measure μ. Then μ is finitely atomic if and only if W_α is recursively generated; i.e., there exist a positive integer k and real numbers $\varphi_0, \cdots, \varphi_{k-1}$ such that*

$$\gamma_{n+k} = \varphi_0\gamma_n + \cdots + \varphi_{k-1}\gamma_{n+k-1} \ (\textit{all } n \geq 0).$$

The property of being recursively generated creates some rigidity on the weights, as follows.

Lemma 5.6 *(cf. [26, Theorem 3.10]) For $0 < \alpha_0 < \alpha_1 < \alpha_2$, let $W_{(\alpha_0,\alpha_1,\alpha_2)^\wedge}$ be the recursively generated weighted shift described by (2.3), (2.4) and (2.5). Let W_η be the restriction of $W_{(\alpha_0,\alpha_1,\alpha_2)^\wedge}$ to $\mathcal{M}$. Then*

$$\alpha_0 = \sup\{x > 0 : \text{shift}(x, \eta_0, \eta_1, \cdots) \equiv \text{shift}(x, \alpha_1, \alpha_2, \cdots) \textit{ is subnormal}\}.$$

We are now ready to briefly discuss the three original classes of examples disproving Conjecture 5.1; what follows is patterned after [48].

2.5.1 The first family of examples

Consider the 2-variable weighted shift whose weight diagram is given in Figure 2.4, and assume that $\max x, y < 1$ and $a < x$.

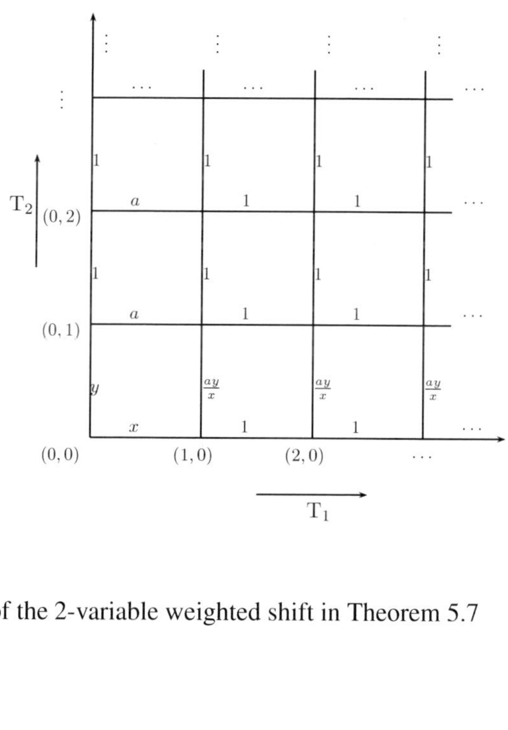

Figure 2.4
Weight diagram of the 2-variable weighted shift in Theorem 5.7

Theorem 5.7 *The 2-variable weighted shift* **T** *given by Figure 2.4 is hyponormal and not subnormal if and only if $x > a$ and $\sqrt{\frac{1-x^2}{1-a^2}} < y \le x\sqrt{\frac{(1-x^2)}{x^2+a^4-2a^2x^2}}$ (see Figure 2.5).*

Subnormality is verified using Theorem 5.4; hyponormality is checked using the Six-point Test. In Figure 2.5 we give a visual representation of Theorem 5.7 in terms of the parameters a, x and y.

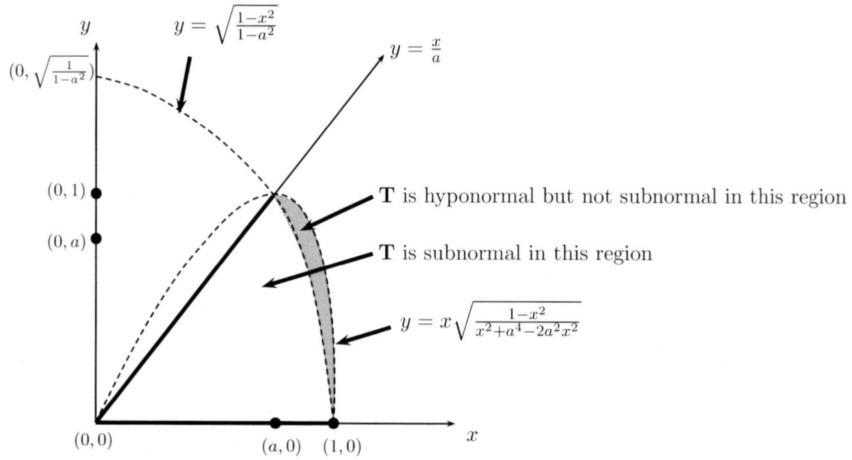

Figure 2.5
Regions of hyponormality and subnormality for the shift in Theorem 5.7

2.5.2 The second family of examples

Let $0 < a,b < 1$ and let $\{\xi_k\}_{k=0}^{\infty}$ and $\{\eta_k\}_{k=0}^{\infty}$ be two strictly increasing weight sequences. Denote by W_ξ and W_η the associated unilateral weighted shifts. Consider the 2-variable weighted shift $\mathbf{T} \equiv (T_1, T_2)$ on $\ell^2(\mathbb{Z}_+^2)$ given by the double-indexed weight sequences

$$\alpha(\mathbf{k}) := \begin{cases} \xi_{k_1} & \text{if } k_1 \geq 1 \text{ or } k_2 \geq 1 \\ a & \text{if } k_1 = 0 \text{ and } k_2 = 0 \end{cases} \tag{2.6}$$

and

$$\beta(\mathbf{k}) := \begin{cases} \eta_{k_2} & \text{if } k_1 \geq 1 \text{ or } k_2 \geq 1 \\ b & \text{if } k_1 = 0 \text{ and } k_2 = 0, \end{cases} \tag{2.7}$$

where W_ξ and W_η are two single-variable subnormal weighted shifts with Berger measures ν and ω, resp., and

$$a\eta_0 = b\xi_0 \tag{2.8}$$

(to guarantee the commutativity of T_1 and T_2, cf. (2.1)). $\mathbf{T}$ can be represented by the following weight diagram (Figure 2.6). It is then clear that T_1 and T_2 are subnormal provided $a \leq \xi_{ext}(\nu_{\mathscr{M}})$ and $b \leq \eta_{ext}(\omega_{\mathscr{M}})$, where, as usual, $\mathscr{M} := \vee\{e_1, e_2, \cdots\}$; in particular, $a < \xi_1$ and $b < \eta_1$.

The 2-variable weighted shift $\mathbf{T} \equiv (T_1, T_2)$ in Figure 2.6 is as close to a doubly commuting subnormal pair as one could wish, since $\mathbf{T}$ is a rank two perturbation of the tensor product $(I \otimes W_\xi, W_\eta \otimes I)$ of two subnormal unilateral weighted shifts.

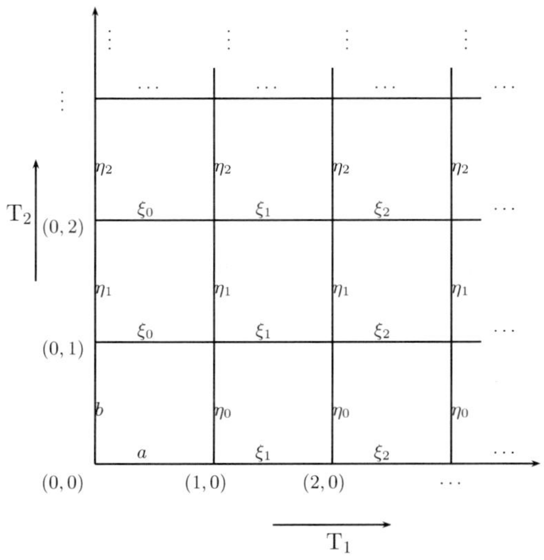

Figure 2.6
Weight diagram needed in the statement of Theorem 5.8

Theorem 5.8 *Let* $\mathbf{T} \equiv (T_1, T_2)$ *be the 2-variable weighted shift defined by (2.6) and (2.7), let*

$$\begin{cases} h := \xi_0 \sqrt{\dfrac{\xi_1^2\eta_1^2-\xi_0^2\eta_0^2}{\xi_0^2\eta_1^2+\xi_1^2\eta_0^2-2\xi_0^2\eta_0^2}}, \\[2ex] s := \sqrt{\dfrac{\xi_0^2\xi_1^2\eta_1^2}{\xi_1^2\eta_0^2+\xi_0^2\eta_1^2-\xi_0^2\eta_0^2}}, \\[2ex] s_2 := \dfrac{\xi_0}{\eta_0}\eta_e, \text{ where } \eta_e \equiv \eta_{ext}(\omega_{\mathcal{M}}), \\[2ex] u := \dfrac{\xi_0^2\eta_e^2\eta_1^2}{\xi_1^2(\eta_1^2-\eta_e^2)+\xi_0^2\eta_e^2}, \text{ and} \\[2ex] v := \dfrac{\xi_1^2(\eta_1^2-\eta_e^2)+2\xi_0^2\eta_e^2-\sqrt{(\eta_1^2-\eta_e^2)(\xi_1^4(\eta_1^2-\eta_e^2)+4\xi_0^2\eta_e^2(\xi_1^2-\xi_0^2))}}{2\xi_0^2}. \end{cases}$$

Assume further that $s_2 = \xi_{ext}(\nu_{\mathcal{M}})$ *and* $\eta_0 \leq \min\{u, v\}$, *and choose* a *such that* $s < a \leq h$. *Then*

(i) $T_1T_2 = T_2T_1$;

(ii) T_1 *is subnormal;*

(iii) T_2 *is subnormal;*

(iv) $\mathbf{T}$ *is hyponormal; and*

(v) $\mathbf{T}$ *is not subnormal.*

As proved in [48], for suitable choices of ξ_0, ξ_1, η_0 and η_1, it is possible to obtain $s < h$, so that any a such that $s < a \leq h$ will yield a 2-variable weighted shift which is hyponormal and not subnormal.

2.5.3 The third family of examples

Let us consider the following 2-variable weighted shift (see Figure 2.7), where

$$\begin{cases} \text{(i) } 0 < \xi_1 < \xi_2 < \cdots < \xi_n \nearrow 1; \\ \text{(ii) } W_\xi := \text{shift}(\xi_1, \xi_2, \cdots) \text{ is subnormal with Berger measure } \nu; \\ \text{(iii) } \frac{1}{s^2} \in L^1(\nu) \text{ (this implies that } \frac{1}{s} \in L^1(\nu) \text{, by Jensen's inequality);} \\ \text{(iv) } \xi_e := (\int \frac{1}{s} d\nu(s))^{-1/2}; \\ \text{(v) } a \leq \frac{1}{\xi_e} (\int \frac{1}{s^2} d\nu(s))^{-1/2}; \\ \text{(vi) } b \leq \xi_e^2 \text{ (this implies the condition } b < \xi_e \text{); and} \\ \text{(vii) } a^2 \leq \frac{b^2 + \xi_e^2}{2}. \end{cases} \tag{2.9}$$

(Recall that ξ_e is the maximum possible value for ξ_0 in Proposition 5.2; cf. Lemma 5.6.)

Observe that $T_1 \cong T_2$ and that $T_1 T_2 = T_2 T_1$. Note that the choice of ξ_e immediately implies that shift$(\xi_e, \xi_1, \xi_2, \cdots)$ is subnormal, with Berger measure $d\nu_e(s) := \frac{\xi_e^2}{s} d\nu(s)$ (cf. Proposition 5.2). Another application of Proposition 5.2 shows that shift$(a, \xi_e, \xi_1, \cdots)$ is subnormal, by (2.9)(iii) and (2.9)(v). This implies that the restriction of T_1 to $\bigvee\{e_{(i,0)} : i \geq 0\}$ is subnormal. Moreover, the subnormality of T_1 on $\bigvee\{e_{(i,j)} : i \geq 0\}$ $(j > 0)$ requires that $b \leq \xi_e$, which holds by (2.9)(vi).

For a concrete numerical example, consider the probability measure $d\nu(s) := 3s^2 ds$ on the interval $[0, 1]$. The measure ν corresponds to a subnormal weighted shift with weights $\xi_1 = \sqrt{\frac{3}{4}}$, $\xi_2 = \sqrt{\frac{4}{5}}$, $\xi_3 = \sqrt{\frac{5}{6}}$, $\cdots$; indeed, in this case W_ξ is the restriction of the Bergman shift B_+ to the invariant subspace $\mathcal{M}_2$ obtained by removing the first two basis vectors in the canonical orthonormal basis of $\ell^2(\mathbb{Z}_+)$. Clearly $\frac{1}{s^2} \in L^1(\nu)$, and $\int \frac{1}{s^2} d\nu(s) = 3$; moreover, $\int \frac{1}{s} d\nu(s) = \frac{3}{2}$, so in this case $\xi_e = \sqrt{\frac{2}{3}}$. Choosing $a = \sqrt{\frac{1}{2}}$ and $b = \sqrt{\frac{1}{3}}$ we see that all conditions in (2.9) are satisfied.

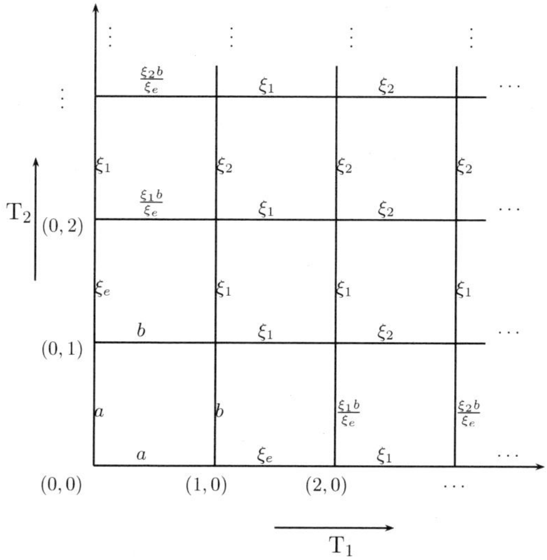

Figure 2.7
Weight diagram of the 2-variable weighted shift in Theorem 5.9

This is the example found independently by M. Dritschel and S. McCullough [56] which we mentioned in the paragraph immediately following Conjecture 5.1.

Theorem 5.9 *Let* $a > 0$ *be such that* $\sqrt{\frac{\xi_e^2}{2}} < a \leq \sqrt{\frac{\xi_e^2+\xi_e^4}{2}}$ *and* $a \leq \frac{1}{\xi_e}(\int \frac{1}{s^2} d\nu(s))^{-1/2}$, *and define* $b := \sqrt{2a^2 - \xi_e^2}$. *Then the 2-variable weighted shift* $\mathbf{T} \equiv (T_1, T_2)$ *satisfies (2.9)(i)-(vii), is hyponormal, and is not subnormal.*

2.5.4 An instance when hyponormality does suffice

Lemma 5.10 *Let* ν *be a probability measure on* $[0,1]$, *and let* $\gamma_n \equiv \gamma_n(\nu) := \int s^n d\nu(s)$ $(n \geq 0)$ *be the moments of* ν. *The sequence* $\{\gamma_n\}_{n=0}^{\infty}$ *is bounded below if and only if* ν *has an atom at* $\{1\}$.

We now consider the 2-variable weighted shift $\mathbf{T}$ given by Figure 2.8), where $W_\xi := \text{shift}(\xi_0, \xi_1, \cdots)$ is a subnormal contraction with associated measure ν, and $y \leq 1$.

It is clear that $T_1 T_2 = T_2 T_1$, and that T_1 is subnormal (being the orthogonal direct sum of W_ξ and copies of U_+). To ensure the subnormality of T_2, we must impose the condition $\frac{y}{\sqrt{\gamma_n}} \leq 1$ (all $n \geq 0$), i.e., $y^2 \leq \gamma_n \equiv \gamma_n(\nu)$ (all $n \geq 0$). Notice that this condition guarantees the boundedness of $\mathbf{T}$.

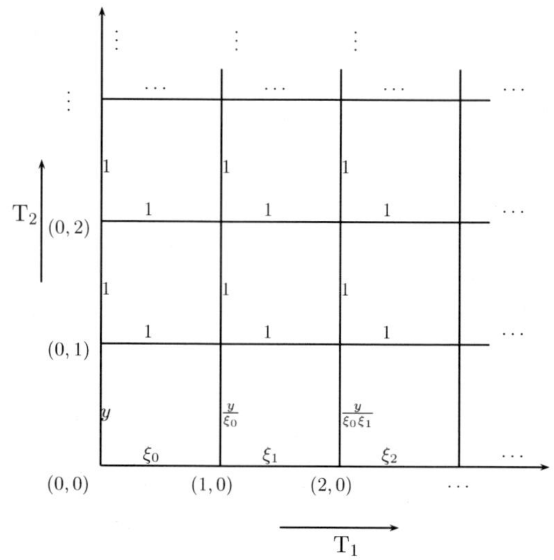

Figure 2.8
Weight diagram of the 2-variable weighted shift in Theorem 5.11

Theorem 5.11 *Let* **T** *be the* 2*-variable weighted shift given by Figure 2.8, and assume that* **T** *is hyponormal. Then* **T** *is subnormal.*

2.6 Propagation in the 2-variable hyponormal case

In this section, we show that if a commuting, (jointly) hyponormal pair $\mathbf{T} \equiv (T_1, T_2)$ with T_1 quadratically hyponormal satisfies $\alpha_{(k_1+1,k_2)} = \alpha_{(k_1,k_2)}$ for some $k_1, k_2 \geq 1$, then $(T_1, T_2(U_+^{k_2-1} \bigotimes I))$ is horizontally flat (see Definition 6.1 below); this is the content of Theorem 6.3. We also prove that Theorem 6.3 is optimal in the following sense: the propagation does not extend either to the left (0-th column) or down (below k_2-th level). We begin with

Definition 6.1 *A* 2*-variable weighted shift* $\mathbf{T} \equiv (T_1, T_2)$ *is horizontally flat (resp. vertically flat) if* $\alpha_{(k_1,k_2)} = \alpha_{(1,1)}$ *for all* $k_1, k_2 \geq 1$ *(resp.* $\beta_{(k_1,k_2)} = \beta_{(1,1)}$ *for all* $k_1, k_2 \geq 1$*). We say that* **T** *is flat if* **T** *is horizontally and vertically flat (cf. Figure 2.1), and we say that* **T** *is symmetrically flat if* **T** *is flat and* $\alpha_{11} = \beta_{11}$.

Remark 6.2 *In connection with propagation, one can simply ask: Can 2-hyponormality imply subnormality? Or more precisely, which 2-hyponormal 2-*

variable weighted shifts are subnormal? It is well known that a 2-hyponormal unilateral weighted shift with two equal weights must be flat, and therefore subnormal [20]. Moreover, a quadratically hyponormal unilateral weighted shift W_α such that $\alpha_n = \alpha_{n+1}$ for some $n \geq 1$ is also flat (cf. [20], [14]); however, if $\alpha_0 = \alpha_1$, flatness may fail [32]. By contrast, a 2-hyponormal 2-variable weighted shift which is both horizontally flat and vertically flat need not be subnormal. On the other hand, in [36] we identified a large class $\mathscr{S}$ of flat 2-variable weighted shifts for which 2-hyponormality is equivalent to subnormality. One measure of the size of $\mathscr{S}$ is given by the fact that within $\mathscr{S}$ there are hyponormal shifts which are not subnormal.

Theorem 6.3 *([50, Theorem 3.3]) Let $\mathbf{T} \equiv (T_1, T_2)$ be a commuting, hyponormal 2-variable weighted shift.*
(i) If T_1 is quadratically hyponormal and $\alpha_{(k_1,k_2)+\varepsilon_1} = \alpha_{(k_1,k_2)}$ for some $k_1, k_2 \geq 1$, then $(T_1, T_2(U_+^{k_2-1} \otimes I)$ is horizontally flat. (That is, the restriction of $\mathbf{T}$ to the subspace generated by orthonormal basis vectors with indices (k_1, k_2-1+n) $(n \geq 0)$ is horizontally flat.)
(ii) If, instead, T_2 is quadratically hyponormal and $\beta_{(k_1,k_2)+\varepsilon_2} = \beta_{(k_1,k_2)}$ for some $k_1, k_2 \geq 1$, then $(T_1(I \otimes U_+^{k_1-1}), T_2)$ is vertically flat.

While propagation works well in the upward direction, the same is not true of the downward direction, as the following result shows.

Theorem 6.4 *([50, Theorem 3.14]) For every $k_2 \geq 1$ and $0 < \alpha_0 < 1$ there exist*
(i) a family $\{B_+^{(\ell_i)}\}_{i=0}^{k_2-1}$ of Bergman-like weighted shifts, and
(ii) a subnormal weighted shift $W_\beta := \mathrm{shift}(\beta_0, \beta_1, \beta_2, \cdots)$ (with $\beta_n < \beta_{n+1}$ for all $n \geq 0$),
such that the commuting 2-variable weighted shift $\mathbf{T} \equiv (T_1, T_2)$ with a weight diagram whose first k_2 rows are $B_+^{(\ell_0)}, \cdots, B_+^{(\ell_{k_2-1})}$, whose remaining rows are S_{α_0}, and whose 0-th column is given by W_β, is (jointly) hyponormal.

The next result shows that (joint) hyponormality does provide some rigidity, in the spirit of Conjecture 5.1.

Theorem 6.5 *([50, Theorem 4.7]) Let $\mathbf{T} \equiv (T_1, T_2)$ be commuting and hyponormal.*
(i) If T_1 is quadratically hyponormal, if T_2 is subnormal, and if $\alpha_{(k_1,k_2)+\varepsilon_1} = \alpha_{(k_1,k_2)}$ for some $k_1, k_2 \geq 0$, then $\mathbf{T}$ is horizontally flat.
(ii) If, instead, T_1 is subnormal, T_2 is quadratically hyponormal, and if $\beta_{(k_1,k_2)+\varepsilon_2} = \beta_{(k_1,k_2)}$ for some $k_1, k_2 \geq 0$, then $\mathbf{T}$ is vertically flat.

Proof (Brief Sketch) (i) The quadratic hyponormality of T_1 implies that the horizontal shift H_{k_2} is flat ([20]; cf. Remark 6.2). The joint hyponormality of (T_1, T_2) leads to L-shaped propagation, as explained in Remark 4.2. As a result, we see that all horizontal shifts H_{k_2+n} are flat for $n \geq 1$. Moreover, for $m \geq 1$ the restrictions of the vertical shifts V_m have identical restrictions to the subspaces generated by the orthonormal basis vectors $e_{m,k_2}, e_{m,k_2+1}, \ldots$ By the subnormality of T_2 and Remark 4.2

we see that all vertical shifts V_m have identical restrictions to the subspaces generated by $e_{m,1}, e_{m,2,\ldots}$; this readily implies that $\mathbf{T}$ is horizontally flat.

Remark 6.6 *Hyponormality alone does not imply flatness. While it is true that in the presence of hyponormality the Six-point Test creates L-shaped propagation (i.e., $\alpha_{\mathbf{k}+\varepsilon_1} = \alpha_{\mathbf{k}} \Longrightarrow \alpha_{\mathbf{k}} = \alpha_{\mathbf{k}+\varepsilon_2}$ and $\beta_{\mathbf{k}} = \beta_{\mathbf{k}+\varepsilon_1}$) (as we mentioned at the end of Section 2.4), without horizontal propagation (as guaranteed by the quadratic hyponormality of T_1) this L-shaped-propagation does not result in vertical propagation, needed to eventually lead to flatness. The same phenomenon arises in one variable, where hyponormality is a very soft condition ($\alpha_k \leq \alpha_{k+1}$ for all $k \geq 0$), while 2-hyponormality is quite rigid. The research in [20] (extending the ideas in [82]) revealed that, for unilateral weighted shifts with two equal weights, 2-hyponormality and subnormality are identical notions. In two variables, however, the analogous result does not hold.*

2.7 A measure-theoretic necessary (but not sufficient!) condition for the existence of a lifting

LPCS for a 2-variable weighted shift $\mathbf{T}$ admits a description in terms of the disintegration of the anticipated Berger measure μ of $\mathbf{T}$. That is, the existence of μ implies that the Berger measures of horizontal slices H_{k_2} of $\mathbf{T}$ must be linearly ordered by absolute continuity. Since T_1 is assumed to be subnormal, all H_{k_2} are subnormal weighted shifts, with Berger measures that will be denoted by ξ_{k_2}; as customary, we also denote ξ_0 by σ. Similarly, the Berger measures of vertical slices V_{k_1} are denoted by η_{k_1}, with η_0 also denoted by τ.

We begin with a brief account of some basic results in the theory of disintegration of measures. In the sequel we always assume that X and Y are compact metric spaces; in case reference is made to weighted shifts, we assume that $X, Y \subseteq \mathbb{R}_+$.

Definition 7.1 *Given a measure μ on $X \times Y$, the marginal measure μ^X is given by $\mu^X := \mu \circ \pi_X^{-1}$, where $\pi_X : X \times Y \to X$ is the canonical projection onto X. Thus, $\mu^X(E) = \mu(E \times Y)$, for every $E \subseteq X$. Observe that if μ is a probability measure, then so is μ^X.*

Example 7.2 *Let $\mu := \xi \times \eta$ be a probability product measure on $X \times Y$. Then $\mu^X = \xi$.*

Lemma 7.3 *(cf. [48, Lemma 3.6]) Let μ be the Berger measure of a 2-variable weighted shift $\mathbf{T} \equiv (T_1, T_2)$, and let σ be the Berger measure of* $\text{shift}(\alpha_{00}, \alpha_{10}, \ldots)$. *Then $\sigma = \mu^X$. As a consequence, $\int f(s)\, d\mu(s,t) = \int f(s)\, d\mu^X(s)$ for all $f \in C(X)$. Similarly, $\tau = \mu^Y$.*

Corollary 7.4 *(cf. [48, Corollary 3.7]) Let μ be the Berger measure of a 2-variable* *weighted shift* $\mathbf{T} \equiv (T_1, T_2)$. *For* $j \geq 0$, *let* $d\mu_j(s,t) := \frac{1}{\gamma_{0j}} t^j d\mu(s,t)$. *Then the Berger measure of* $\text{shift}(\alpha_{0j}, \alpha_{1j}, ...)$ *is* $\xi_j \equiv \mu_j^X$.

Recall that given two positive regular Borel measures κ and ω, κ is said to be absolutely continuous with respect to ω (in symbols, $\kappa \ll \omega$) if for every Borel set E, $\omega(E) = 0 \Rightarrow \kappa(E) = 0$. It follows at once that $\kappa \ll \omega \Rightarrow \text{supp}\kappa \subseteq \text{supp}\omega$.

Lemma 7.5 *Let* κ *and* ω *be two measures on* $X \times Y$, *and assume that* $\kappa \ll \omega$. *Then* $\kappa^X \ll \omega^X$ *and* $\kappa^Y \ll \omega^Y$.

Let μ be the Berger measure of a 2-variable weighted shift $\mathbf{T} \equiv (T_1, T_2)$. Although Corollary 7.4 indicates how to obtain the Berger measure ξ_j of the horizontal j-th slice of T_1 in terms of μ, the description is not completely satisfactory, in that it may not be easy to find the measures μ_j $(j \geq 0)$. We will now employ disintegration of measure techniques to give a precise description of ξ_j. First, we need to review some basic concepts and general results about disintegration of measures; most of the discussion in the rest of this section is is taken from [15, VII.2, pp. 317-319].

Let X and Z be compact metric spaces and let μ be a positive regular Borel measure on Z. For $\varphi : Z \to X$ a Borel mapping, let ν be the Borel measure $\mu \circ \varphi^{-1}$ on X; that is,

$$\nu(A) := \mu(\varphi^{-1}(A)) \tag{2.10}$$

for all Borel sets A contained in X. Let $\pounds^1(\mu) := \{f : f \text{ is Borel function on } Z \text{ such that } \int |f|\, d\mu < \infty\}$, and let $L^1(\mu) := \{[f] : f \in \pounds^1(\mu)\}$, where $[f] := \{g \in \pounds^1(\mu) : \int |f - g|\, d\mu = 0\}$. The map

$$\psi \to \int_Z (\psi \circ \varphi) f d\mu \tag{2.11}$$

defines a bounded linear functional on $L^\infty(\nu)$. If attention is restricted to characteristic functions χ_A in $L^\infty(\nu)$, then the map

$$A \to \int_Z (\chi_A \circ \varphi) f d\mu = \int_{\varphi^{-1}(A)} f d\mu \tag{2.12}$$

is a countably additive measure defined on Borel sets in X, that is absolutely continuous with respect to ν. Hence there is a unique element $E(f)$ in $L^1(\nu)$ such that

$$\int_Z (\chi_A \circ \varphi) f d\mu = \int_X \chi_A E(f) d\nu \tag{2.13}$$

for all Borel subsets A of X. By an an approximation argument one can show that

$$\int_Z (\psi \circ \varphi) f d\mu = \int_X \psi E(f) d\nu \tag{2.14}$$

for all ψ in $L^\infty(\nu)$. This defines a linear map

$$E : \pounds^1(\mu) \to L^1(\nu) \tag{2.15}$$

called the *expectation operator*.

Notation 7.6 *The space of all Borel measures on Z will be denoted by* $M(Z)$.

Definition 7.7 *A disintegration of the measure* μ *with respect to* φ *is a function* $x \mapsto \Phi_x$ *from* X *to* $M(Z)$, *such that*
(i) for each x *in* X, Φ_x *is a probability measure;*
(ii) if $f \in \pounds^1(\mu)$, $E(f)(x) = \int_Z f d\Phi_x$ *a.e.* $[\nu]$.

We now list the main theorem on existence and uniqueness of disintegration of measures.

Theorem 7.8 *([15, Theorem VII.2.11]) Given a regular Borel measure* μ *on a compact metric space* Z, *and a Borel function* φ *from* Z *into a compact metric space* X, *there is a disintegration* $x \mapsto \Phi_x$ *of* μ *with respect to* φ. *If* $x \mapsto \Phi_x^{'}$ *is another disintegration of* μ *with respect to* φ, *then* $\Phi_x = \Phi_x^{'}$ *a.e.* $[\nu]$.

We are now ready to calculate explicitly the measures ξ_j $(j \geq 0)$. Fix $j \geq 0$ and observe that the moments of ξ_j are given by

$$\int_X s^i \, d\xi_j(s) = \alpha_{0j}^2 \cdot ... \alpha_{i-1,j}^2 = \frac{\gamma_{ij}}{\gamma_{0j}} = \frac{1}{\gamma_{0j}} \int \int_R s^i t^j \, d\mu(s,t) \; (\text{all } i \geq 0), \tag{2.16}$$

where $R := X \times Y \equiv [0,a_1] \times [0,a_2]$. Since μ is regular and Borel, we can use Theorem 7.8 to disintegrate μ with respect to $\varphi \equiv \pi_X$ and obtain

$$\mu(A) = \int_X \Phi_x(A) \, d\mu^X(x),$$

where as above $\mu^X = \mu \circ \pi_X^{-1}$.

Theorem 7.9 *([49, Theorem 3.1]) Let* μ *be the Berger measure of a subnormal* 2-*variable weighted shift, and for* $j \geq 0$ *let* ξ_j *be the Berger measure of the associated* j*-th horizontal* 1*-variable weighted shift* $W_{\alpha^{(j)}}$. *Then* $\xi_j = \mu_j^X$ *(cf. Definition 7.1), where* $d\mu_j(s,t) := \frac{1}{\gamma_{0j}} t^j d\mu(s,t)$; *more precisely,*

$$d\xi_j(s) = \{\frac{1}{\gamma_{0j}} \int_Y t^j \, d\Phi_s(t)\} \, d\mu^X(s), \tag{2.17}$$

where $d\mu(s,t) \equiv d\Phi_s(t) \, d\mu^X(s)$ *is the disintegration of* μ *by vertical slices. A similar result holds for the Berger measure* η_i *of the associated* i*-th vertical* 1*-variable weighted shifts* $W_{\beta^{(i)}}$ $(i \geq 0)$.

We are finally ready to establish a necessary condition for the existence of a lifting for two commuting subnormal weighted shifts, that is, the sequences of Berger measures of horizontal and vertical slices form non-increasing families relative to the order given by absolute continuity.

Theorem 7.10 *(cf. [49, Theorem 3.3], [55, Theorem 5.11]) Let μ, ξ_j and η_i be as in Theorem 7.9. For every $i, j \geq 0$ we have*

$$\xi_{j+1} \ll \xi_j \tag{2.18}$$

and

$$\eta_{i+1} \ll \eta_i. \tag{2.19}$$

More is true, as it follows from (2.17): for $i, j \geq 1$, one actually has

$$\xi_{j+1} \approx \xi_j$$

and

$$\eta_{i+1} \approx \eta_i;$$

that is, the respective measures are mutually absolutely continuous.

Remark 7.11 *(The Necessary Condition is Not Sufficient.) Consider the 2-variable weighted shift whose weight diagram is given by Figure 2.4. Assume that $y = x$ and that $x > a$. Observe that all horizontal weighted shifts have the 2-atomic Berger measure $(1-a^2)\delta_0 + a^2\delta_1$; thus, the necessary condition is clearly satisfied. However, for $\mathbf{T}$ to be subnormal the necessary and sufficient condition is $1 - 2x^2 + a^2x^2 \geq 0$, while hyponormality requires $1 - 2x^2 + a^2 \geq 0$. It is thus sufficient to find a pair (x, a) satisfying the second condition but not the first to conclude that the necessary condition in Theorem 7.10 is not sufficient.*

2.8 Reconstruction of the Berger measure for 2-variable weighted shifts whose core is of tensor form

In this section we focus on the case when the 2-variable weighted shift $\mathbf{T}$ has a core of tensor form (cf. Figure 2.9). That is, the restriction of $\mathbf{T}$ to the intersection $\mathcal{M} \bigcap \mathcal{N}$ is a tensor product of two 1-variable subnormal weighted shifts, with Berger measures ξ and η. Of course we also assume that T_1 and T_2 are subnormal, so that the horizontal shift H_0 has Berger measure σ and the vertical shift V_0 has Berger measure τ. Due to commutativity, the entire 2-variable weighted shift $\mathbf{T}$ is completely determined once we also know α_{01}.

Problem 8.1 *Given the measures ξ, η, σ and τ, and the positive real number α_{01}, characterize the subnormality of $\mathbf{T}$ in terms of these five parameters.*

(For the terminology from disintegration of measures, we refer the reader to Section 2.7.)

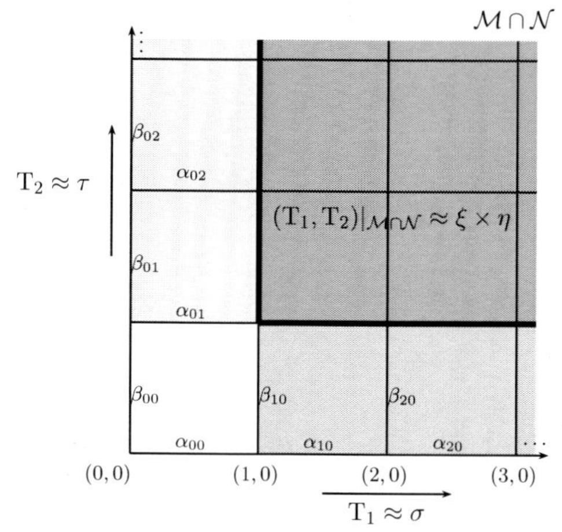

Figure 2.9
A 2-variable weighted shift with core of tensor form.

Motivated by Theorem 5.4, we let

$$\psi := \tau_1 - a^2 \left\| \frac{1}{s} \right\|_{L^1(\xi)} \eta \tag{2.20}$$

and

$$\varphi := \sigma - \beta_{0,0}^2 \left\| \frac{1}{t} \right\|_{L^1(\psi)} \delta_0 - a^2 \beta_{0,0}^2 \left\| \frac{1}{s} \right\|_{L^1(\xi)} \left\| \frac{1}{t} \right\|_{L^1(\eta)} \widetilde{\xi}, \tag{2.21}$$

where τ_1 is the Berger measure of the subnormal shift shift$(\beta_{0,1}, \beta_{0,2}, \cdots)$ and $d\widetilde{\xi}(s) := \frac{1}{s\left\|\frac{1}{s}\right\|_{L^1(\xi)}} d\xi(s)$. Trivially, ψ and φ are measures, but they may or may not be *positive* measures. The following result gives a complete solution to Problem 8.1.

Theorem 8.2 *([34, Theorem 2.3] Let* $\mathbf{T} \equiv (T_1, T_2)$ *be as above. Then* $\mathbf{T}$ *is subnormal if and only if* ψ *and* φ *are positive measures.*

The Proof of Theorem 8.2 involves careful applications of the ideas in Section 2.7 together with Theorem 5.4. As an application of Theorem 8.2 we have

Theorem 8.3 *([38, Theorem 8.3]) (The case when* $W_{(\alpha,\beta)}$ *has a core of tensor form; see Figure 2.9.) Assume that* $W_{(\alpha,\beta)}|_{\mathcal{M}}$ *and* $W_{(\alpha,\beta)}|_{\mathcal{N}}$ *are subnormal with Berger measures* $\mu_{\mathcal{M}}$ *and* $\mu_{\mathcal{N}}$*, respectively, and let* $\rho := \mu^X_{\mathcal{M}}$*; i.e.,* ρ *is the Berger measure of* shift$(\alpha_{01}, \alpha_{11}, \cdots)$*. Also assume that* $\mu_{\mathcal{M}\cap\mathcal{N}} = \xi \times \eta$ *for some* 1*-variable probability measures* ξ *and* η*. Then* $W_{(\alpha,\beta)}$ *is subnormal if and only if the following conditions hold:*

(i) $\frac{1}{t} \in L^1(\mu_{\mathscr{M}})$;
(ii) $\beta_{00}^2 \left\|\frac{1}{t}\right\|_{L^1(\mu_{\mathscr{M}})} \leq 1$;
(iii) $\left(\beta_{00}^2 \left\|\frac{1}{t}\right\|_{L^1(\tau_1)}\right) \rho = \left(\beta_{00}^2 \left\|\frac{1}{t}\right\|_{L^1(\mu_{\mathscr{M}})}\right) \rho \leq \sigma$.

2.9 The subnormal completion problem for 2-variable weighted shifts

In this section we discuss the Subnormal Completion Problem (SCP) for 2-variable weighted shifts. We use tools and techniques from the theory of truncated moment problems to give a general strategy to solve SCP in the foundational case of six prescribed initial weights; these weights give rise to the quadratic moments. For this case, the natural necessary conditions for the existence of a subnormal completion are also sufficient. To calculate explicitly the associated Berger measure, we compute the algebraic variety of the associated truncated moment problem; it turns out that this algebraic variety is precisely the support of the Berger measure of the subnormal completion.

Definition 9.1 *Given $m \geq 0$ and a finite family of positive numbers $\Omega_m \equiv \{(\alpha_{\mathbf{k}}, \beta_{\mathbf{k}})\}_{|\mathbf{k}| \leq m}$, we say that a 2-variable weighted shift $\mathbf{T} \equiv (T_1, T_2)$ with weight sequences $\alpha_{\mathbf{k}}^{\mathbf{T}}$ and $\beta_{\mathbf{k}}^{\mathbf{T}}$ is a subnormal completion of Ω_m if (i) $\mathbf{T}$ is subnormal, and (ii) $(\alpha_{\mathbf{k}}^{\mathbf{T}}, \beta_{\mathbf{k}}^{\mathbf{T}}) = (\alpha_{\mathbf{k}}, \beta_{\mathbf{k}})$ whenever $|\mathbf{k}| \leq m$.*

Remark 9.2 *Note that since a subnormal 2-variable weighted shift is necessarily commuting, Ω_m in Definition 9.1 must necessarily satisfy the commutativity condition in (2.1). When a family Ω_m of positive numbers has this property, we say that Ω_m is commutative.*

Definition 9.3 *Given $m \geq 0$ and a finite family of positive numbers $\Omega_m \equiv \{(\alpha_{\mathbf{k}}, \beta_{\mathbf{k}})\}_{|\mathbf{k}| \leq m}$, we say that $\hat{\Omega}_{m+1} \equiv \{(\hat{\alpha}_{\mathbf{k}}, \hat{\beta}_{\mathbf{k}})\}_{|\mathbf{k}| \leq m+1}$ is an extension of Ω_m if $(\hat{\alpha}_{\mathbf{k}}, \hat{\beta}_{\mathbf{k}}) = (\alpha_{\mathbf{k}}, \beta_{\mathbf{k}})$ whenever $|\mathbf{k}| \leq m$. The degree of Ω_m,* deg Ω_m, *is $m+1$. When $m = 1$, we say that Ω_1 is quadratic. For $m = 2\ell + 1$, the moment matrix of Ω_m is*

$$M(\ell) \equiv M(\Omega_m) \equiv M_{\mathbf{0}}(\Omega_m) := (\gamma_{(i,j)+(p,q)})_{\substack{0 \leq i+j \leq m \\ 0 \leq p+q \leq m}}.$$

Observe that if $\hat{\Omega}_{m+1}$ is commutative, then so is Ω_m. For m odd, $M(\hat{\Omega}_{m+2})$ is an extension of $M(\Omega_m)$.

Notation 9.4 *When $m = 1$, we shall let $a := \alpha_{00}^2$, $b := \beta_{00}^2$, $c := \alpha_{10}^2$, $d := \beta_{01}^2$, $e := \alpha_{01}^2$ and $f := \beta_{10}^2$. To be consistent with the commutativity of a 2-variable weighted* *shifts whose weight sequences satisfy (2.1), we shall always assume $af = be$. The moments of Ω_1 are*

$$\begin{cases} \gamma_{00} := 1 & & \\ \gamma_{01} := a & \gamma_{10} := b & \\ \gamma_{02} := ac & \gamma_{11} := be & \gamma_{20} := bd \end{cases},$$

and the associated moment matrix is

$$M(\Omega_1) := \begin{pmatrix} 1 & a & b \\ a & ac & be \\ b & be & bd \end{pmatrix}. \tag{2.22}$$

In this case, solving the SCP consists of finding a probability measure μ supported on $\mathbb{R}_+^2$ such that $\int_{\mathbb{R}_+^2} y^i x^j \, d\mu(x,y) = \gamma_{ij}$ $(i, j \geq 0,\ i+j \leq 2)$.

Notation 9.5 *We also need the notion of localizing matrix. Rather than giving the general definition from [29], we will content ourselves with understanding the two key localizing matrices we need here; that is, $M_x(2)$ and $M_y(2)$. Given a 6×6 moment matrix $M(2)$, with rows and columns labeled 1, X, Y, X^2, XY and Y^2, the localizing matrix $M_x(2)$ is a 3×3 matrix obtained from $M(2)$ by crossing out the three columns whose monomial labels do not include X, and by crossing out the three bottom rows. The localizing matrix $M_y(2)$ is defined similarly.*

2.9.1 Localizing matrices as flat extension builders

In what follows, and for simplicity, we will specialize to the case $m = 1$ in two variables, and show that the condition $M(\Omega_1) \geq 0$ is sufficient for the existence of a subnormal completion.

Theorem 9.6 *([37, Theorem 5.1]) Let Ω_1 be a quadratic, commutative, initial set of positive weights, and assume $M(\Omega_1) \geq 0$. Then there always exists a quartic commutative extension $\hat{\Omega}_3$ of Ω_1 such that $M(\hat{\Omega}_3)$ is a flat extension of $M(\Omega_1)$, and $M_x(\hat{\Omega}_3) \geq 0$ and $M_y(\hat{\Omega}_3) \geq 0$. As a consequence, Ω_1 admits a subnormal completion $\mathbf{T}_{\hat{\Omega}_\infty}$. (The family Ω_1 is shown in Figure 2.10.)*

Proof (Sketch) We first need to show that six new weights, $\hat{\alpha}_{20}, \hat{\beta}_{20}, \hat{\alpha}_{11}, \hat{\beta}_{11}, \hat{\alpha}_{02}$ and $\hat{\beta}_{02}$ can be chosen in such a way that $M_x(\hat{\Omega}_3) \geq 0$ and $M_y(\hat{\Omega}_3) \geq 0$. Once this is done, the next step is to employ techniques from truncated moment problems to establish the existence of a flat extension $M(\hat{\Omega}_3)$ of $M(\Omega_1)$. After that, one appeals to the main result in [29]; the existence of a flat extension will readily imply the existence of a representing measure μ for $M(1)$, and the positivity of the localizing matrices $M_x(2)$ and $M_y(2)$ means that supp $\mu \subseteq \mathbb{R}_+^2$. Thus, μ will be the Berger measure of a subnormal 2-variable weighted shift $\mathbf{T}_{\Omega_\infty}$, which will be the desired subnormal completion of Ω_1.

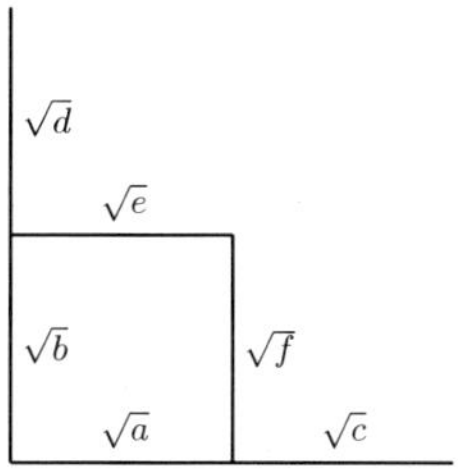

Figure 2.10
The initial family of weights Ω_1.

We now build $M(2)$.

Since $M(1) \equiv M(\Omega_1) \geq 0$, it follows from (2.22) that $\det \begin{pmatrix} ac & be \\ be & bd \end{pmatrix} \geq 0$, i.e.,

$$acd \geq be^2. \tag{2.23}$$

By the commutativity of Ω_1, we have

$$af = be, \tag{2.24}$$

and therefore

$$cd \geq ef. \tag{2.25}$$

A straightforward calculation shows that

$$\det M(1) = acbd - b^2e^2 - a^2bd + 2ab^2e - b^2ac$$

and that

$$\det M(1) > 0 \Longrightarrow cd - ef > 0. \tag{2.26}$$

Without loss of generality, we shall assume that $c \geq e$. We also assume that $a < c$, since otherwise a trivial solution exists. To build $M(2) \equiv M(\hat{\Omega}_3)$, we first need six new weights (the quadratic weights). Since the extension $\hat{\Omega}_3$ will also be commutative, two of these weights will be expressible in terms of other weights. We thus denote $\hat{\alpha}_{20}$ by $\sqrt{p}$, $\hat{\alpha}_{11}$ by $\sqrt{q}$, $\hat{\alpha}_{02}$ by $\sqrt{r}$, and $\hat{\beta}_{02}$ by $\sqrt{s}$ ($\hat{\beta}_{20}$ and $\hat{\beta}_{11}$ can be written in terms of the other four new weights). It follows that

$$M(2) = \begin{pmatrix} 1 & a & b & ac & be & bd \\ a & ac & be & acp & beq & bdr \\ b & be & bd & beq & bdr & bds \\ ac & acp & beq & & & \\ be & beq & bdr & & & \\ bd & bdr & bds & & & \end{pmatrix} \tag{2.27}$$

(with the lower right-hand 3×3 corner yet undetermined) and

$$M_x(2) = \begin{pmatrix} a & ac & be \\ ac & acp & beq \\ be & beq & bdr \end{pmatrix} \text{ and } M_y(2) = \begin{pmatrix} b & be & bd \\ be & beq & bdr \\ bd & bdr & bds \end{pmatrix}.$$

Now, since the zero-th row of a subnormal completion of Ω_1 will be a subnormal completion of the zero-th row of Ω_1, which is given by the weights $a < c$, we let $p := c$. By L-shaped propagation (Remark 4.2), having $\alpha_{10} = \hat{\alpha}_{20}$ immediately implies that $\hat{\alpha}_{11} = \sqrt{c}$, that is, $q := c$. Thus,

$$M_x(2) = \begin{pmatrix} a & ac & be \\ ac & ac^2 & bce \\ be & bce & bdr \end{pmatrix}.$$

By Choleski's Algorithm [8] (or its generalization, proved by J.L. Smul'jan [81]), $M_x(2) \geq 0$ if and only if $bdr \geq \frac{(be)^2}{a}$, so that we need $r \geq \frac{ef}{d}$. Thus, provided we take $r \geq \frac{ef}{d}$, the positivity of $M_x(2)$ is guaranteed. It remains to show that we can choose s in such a way that $s \geq d$ and $M_y(2) \equiv M_y(2)(s) \geq 0$. This can certainly be done. To complete the proof, we need to define the 3×3 lower right-hand corner of $M(2)$, and then show that $M(2)$ is a flat extension of $M(1)$, and therefore $M(2) \geq 0$. This is done by examining the rank of $M(1)$.

2.9.2 Description of the representing measure

In this subsection we provide a concrete description of the Berger measure for the subnormal completion in Theorem 9.6. The cases when rank$M(1)$ is 1 or 2 are straightforward, so we focus on the case rank $M(1) = 3$.

Since $M(1)$ is invertible, the last three columns of the flat extension $M(2)$ can be written in terms of the first three columns; that is, the columns labeled X^2, YX and Y^2 are linear combinations of 1, X and Y; moreover, $e < c$. Each of these column relations is associated with a quadratic polynomial in x and y, whose zero sets give rise to the so-called *algebraic variety* of $\hat{\Omega}_3$ [31]; concretely, $\mathscr{V}(\hat{\Omega}_3) := \bigcap_{p(X,Y)=0,\ \deg p \leq 2} \mathscr{Z}(p)$, where $\mathscr{Z}(p)$ denotes the zero set of p. In our case, the three column relations are

$$\begin{aligned} X^2 &= cX \\ YX &= fX \\ Y^2 &= \frac{be(f-d)}{a(c-e)}X + \frac{cd-ef}{c-e}Y. \end{aligned}$$

The associated zero sets are

$$\begin{aligned} \{(x,y) &: x = 0 \text{ or } x = c\} \\ \{(x,y) &: x = 0 \text{ or } y = f\} \\ \{(x,y) &: y^2 = \frac{be(f-d)}{a(c-e)}x + \frac{cd-ef}{c-e}y\}. \end{aligned}$$

Let $z := \frac{cd-ef}{c-e}$ and observe that $z > 0$ by (2.26). The algebraic variety of $\hat{\Omega}_3$ is then $\mathscr{V}(\hat{\Omega}_3) = \{(0,0),(0,z),(c,f)\}$ and these are the three atoms of the unique representing measure for $M(2)$. To find the densities, we use the first three moments, γ_{00}, γ_{01} and γ_{10}:

$$\begin{cases} \rho_{(0,0)} + \rho_{(0,z)} + \rho_{(c,f)} & = & 1 \\ \rho_{(c,f)} c & = & a \\ \rho_{(0,z)} z + \rho_{(c,f)} f & = & b. \end{cases}$$

Remark 9.7 *As shown in [37], flat extensions may not exist in the general case $m > 1$; that is, one can build an example of a set Ω_3 for which the associated moment matrix $M(2)$ admits a representing measure, but such that $M(2)$ has no flat extension $M(3)$.*

For more on the SCP, the reader is referred to [37], [74] and [75].

2.10 Spectral picture of hyponormal 2-variable weighted shifts

As stated in Lemma 1.1, the spectral picture of a norm-one hyponormal weighted shift W_α is easy to describe: the spectrum is the closed unit disk, the essential spectrum is the unit circle, and the Fredholm index is -1 in the open unit disk. Thus, from a spectral perspective all norm-one hyponormal shifts are equivalent to the unilateral shift U_+ (which is indeed subnormal). The associated C*-algebras, on the other hand, are not necessarily isomorphic, since the $*$-homomorphic image of a subnormal operator is again subnormal. But they do give rise to the same element of $Ext(\mathbb{T})$, since W_α and U_+ differ by a compact.

For 2-variable weighted shifts the situation is quite different; a complete description of the spectral picture in the subnormal case was given in [46] and [47]. We now briefly review the main ideas, in preparation for the study of the spectral picture of *hyponormal* 2-variable weighted shifts. (For basic information on the notion of Taylor spectrum and related results, the reader is referred to [17], [18], [22], [83], [84].)

The spectral picture in the subnormal case was obtained using the groupoid machinery introduced in [79] and [78], and refined in [41]. If μ is a *Reinhardt* measure on $\mathbb{C}^n$ (i.e., invariant under the action of the n-torus), the analysis of the C^*-algebra associated with $M_z \equiv (M_{z_1}, \cdots, M_{z_n})$ acting on the Hilbert space $P^2(\mu) := \overline{\mathbb{C}[z_1, \cdots, z_n]}^{L^2(\mu)}$ is related to the theory of multivariable weighted shifts. Indeed, each M_{z_i} is an operator weighted shift, M_z is subnormal, and all powers $M_z^\beta := M_{z_1}^{\beta_1} \cdots M_{z_n}^{\beta_n}$ are associated with weight sequences $w_\beta(\cdot)$, $\beta \in \mathbb{Z}_n^+$. We extend w_β to all of $\mathbb{Z}^n$ via $w_\beta(\alpha) := 0$ $(\alpha \notin \mathbb{Z}_+^n)$, and we let $\mathscr{A}$ denote the closed translation-invariant subalgebra of $\ell^\infty(\mathbb{Z}^n)$ generated by $\{w_\beta\}_{\beta \in \mathbb{Z}_+^n}$, not including the constants.

The maximal ideal space of $\mathscr{A}$, denoted by Y, is a non-compact, locally compact, Hausdorff space on which $\mathbb{Z}^n$ acts by translation.

When M_z is (jointly) bounded below, the map $\varphi : \mathbb{Z}^n \to Y$ given by $\varphi(\alpha)(a) := a(\alpha)$ $(\alpha \in \mathbb{Z}^n, a \in Y)$ is injective and open, and $X := \varphi(\mathbb{Z}^n_+)^- \subseteq Y$ is compact. Thus, X is a suitable *compactification* of $\mathbb{Z}^n_+$ [41, Lemmas 2.1 and 2.3]. If $\mathscr{G} := Y \times \mathbb{Z}^n \mid_X := \{(y, \alpha) \in Y \times \mathbb{Z}^n : y \in X \text{ and } y + \alpha \in X\}$, we see that $\mathscr{G}$ is the *groupoid* obtained by reducing the transformation group $Y \times \mathbb{Z}^n$ to X (X then becomes the *unit space* of $\mathscr{G}$). Analysis of X leads to a description of the ideal structure of $C^*(M_z)$, based on the correspondence between open invariant subsets of X and closed ideals in $C^*(M_z)$.

Since X is obtained from $\varphi(\mathbb{Z}^n_+)$ by adding suitable limit points at infinity, the *asymptotic behavior* of the weight sequences is crucial. For simplicity we assume $n = 2$. One is led to consider the sets

$$\infty_{\vec{u}} := \{x \in X : \exists \, \{k^{(j)}\}_{j=1}^{\infty} \text{ with } \varphi(k^{(j)}) \overset{w^*}{\to} x, k^{(j)} = p_j \vec{u} + q_j \vec{u}^{\perp}, \text{ and } q_j/p_j \to 0\},$$

where $\vec{u} \in \mathbb{R}^2_+$ is a unit vector, called a *direction.* This description works equally well for *hyponormal* 2-variable weighted shifts. Moreover, as we have described, one needs knowledge not only of the shift but also of all of its powers [35]; this also explains the relevance of invariance of hyponormality under powers [52]. However, in the subnormal case, more can be said. Indeed, our research with K. Yan ([46], [47]) exploited some special features of $\infty_{\vec{u}}$ when a Berger measure μ exists, and we were able to fully describe the spectral picture of M_z on $P^2(\mu)$: the Taylor spectrum $\sigma_T(M_z)$ is the polynomially convex hull of supp μ, the Taylor essential spectrum is the boundary of $\sigma_T(M_z)$, and the Fredholm index is 1 in the interior of $\sigma_T(M_z)$. This establishes the perfect analog of the above mentioned one-variable results.

The presence of the Berger measure was essential in the study of the asymptotic behavior of the weight sequences (e.g., we used the Lebesgue Dominated Convergence Theorem), and led to concrete Taylor spectral results. Consider now the class

$$\mathfrak{H}_0 := \{\mathbf{T} \equiv (T_1, T_2) : T_1, T_2 \text{ commuting subnormals}\}.$$

For *hyponormal* weighted shifts $\mathbf{T} \in \mathfrak{H}_0$, the study of the spectral properties of $\mathbf{T}$ requires new techniques, since no (joint) Berger measure is present. On the other hand, the Six-Point Test (Theorem 4.1) does provide a certain rigidity that may be enough to describe the asymptotic behavior of the weight sequences. We naturally arrive at the following problem.

Problem 10.1 *Describe in detail the spectral picture of hyponormal* $\mathbf{T} \in \mathfrak{H}_0$.

One famous example of a commuting pair of subnormal operators is the Drury-Arveson 2-shift $M_{z,1}$; however, it is *not* hyponormal (much less subnormal), as a simple application of the Six-Point Test reveals. In recent work, we have obtained some partial results, which give strong indications that a solution to Problem 10.1 is within reach. For instance, in [51, Theorem 2.2] we find a sufficient condition that guarantees the *disconnectedness* of the Taylor essential spectrum, namely,

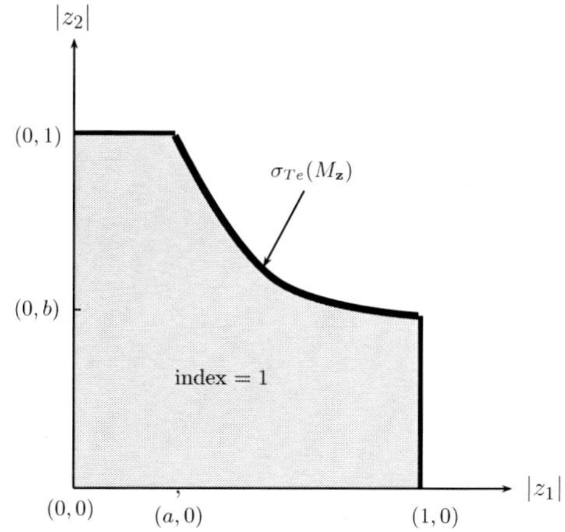

Figure 2.11
Spectral picture of a typical subnormal 2-variable weighted shift.

$\left\|W_{\alpha^{(1)}}\right\| < \left\|W_{\alpha^{(0)}}\right\|$, where $W_{\alpha^{(j)}}$ denotes the j-th horizontal slice of T_1; moreover, in [51, Example 1] we show that this condition can be present in hyponormal pairs with mutually absolutely continuous horizontal slices $W_{\alpha^{(j)}}$ $(j \geq 1)$. We improve this by showing in [51, Theorem 2.5] that more connected components of the Taylor essential spectrum can be formed, still preserving hyponormality, if we use Bergman-like weighted shifts (introduced and studied in [43]) on each horizontal level of T_1. This fact is quite surprising, in view of the well known one-variable results, and especially if we observe that in this case the Taylor spectrum is contained in the set $\{(z_1,z_2) \in \mathbb{C}^2 : z_1 z_2 = 0\}$. Along the way we prove that, for subnormal weighted shifts, the Berger measures ξ_j of the horizontal slices $\{W_{\alpha^{(j)}}\}_{j=1}^{\infty}$ are all mutually absolutely continuous (as expected from Theorem 7.10) and, as a result, $\left\|W_{\alpha^{(j)}}\right\| = \left\|W_{\alpha^{(1)}}\right\|$ (all $j \geq 1$) (and similarly for the vertical slices). This is relevant to the study of the asymptotic behavior of the weight sequences. Moreover, this new necessary condition is easily computable, and complements the necessary condition found in [49, Theorem 3.3].

Problem 10.2 *Let* $\mathbf{T} \equiv (T_1,T_2)$ *be a 2-variable weighted shift. When* $\mathbf{T}$ *is subnormal, we have* $\left\|W_{\alpha^{(j)}}\right\| = \left\|W_{\alpha^{(1)}}\right\|$ *(all* $j \geq 1$*). What is the analog of this condition for hyponormal* $\mathbf{T}$ *in* $\mathfrak{H}_0$*? Within the class* $\mathfrak{H}_0$*, is there a suitable analog for having the* ξ_j*'s mutually absolutely continuous?*

In the case of the Drury-Arveson 2-shift $M_{z,1}$, the measures of the horizontal slices are given by $\xi_0 = \delta_1$ (the Dirac measure at 1) and $d\xi_j(s) = j(1-s)^{j-1}ds$ $(j \geq 1)$. Since ξ_1 is not absolutely continuous with respect to σ, $M_{z,1}$ cannot be subnormal (by Theorem 7.10). However, the restriction of $M_{z,1}$ to the subspace $\mathscr{C} := \{(k_1,k_2) \in \ell^2(\mathbb{Z}_+^2) : k_1,k_2 \geq 1\}$ enjoys good asymptotic behavior for the weights, which we use to analyze its spectral picture. It is this kind of regularity, often present in nonsubnormal weighted shifts, that one tries to exploit.

Problem 10.2 is intimately related to Problem 10.1. For instance, we have shown that the joint k-hyponormality [33] of $\mathbf{T}$ (for a fixed but arbitrary $k \geq 1$) is not sufficient to force the equality $|\sigma_{Te}(\mathbf{T})| = \partial\, |\sigma_T(\mathbf{T})|$. Thus, we begin to see how elements of the spectral picture can be used to distinguish subnormality from hyponormality, something that cannot be done in one variable. Figure 2.12 below exhibits three distinct spectral pictures of *nonsubnormal* 2-variable weighted shifts, giving us a glimpse of the type of description we seek.

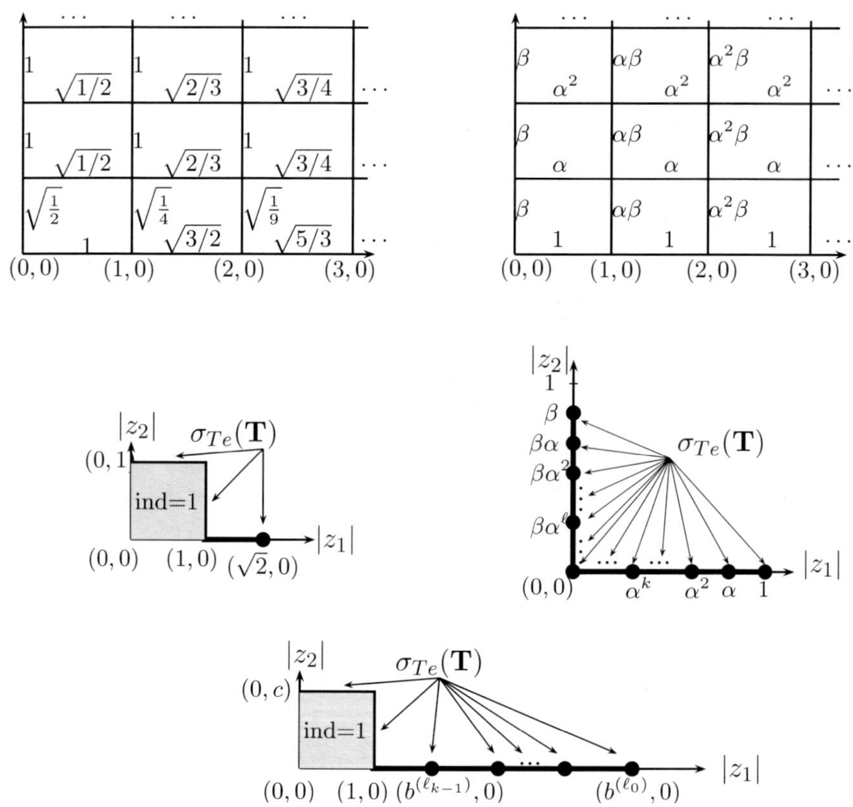

Figure 2.12

(top left and center left) Weight diagram and spectral picture of a nonsubnormal hyponormal 2-variable weighted shift (σ_{Te} is *disconnected*, although for all $k_2 \geq 1$, $W_{\alpha_{k_1,k_2}}$ is the Bergman shift); (top right and center right) weight diagram and spectral picture of a non-hyponormal commuting pair of subnormals (int. $\sigma_T = \emptyset$, and σ_{Te} has infinitely many connected components); (bottom) spectral picture of a nonsubnormal hyponormal 2-variable weighted shift with finitely many Bergman-like shifts as rows (int. $\sigma_T \neq \emptyset$ and σ_{Te} has finitely many connected components).

2.11 A bridge between 2-variable weighted shifts and shifts on directed trees

In this section we propose a link between two areas of multivariable operator theory that have developed independently. Following the notation in [66] and [11], let $\mathfrak{T} \equiv (V,E)$ be a directed tree with root. If $u,v \in V$ are two vertices such that $(u,v) \in E$, u is called a parent of v and v is called a child of u. Thus, the root is a vertex that has no parent. A weighted shift on $\mathfrak{T}$ is an operator S_λ defined by $S_\lambda e_u := \sum_{v \in Chi(u)} \lambda_v e_v \ (u \in V)$, where $\{e_v\}_{v \in V}$ is the standard orthonormal basis of $\ell^2(V)$, $Chi(u)$ is the set of all children of u, and $\{\lambda_v\}$ is a system of complex numbers (called the weights of S_λ) indexed by the non-root vertices in V. The research monograph [66] is an extensive study of weighted shifts on directed trees, which includes structural results (e.g., polar decomposition and Fredholmness), as well as various characterizations of properties such as hyponormality and subnormality. For example, S_λ is hyponormal if and only if (i) $v \in Chi(u), S_\lambda e_v = 0 \Rightarrow \lambda_v = 0$, and (ii) $\sum_{v \in Chi_\lambda^+(u)} \frac{|\lambda_v|^2}{\|S_\lambda e_v\|^2} \leq 1 \ (u \in V)$, where $Chi_\lambda^+(u) := \left\{ v \in Chi(u) : \sum_{w \in Chi(v)} |\lambda_w|^2 > 0 \right\}$.

2.11.1 From 2-variable weighted shifts to directed trees.

We have recently discovered a link between weighted shifts S_λ on directed tress and n-variable commuting weighted shifts. To describe the idea, assume for simplicity that $n = 2$. On one hand, any commuting 2-variable weighted shift (T_1, T_2) with weight sequences α and β can be regarded as a weighted shift on a directed tree, if we let $V := \mathbb{Z}_+^2$ and for every $\mathbf{k} \in V$ we let $Chi(\mathbf{k}) := \{\mathbf{k} + \varepsilon_1, \mathbf{k} + \varepsilon_2\}$. The root is $(0,0)$ and the tree (with weights) consists of the weight diagram of (T_1, T_2), when we regard the points $\mathbf{k} \in \mathbb{Z}_+^2$ with $k_1, k_2 > 0$ as *double points*, labeled $\mathbf{k}^{(1)}$ and $\mathbf{k}^{(2)}$, with the parent of $\mathbf{k}^{(i)}$ being $\mathbf{k}^{(i)} - \varepsilon_i \ (i = 1,2)$. Similarly, for $k_1 > 0$ the parent of $(k_1, 0)$ is $(k_1 - 1, 0)$ and for $k_2 > 0$ the parent of $(0, k_2)$ is $(0, k_2 - 1)$. With this notation in hand, the definition of the weights is straightforward; for example, $\lambda_{(3,2)}^{(1)} := \alpha_{(2,2)}$.

Problem 11.1 *Describe in detail how structural and spectral properties of the 2-variable weighted shift (T_1, T_2) are transferred to the weighted shift S_λ.*

Here's a sample result, which we have recently obtained in joint work with J. Yoon: if (T_1, T_2) is jointly hyponormal, then S_λ is hyponormal.

Problem 11.2 *For $k \geq 2$, assume that (T_1, T_2) is k-hyponormal. Does it follow that S_λ is k-hyponormal? (Recall that T is k-hyponormal if $(T^{*j}T^i - T^iT^{*j})_{i,j=1}^k \geq 0$.)*

2.11.2 From directed trees to 2-variable weighted shifts.

To describe the link in the opposite direction, start with a directed tree with root and assume for simplicity that the cardinality of $Chi(u)$ is at most 2 for every vertex u. We then proceed to build a commutative 2-variable weighted shift, as shown in Figure 2.13 below. The tree is represented by the **solid black lines**, and each edge (u,v) is associated with a weight λ_v. For the unit square containing the root $(0,0)$, we know the weights $\alpha_{(0,0)} := \lambda_{(1,0)}$, $\beta_{(0,0)} := \lambda_{(0,1)}$ and $\beta_{(1,0)} := \lambda_{(1,1)}$, so that we can use the commutative property to define $\alpha_{(0,1)} := \alpha_{(0,0)}\beta_{(1,0)}/\beta_{(0,0)}$. In Figure 2.13 we record this with the label A. Having found the weight associated with A, we then repeat the process to find the weight associated with the edge B; and so on. This process will not necessarily produce every weight in the 2-variable weighted shift, so a completion will generally be needed.

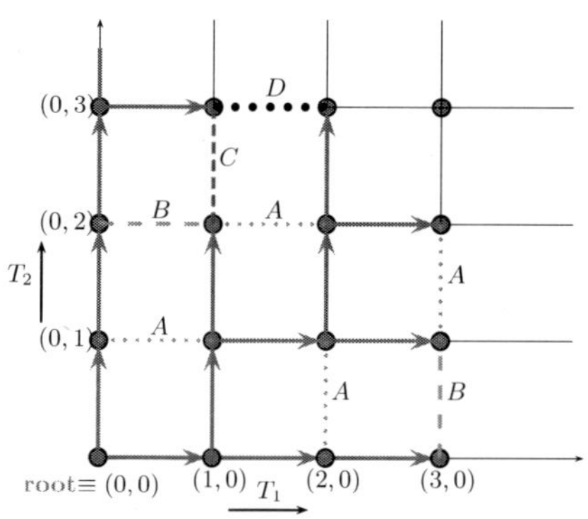

Figure 2.13
Conversion of a directed tree with root to a weight diagram for a 2-variable weighted shift. Shown in **solid black** is the original directed tree; shown in dotted lines (labeled A) are the new weights that can be defined using commutativity and the original tree. Shown in long-dashed lines (labeled B) are weights that require the original tree and the weights labeled A. Next, shown in short-dashed lines (labeled C) are weights that require the original tree and weights labeled B (and possibly weights labeled A). Finally, shown in thick dotted lines (labeled D) are weights that require the original tree and weights labeled C (and possibly weights labeled A or B).

Problem 11.3 *How are the structural and spectral properties of S_λ captured by suitable completions of the partially defined 2-variable weighted shift (T_1, T_2)?*

Recall that a subnormal unilateral weighted shift W_α is said to be recursively generated if there exists $N > 0$ such that the moments γ_n satisfy the recursive relation $\gamma_{N+k} = \varphi_0\gamma_k + \cdots + \varphi_{N-1}\gamma_{N+k-1}$ for all $k \geq 0$, where $\varphi_0, \cdots, \varphi_{N-1} \in \mathbb{R}$. In [25], we proved that W_α is recursively generated if and only if its Berger measure is finitely atomic. This characterization can be used to define recursiveness for 2-variable weighted shifts, by requiring that the Berger measure be finitely atomic. Consider now the homogeneous subspace $\mathscr{P}_n$ of degree n in $\ell^2(\mathbb{Z}^2_+)$. Then a subnormal 2-variable weighted shift (T_1,T_2) is recursively generated if and only if there exists $N>0$ such that all the moments of degree $N+k$ can be written as linear combinations, with fixed coefficients $\varphi_{(0,0)},\varphi_{(1,0)},\varphi_{(0,1)},\cdots,\varphi_{(0,N)}$, of moments of degree $k,k+1,\cdots,N+k-1$ (cf. [37]).

Problem 11.4 *Using the above mentioned link between 2-variable weighted shifts and weighted shifts on directed trees, define recursiveness for weighted shifts on directed trees.*

2.12 The spherical Aluthge transform

Recall that a Hilbert space operator $T\in\mathscr{B}(\mathscr{H})$ is *quasinormal* if T commutes with T^*T. It is well known that

$$\text{normal} \Longrightarrow \text{quasinormal} \Longrightarrow \text{subnormal} \Longrightarrow \text{hyponormal}.$$

For $T\in\mathscr{B}(\mathscr{H})$, consider now the *canonical polar decomposition* of T, $T\equiv VP$, where V is a partial isometry, $P:=(T^*T)^{\frac{1}{2}}$, and $\ker T=\ker V=\ker P$. The *Aluthge transform* $\widehat{T}$ is $\widehat{T}:=P^{\frac{1}{2}}VP^{\frac{1}{2}}$. The Aluthge transform was first introduced in [2] and it has attracted considerable attention over the last two decades (see [3], [4], [10], [13], [39], [57], [59], [60], [68], [69], [70], [71], [72]), [73], [76], [77], [86]).

It is well known that $T\in\mathscr{B}(\mathscr{H})$ is quasinormal if and only if T commutes with P in the canonical polar decomposition $T\equiv VP$; equivalently, if V commutes with P. It follows easily that T is quasinormal if and only if $T=\widehat{T}$, that is, if and only if T is a fixed point for the Aluthge transform.

We next consider a suitable polar decomposition (and corresponding Aluthge transform) for commuting pairs $\mathbf{T}\equiv(T_1,T_2)$. Given such a $\mathbf{T}$, let

$$Q:=(T_1^*T_1+T_2^*T_2)^{\frac{1}{2}}. \tag{2.28}$$

Clearly, $\ker Q=\ker T_1\bigcap\ker T_2$. For $x\in\ker Q$, let $V_ix:=0$, and for $y\in\operatorname{Ran}Q$, say $y=Qx$, let $V_iy:=T_ix$ $(i=1,2)$. It is easy to see that V_1 and V_2 are well defined. We then have

$$\begin{pmatrix} T_1 \\ T_2 \end{pmatrix}=\begin{pmatrix} V_1 \\ V_2 \end{pmatrix}Q,$$

as operators from $\mathscr{H}$ to $\mathscr{H}\oplus\mathscr{H}$. Moreover, this is the unique canonical polar decomposition of $\begin{pmatrix} T_1 \\ T_2 \end{pmatrix}$. It follows that $\begin{pmatrix} V_1 \\ V_2 \end{pmatrix}$ is a partial isometry from $(\ker Q)^{\perp}$ onto $\overline{\operatorname{Ran}\begin{pmatrix} T_1 \\ T_2 \end{pmatrix}}$.

Following [7] and [63], we say that $\mathbf{T}$ is *(jointly) quasinormal* if T_i commutes with $T_j^* T_j$ for all $i, j = 1,2$; and *spherically quasinormal* if T_i commutes with Q for $i = 1,2$. By [7], for all $k \geq 1$, one has

$$\begin{aligned} \text{normal} &\implies \text{(jointly) quasinormal} \implies \text{spherically quasinormal} \\ &\implies \text{subnormal} \implies k\text{-hyponormal}. \end{aligned} \tag{2.29}$$

On the other hand, the results in [33] and [63] show that the reverse implications in (2.29) do not necessarily hold.

We are now ready to introduce a bivariate operator transform.

Definition 12.1 *(cf. [53], [54]) With* $\mathbf{T}$*,* V_1*,* V_2 *and* Q *as above, the spherical Aluthge transform of* $\mathbf{T}$ *is*

$$\widehat{\mathbf{T}} \equiv (\widehat{T_1}, \widehat{T_2}),$$

where

$$\widehat{T_i} := Q^{\frac{1}{2}} V_i Q^{\frac{1}{2}} \ (i = 1,2). \tag{2.30}$$

Lemma 12.2 *([10], [54])* $\widehat{\mathbf{T}}$ *is commutative.*

The spherical Aluthge transform was introduced in [53]; its general theory was developed in [54] and [55]. We will now focus on the spherical quasinormal pairs, which are fixed points of the spherical Aluthge transform.

Given a commuting 2-variable weighted shift $\mathbf{T} \equiv (T_1, T_2) = W_{(\alpha,\beta)}$, and given $k_1, k_2 \geq 0$, we let

$$H_{k_2} := \text{shift}(\alpha_{(0,k_2)}, \alpha_{(1,k_2)}, \cdots) \tag{2.31}$$

be the k_2-th horizontal slice of T_1; similarly we let

$$V_{k_1} := \text{shift}(\beta_{(k_1,0)}, \beta_{(k_1,1)}, \cdots) \tag{2.32}$$

be the k_1-th vertical slice of T_2. By the commutativity condition (2.1), we note that

$$\gamma_{(k_1,k_2)}(W_{(\alpha,\beta)}) = \frac{\gamma_{k_1}(H_{k_2})}{\beta^2_{(0,0)} \cdots \beta^2_{(0,k_2-1)}}, \tag{2.33}$$

where $\gamma_{k_1}(H_{k_2})$ is given by (2.2). A similar identity holds for V_{k_1}.

We next consider the structure of commuting pairs which are fixed points of the spherical Aluthge transform. It is well known that an operator T is quasinormal if and only if $T = \widehat{T}$. We will extend this result to the case of commuting pairs $\mathbf{T} \equiv (T_1, T_2)$. First, we need an auxiliary result.

Lemma 12.3 *For* $i = 1,2$*,* T_i *commutes with* Q *if and only if* V_i *commutes with* Q*.*

We now consider spherical quasinormality for commuting pairs. Suppose a commuting pair $\mathbf{T}$ is spherically quasinormal. Since for $i = 1,2$, T_i commutes with

$T_1^*T_1 + T_2^*T_2$, then for $i = 1,2$ T_i commutes with Q (by the continuous functional calculus for Q). Observe now that

$$\widehat{(T_1,T_2)}\sqrt{Q} = \left(\sqrt{Q}V_1\sqrt{Q}, \sqrt{Q}V_2\sqrt{Q}\right)\sqrt{Q} = \left(\sqrt{Q}T_1, \sqrt{Q}T_2\right) = (T_1,T_2)\sqrt{Q},$$

so that

$$\widehat{(T_1,T_2)} = (T_1,T_2) \text{ on } \overline{\operatorname{Ran}\sqrt{Q}}. \tag{2.34}$$

On the other hand, since $\ker Q = \ker T_1 \bigcap \ker T_2$, it follows easily that

$$\widehat{(T_1,T_2)} = (T_1,T_2) \text{ on } \ker Q. \tag{2.35}$$

Since $\mathcal{H} = \left(\overline{\operatorname{Ran} Q}\right) \oplus \ker Q$, we can combine (2.34) and (2.35) to prove that $\widehat{(T_1,T_2)} = (T_1,T_2)$.

Theorem 12.4 *([53]) Let $\mathbf{T} \equiv (T_1,T_2)$ be a commuting 2-variable weighted shift. The following statements are equivalent.*
(i) $\mathbf{T}$ is spherically quasinormal.
(ii) $\widehat{(T_1,T_2)} = (T_1,T_2)$.

2.12.1 Spherically quasinormal 2-variable weighted shifts

In this Subsection we present a characterization of spherical quasinormality for 2-variable weighted shifts. Before we state it, we list a simple fact about quasinormality for 2-variable weighted shifts.

Remark 12.5 *A 2-variable weighted shift $\mathbf{T} \equiv (T_1,T_2) = W_{(\alpha,\beta)}$ is (jointly) quasinormal if and only if $\alpha_{(k_1,k_2)} = \alpha_{(0,0)}$ and $\beta_{(k_1,k_2)} = \beta_{(0,0)}$ for all $k_1,k_2 \geq 0$. As a result, up to a scalar multiple in each component, a quasinormal 2-variable weighted shift is identical to the so-called Helton-Howe shift, that is, the shift that corresponds to the pair of multiplications by the coordinate functions in the Hardy space $H^2(\mathbb{T}\times\mathbb{T})$ of the 2-torus, with respect to arclength measure on each circle $\mathbb{T}$ (cf. [63]). This fact is consistent with the one-variable result: a unilateral weighted shift W_ω is quasinormal if and only if $W_\omega = cU_+$ for some $c > 0$.*

Theorem 12.6 *Let $\mathbf{T} \equiv (T_1,T_2) = W_{(\alpha,\beta)}$ be a commuting 2-variable weighted shift. Then the following statements are equivalent.*
(i) $\mathbf{T} \equiv (T_1,T_2)$ is spherically quasinormal.
(ii) There exists a constant $c > 0$ such that for all $\mathbf{k} \equiv (k_1,k_2) \in \mathbb{Z}_+^2$,

$$\alpha_{(k_1,k_2)}^2 + \beta_{(k_1,k_2)}^2 = c.$$

(iii) $T_1^*T_1 + T_2^*T_2 = c\,I$.

We now describe the weight diagrams of $\widehat{\mathbf{T}}$ for a 2-variable weighted shift $\mathbf{T} \equiv (T_1,T_2) = W_{(\alpha,\beta)}$.

Proposition 12.7 *Let* $\mathbf{T} \equiv (T_1, T_2) = W_{(\alpha,\beta)}$ *be a 2-variable weighted shift. Then*

$$\widehat{T}_1 e_{\mathbf{k}} = \alpha_{\mathbf{k}} \frac{(\alpha^2_{\mathbf{k}+\varepsilon_1} + \beta^2_{\mathbf{k}+\varepsilon_1})^{1/4}}{(\alpha^2_{\mathbf{k}} + \beta^2_{\mathbf{k}})^{1/4}} e_{\mathbf{k}+\varepsilon_1}; \ \widehat{T}_2 e_{\mathbf{k}} = \beta_{\mathbf{k}} \frac{(\alpha^2_{\mathbf{k}+\varepsilon_2} + \beta^2_{\mathbf{k}+\varepsilon_2})^{1/4}}{(\alpha^2_{\mathbf{k}} + \beta^2_{\mathbf{k}})^{1/4}} e_{\mathbf{k}+\varepsilon_2} \quad (2.36)$$

for all $\mathbf{k} \in \mathbb{Z}^2_+$.

Recall now the class of spherically isometric commuting pairs of operators (cf. [6], [7], [9], [58], [63]).

Definition 12.8 *A commuting pair* $\mathbf{T} \equiv (T_1, T_2)$ *is a spherical isometry if* $T_1^* T_1 + T_2^* T_2 = I$.

The following result is a straightforward application of Definition 12.8.

Lemma 12.9 *A 2-variable weighted shift* $\mathbf{T} \equiv (T_1, T_2) = W_{(\alpha,\beta)}$ *is a spherical isometry if and only if*

$$\alpha^2_{\mathbf{k}} + \beta^2_{\mathbf{k}} = 1 \ (\textit{for all } \mathbf{k} \in \mathbb{Z}^{\mathbf{2}}_+).$$

By Theorem 12.6, we have:

Corollary 12.10 *A 2-variable weighted shift* $\mathbf{T} \equiv (T_1, T_2)$ *is spherically quasinormal if and only if there exists* $c > 0$ *such that* $\frac{1}{\sqrt{c}}\mathbf{T}$ *is a spherical isometry, that is,* $T_1^* T_1 + T_2^* T_2 = I$.

We pause to recall an important result about spherical isometries.

Lemma 12.11 *[58] Any spherical isometry is subnormal.*

Combining Corollary 12.10 and Lemma 12.11, we easily obtain the following result.

Theorem 12.12 *Any quasinormal 2-variable weighted shift is subnormal.*

Remark 12.13 *(cf. [53, Remark 2.14])*
(i) A. Athavale and S. Poddar have recently proved that a commuting spherically quasinormal pair is always subnormal [7, Proposition 2.1]; this provides a different proof of Theorem 12.12.
(ii) In a different direction, let $Q_{\mathbf{T}}(X) := T_1^* X T_1 + T_2^* X T_2$. *By induction, it is easy to prove that if* $\mathbf{T}$ *is spherically quasinormal, then* $Q^n_{\mathbf{T}}(I) = (Q_{\mathbf{T}}(I))^n$ $(n \geq 0)$*; by [12, Remark 4.6],* $\mathbf{T}$ *is subnormal.*

2.12.2 Construction of spherically quasinormal 2-variable weighted shifts

As observed in [54], within the class of 2-variable weighted shifts there is a simple description of spherical isometries, in terms of the weight sequences $\alpha \equiv \{\alpha_{(k_1,k_2)}\}$ and $\beta \equiv \{\beta_{(k_1,k_2)}\}$. Indeed, since spherical isometries are (jointly) subnormal, we know that the unilateral weighted shift associated with the 0-th row in the weight diagram must be subnormal. Thus, without loss of generality, we can always assume that the 0-th row corresponds to a subnormal unilateral weighted shift, and denote its weights by $\{\alpha_{(k,0)}\}_{k=0,1,2,\cdots}$. Also, in view of Corollary 12.10 we can assume that $c = 1$. Using the identity

$$\alpha_{\mathbf{k}}^2 + \beta_{\mathbf{k}}^2 = 1 \quad (\mathbf{k} \in \mathbb{Z}_+^2) \tag{2.37}$$

and the above mentioned 0-th row, we can compute $\beta_{(k,0)} := \sqrt{1-\alpha_{k,0}^2}$ for $k = 0,1,2,\cdots$. With these new values at our disposal, we can use the commutativity property (2.1) to generate the values of α in the first row; that is,

$$\alpha_{(k,1)} := \alpha_{(k,0)}\beta_{(k+1,0)}/\beta_{(k,0)}.$$

We can now repeat the algorithm, and calculate the weights $\beta_{(k,1)}$ for $k = 0,1,2,\cdots$, again using the identity (2.37). This in turn leads to the α weights for the second row, and so on.

This simple construction of spherically isometric 2-variable weighted shifts will allow us to study properties like recursiveness (tied to the existence of finitely atomic Berger measures) and propagation of recursive relations. We pursue this in Subsection 2.12.4 below.

We briefly pause to show that one can show directly that having a strictly increasing sequence of weights on the 0-th row is a necessary condition for the above algorithm to proceed. Concretely, we establish the following result.

2.12.3 The above construction may stall if $\{\alpha_{(k,0)}\}_{k\geq 0}$ is not strictly increasing.

Proposition 12.14 *Let $\alpha_{(0,0)} := \sqrt{p}$, $\alpha_{(1,0)} := \sqrt{q}$, $\alpha_{(2,0)} := \sqrt{r}$ and $\alpha_{(3,0)} := \sqrt{r}$, and assume that $0 < p < q < r < 1$. Then the algorithm described in this section fails at some stage. As a consequence, there does not exist a spherical isometry interpolating these initial data.*

Proof Assume that the algorithm works, and let $\beta_{(0,0)} := \sqrt{1-\alpha_{(0,0)}^2} = \sqrt{1-p}$, $\beta_{(1,0)} := \sqrt{1-\alpha_{(1,0)}^2} = \sqrt{1-q}$, $\beta_{(2,0)} := \sqrt{1-\alpha_{(2,0)}^2} = \sqrt{1-r}$ and $\beta_{(3,0)} := \sqrt{1-\alpha_{(3,0)}^2} = \sqrt{1-r}$. Also, let $\alpha_{(0,1)} := \frac{\alpha_{(0,0)}\beta_{(1,0)}}{\beta_{(0,0)}} = \sqrt{\frac{p(1-q)}{1-p}}$. We recursively define two sequences a and b of real numbers by

$$a_0 := p \tag{2.38}$$

$$a_1 := \frac{p(1-q)}{1-p} \tag{2.39}$$

$$b_0 := q \tag{2.40}$$

$$b_1 \equiv \alpha^2_{(1,1)} := \frac{\alpha^2_{(1,0)}\beta^2_{(2,0)}}{\beta^2_{(1,0)}} = \frac{q(1-r)}{1-q} \tag{2.41}$$

$$a_{n+1} := \frac{a_n(1-b_n)}{1-a_n} \quad (n \geq 1) \tag{2.42}$$

$$b_{n+1} := \frac{b_n(1-r)}{1-b_n} \quad (n \geq 1). \tag{2.43}$$

Claim 1. For all $n \geq 0$, we have $0 < b_{n+1} < b_n < r$.

Proof of Claim 1. It is clear that $0 < b_0 = q < r$. If $b_1 \geq b_0$ then $\frac{q(1-r)}{1-q} \geq q$ and therefore $q \geq r$, a contradiction. Thus, $b_1 < b_0$. Assume now that Claim 1 is true for $n = k$. If $b_{k+1} \geq b_k$ then $\frac{b_k(1-r)}{1-b_k} \geq b_k$, which implies that $b_k \geq r$, a contradiction. By mathematical induction, Claim 1 is thus established.

Claim 2. The sequence b converges to 0.

Proof of Claim 2. By Claim 1, b is decreasing; let b_∞ be its limit. Observe that $b_\infty < r < 1$. Then

$$b_\infty = \frac{b_\infty(1-r)}{1-b_\infty}.$$

It follows that

$$b_\infty r = b_\infty^2,$$

so that $b_\infty(r - b_\infty) = 0$, and a fortiori $b_\infty = 0$.

Using *Mathematica*, one conjectures that

$$b_n = \frac{qr(1-r)^n}{q[(1-r)^n - 1] + r}$$

and

$$a_n = \frac{pr[q(1-r)^n + r - q]}{pq[(1-r)^n - 1] + r(r + pqn - prn)} =: \frac{K_n}{L_n},$$

for all $n \geq 0$. Both formulas can be proved by mathematical induction.

Claim 3. The inequality $0 < a_n < 1$ is not true for every $n \geq 0$; that is, for some $n > 1$ the algorithm that constructs spherical isometries fails.

Proof of Claim 3. Assume to the contrary that $0 < K_n < L_n$ for all $n \geq 0$. It follows that

$$pr[q(1-r)^n + r - q] < pq[(1-r)^n - 1] + r(r + pqn - prn),$$

for all $n \geq 0$. Then

$$(pqr - pq)(1-r)^n + pr^2 - pqr < -pq + r^2 + pqrn - pr^2 n,$$

so that

$$-pq(1-r)^{n+1} < -pr(r-q)n - pr(r-q) + r^2 - pq,$$

and therefore

$$pr(r-q)(n+1) < r^2 - pq + pq(1-r)^{n+1}, \tag{2.44}$$

for all $n \geq 0$. When $n \to \infty$, the left-hand side of (2.44) converges to ∞, while the right-hand side converges to $r^2 - pq$. This is a contradiction, and therefore Claim 3 has been established.

Putting together Claims 1, 2 and 3, we complete the proof of Proposition 12.14.

2.12.4 Recursively generated spherically quasinormal two-variable weighted shifts

We begin by recalling some terminology and basic results from [25] and [26]. A subnormal unilateral weighted shift W_ω is said to be *recursively generated* if the sequence of moments γ_n admits a finite-step recursive relation; that is, if there exists an integer $k \geq 1$ and real coefficients $\varphi_0, \varphi_1, \cdots, \varphi_{k-1}$ such that

$$\gamma_{n+k} = \varphi_0 \gamma_n + \varphi_1 \gamma_{n+1} + \cdots + \varphi_{k-1} \gamma_{n+k-1} \ \text{(all } n \geq 0). \tag{2.45}$$

In conjunction with (2.45) we consider the generating function

$$g_\omega(s) := s^k - (\varphi_0 + \varphi_1 s + \cdots + \varphi_{k-1} s^{k-1}). \tag{2.46}$$

The following result characterizes recursively generated subnormal unilateral weighted shifts.

Lemma 12.15 *[27] Let W_ω be a subnormal unilateral weighted shift. The following statements are equivalent.*
(i) W_ω is recursively generated.
(ii) The Berger measure μ of W_ω is finitely atomic, and $\operatorname{supp}\mu \subseteq \mathscr{Z}(g_\omega)$, *where $\mathscr{Z}(g_\omega)$ denotes the zero set of g_ω, that is, the set of roots of the equation* $g_\omega = 0$.

In [55], J. Yoon and the author have proved that for spherically quasinormal 2-variable weighted shifts a recursive relation in the 0-th row automatically propagates to the first row, preserving the coefficients of recursion. Given a 2-variable weighted shift $\mathbf{T} \equiv (T_1, T_2) = W_{(\alpha,\beta)}$, recall from (2.31) the notation H_0 and H_1.

Theorem 12.16 *([55]) Let $\mathbf{T}$ be a spherically quasinormal 2-variable weighted shift, and assume that H_0 is recursively generated, with coefficients $\varphi_0, \varphi_1, \ldots, \varphi_{n-1}$; that is,*

$$\gamma_{n+k}(H_0) = \varphi_0 \gamma_k(H_0) + \varphi_1 \gamma_{k+1}(H_0) + \cdots + \varphi_{n-1} \gamma_{n+k-1}(H_0) \ \ (\mathit{all}\ k \geq 0). \tag{2.47}$$

Then H_1 is recursively generated, with the same recursion coefficients.

A straightforward induction argument yields the following result.

Corollary 12.17 *Let* $\mathbf{T}$ *be a spherically quasinormal* 2*-variable weighted shift, and assume that* H_0 *is recursively generated, with coefficients* $\varphi_0, \varphi_1, \ldots, \varphi_{n-1}$*, and let* $k_2 > 1$*. Then* H_{k_2} *is recursively generated, with the same recursion coefficients.*

In view of Theorem 12.16 and (2.37), one is naturally led to the following question. If H_0 is recursively generated, is it also the case that V_0 is recursively generated? To study this question, we will take advantage of the theory of truncated moment problems in two real variables. (The reader is referred to [28], [29] and [30] for terminology and basic results.) Here we will only make use of the moment matrix associated with $W_{(\alpha,\beta)}$; that is, the infinite matrix $M(\alpha,\beta)$ whose rows and columns are indexed by $\mathbf{k} \in \mathbb{Z}_+^2$ and whose $(\mathbf{i},\mathbf{j})$-entry is given by $\gamma_{\mathbf{i}+\mathbf{j}}$. As typically done in the theory of truncated real moment problems, it is natural to label the rows and columns of $M(\alpha,\beta)$ using the homogenous monomials of ascending degree $1, S, T, S^2, ST, T^2, S^2, S^2T, ST^2, T^3, \cdots$. For instance, when we refer to the entry in the position $((1,2),(0,1))$, we mean the entry corresponding to row $(1,2)$ and column $(0,1)$, that is, the row labeled by the monomial ST^2 and the column labeled by the monomial T.

The proof of the following result is straightforward.

Lemma 12.18 *Let* $W_{(\alpha,\beta)}$ *be a* 2*-variable weighted shift, let* $c > 0$ *and fix* $\mathbf{k} \in \mathbb{Z}_+^2$*. The following statements are equivalent.*

(i) $\alpha_{\mathbf{k}}^2 + \beta_{\mathbf{k}}^2 = c$.

(ii) $\gamma_{\mathbf{k}+\varepsilon_1} + \gamma_{\mathbf{k}+\varepsilon_2} = c\gamma_{\mathbf{k}}$.

Corollary 12.19 *Let* $W_{(\alpha,\beta)}$ *be a spherically quasinormal* 2*-variable weighted shift, with constant* $c > 0$*. Then the columns of the moment matrix* $M(\alpha,\beta)$ *satisfy the linear relation* $S + T = c\,1$.

Corollary 12.20 *([55]) Let* $W_{(\alpha,\beta)}$ *be a spherically quasinormal* 2*-variable weighted shift, with constant* $c > 0$*, and let* σ *and* τ *be the Berger measures of* H_0 *and* V_0*, respectively. Then* $\operatorname{supp}\tau = c - \operatorname{supp}\sigma := \{c - s : s \in \operatorname{supp}\sigma\}$.

We are now ready to state that for spherically quasinormal 2-variable weighted shifts the property of being recursively generated transfers from the 0-th row in the weight diagram to the 0-th column.

Theorem 12.21 *([55]) Let* $W_{(\alpha,\beta)}$ *be a spherically quasinormal* 2*-variable weighted shift, with constant* $c > 0$*, and assume that the unilateral weighted shift* H_0 *(which corresponds to the* 0*-th row in the weight diagram of* $W_{(\alpha,\beta)}$*) is recursively generated. Then the unilateral weighted shift* V_0 *(which corresponds to the* 0*-th column) is also recursively generated.*

Corollary 12.22 *Let* $W_{(\alpha,\beta)}$ *be a spherically quasinormal* 2*-variable weighted shift, with constant* $c > 0$*, and assume that the unilateral weighted shift* H_0 *(which corresponds to the* 0*-th row in the weight diagram of* $W_{(\alpha,\beta)}$*) is recursively generated. Let*

σ be the Berger measure of H_0, and let μ be the Berger measure of $W_{(\alpha,\beta)}$. Then

(i) $\operatorname{supp}\mu \subseteq \operatorname{supp}\sigma \times (c - \operatorname{supp}\sigma)$*; and*

(ii) μ *is finitely atomic.*

We conclude this section with an intriguing question.

Question 12.23 *Let H_0 be the Bergman shift,* $\operatorname{shift}(\sqrt{\frac{1}{2}}, \sqrt{\frac{2}{3}}, \sqrt{\frac{3}{4}}, \cdots)$*, and use Subsection 2.12.2 to build a spherically quasinormal 2-variable weighted shift W (cf. Figure 2.14). For this shift the j-th row is identical to the j-th column (all $j \geq 0$). Also, W is a close relative of the Drury-Arveson 2-variable weighted shift, in that the j-th row of W is the Agler A_{j+2} shift [1]. What is the Berger measure of W?*

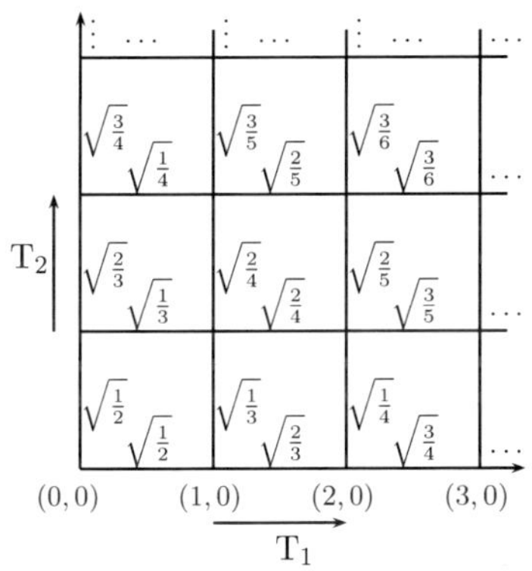

The 2-variable weighted shift $W_{(\alpha,\beta)}$ whose weight diagram is shown on the left has the following properties:

(i) $\alpha^2_{(k_1,k_2)} + \beta^2_{(k_1,k_2)} = 1$ for all $(k_1, k_2) \in \mathbb{Z}^2_+$.

(ii) $\widehat{W_{(\alpha,\beta)}} = W_{(\alpha,\beta)}$.

(iii) $W_{(\alpha,\beta)}$ is a spherical isometry.

Figure 2.14
The weight diagram on the left corresponds to the 2-variable weighted shift in Question 12.23.

Acknowledgment

All the examples, and the basic construction in Subsection 2.5.3 were obtained using calculations with the software tool *Mathematica [85]*.

References

[1] J. Agler, Hypercontractions and subnormality, *J. Operator Theory* **13** (1985), 203–217.

[2] A. Aluthge, On p-hyponormal operators for $0 < p < 1$, *Integral Equations Operator Theory* **13** (1990), 307–315.

[3] I. Amemiya, T. Ito, and T.K. Wong, On quasinormal Toeplitz operators, *Proc. Amer. Math. Soc.* **50** (1975), 254–258.

[4] T. Ando, Aluthge transforms and the convex hull of the spectrum of a Hilbert space operator, in *Recent advances in operator theory and its applications*, Oper. Theory Adv. Appl. **160** (2005), 21–39.

[5] A. Athavale, On joint hyponormality of operators, *Proc. Amer. Math. Soc.* **103**(1988), 417–423.

[6] A. Athavale, On the intertwining of joint isometries, *J. Operator Theory* **23** (1990), 339–350.

[7] A. Athavale and S. Poddar, On the reflexivity of certain operator tuples, *Acta Math. Sci. (Szeged)* **81** (2015), 285–291.

[8] K. Atkinson, *Introduction to Numerical Analysis*, 2nd Edition, John Wiley and Sons, New York, 1989.

[9] K.R.M. Attele and A.R. Lubin, Commutant lifting for jointly isometric operators-a geometrical approach, *J. Funct. Anal.* **140** (1996), 300–311.

[10] C. Benhida and R. Curto, Spectral properties of the spherical Aluthge transform, in final stages of preparation.

[11] P. Budzyński, Z. J. Jabłoński, I. B. Jung, J. Stochel, A subnormal weighted shift on a directed tree whose nth power has trivial domain, *J. Math. Anal. Appl.* **435** (2016), 302–314.

[12] S. Chavan and V. Sholapurkar, Rigidity theorems for spherical hyperexpansions, *Complex Anal. Oper. Theory* **7** (2013), 1545-1568.

[13] M. Chō, I.B. Jung and W. Y. Lee, On Aluthge Transforms of p-hyponormal Operators, *Integral Equations Operator Theory* **53** (2005), 321–329.

[14] Y.B. Choi, A propagation of quadratically hyponormal weighted shifts, *Bull. Korean Math. Soc.* **37** (2000), 347–352.

[15] J. Conway, *The Theory of Subnormal Operators*, Mathematical Surveys and Monographs, vol. 36, Amer. Math. Soc., Providence, 1991.

[16] J.B. Conway and W. Szymanski, Linear combination of hyponormal operators, *Rocky Mountain J.* **18** (1988), 695–705.

[17] R. Curto, On the connectedness of invertible n-tuples, *Indiana Univ. Math. J.* **29** (1980), 393–406.

[18] R. Curto, Applications of several complex variables to multiparameter spectral theory, in *Surveys of some recent results in operator theory*, Vol. II, 25–90, Pitman Res. Notes Math. Ser., 192, Longman Sci. Tech., Harlow, 1988.

[19] R. Curto, Joint hyponormality: A bridge between hyponormality and subnormality, *Proc. Symposia Pure Math.* **51** (1990), 69–91.

[20] R. Curto, Quadratically hyponormal weighted shifts, *Integral Equations Operator Theory* **13** (1990), 49–66.

[21] R. Curto, Polynomially hyponormal operators on Hilbert space, in *Proceedings of ELAM VII*, Revista Unión Mat. Arg. **37** (1991), 29–56.

[22] R. Curto, Spectral theory of elementary operators, in *Elementary Operators and Applications*, M. Mathieu, ed., World Sci. Publishing, River Edge, NJ, 1992; pp. 3–52.

[23] R. Curto, An operator-theoretic approach to truncated moment problems, in *Linear Operators*, Banach Center Publ., vol. 38, 1997, pp. 75–104.

[24] R. Curto and L. Fialkow, Recursiveness, positivity, and truncated moment problems, *Houston J. Math.* **17** (1991), 603–635.

[25] R. Curto and L. Fialkow, Recursively generated weighted shifts and the subnormal completion problem, *Integral Equations Operator Theory* **17** (1993), 202–246.

[26] R. Curto and L. Fialkow, Recursively generated weighted shifts and the subnormal completion problem, II, *Integral Equations Operator Theory* **18** (1994), 369–426.

[27] R. Curto and L. Fialkow, Solution of the truncated complex moment problem with flat data, *Memoirs Amer. Math. Soc.* no. 568, Amer. Math. Soc., Providence, 1996.

[28] R. Curto and L. Fialkow, Flat extensions of positive moment matrices: Recursively generated relations, *Memoirs Amer. Math. Soc.* no. 648, Amer. Math. Soc., Providence, 1998.

[29] R. Curto and L. Fialkow, The truncated complex K-moment problem, *Trans. Amer. Math. Soc.* **352** (2000), 2825–2855.

[30] R. Curto and L. Fialkow, Solution of the singular quartic moment problem, *J. Operator Theory* **48** (2002), 315–354.

[31] R. Curto and L. Fialkow, Solution of the truncated parabolic moment problem, *Integral Equations Operator Theory* **50** (2004), 169–196.

[32] R. Curto and I.B. Jung, Quadratically hyponormal shifts with two equal weights, *Integral Equations Operator Theory* **37** (2000), 208–231.

[33] R. Curto, S.H. Lee and J. Yoon, k-hyponormality of multivariable weighted shifts, *J. Funct. Anal.* **229** (2005), 462–480.

[34] R. Curto, S.H. Lee and J. Yoon, Reconstruction of the Berger measure when the core is of tensor form, *Bibl. Rev. Mat. Iberoamericana* (Actas del XVI Coloquio Latinoamericano de Álgebra, Colonia, Uruguay, 2005), 317–331.

[35] R. Curto, S.H. Lee and J. Yoon, Hyponormality and subnormality for powers of commuting pairs of subnormal operators, *J. Funct. Anal.* **245** (2007), 390–412.

[36] R. Curto, S.H. Lee and J. Yoon, Which 2-hyponormal 2-variable weighted shifts are subnormal?, *Linear Algebra Appl.* **429** (2008), 2227–2238.

[37] R. Curto, S.H. Lee and J. Yoon, A new approach to the 2-variable subnormal completion problem, *J. Math. Anal. Appl.* **370** (2010), 270–283.

[38] R. Curto, S.H. Lee and J. Yoon, One-step extensions of subnormal 2-variable weighted shifts, *Integral Equations Operator Theory* **78** (2014), 415–426.

[39] R. Curto, S.H. Lee and J. Yoon, Quasinormality of powers of commuting pairs of bounded operators, in final stages of preparation.

[40] R. Curto and W.Y. Lee, Towards a model theory for 2-hyponormal operators, *Integral Equations Operator Theory* **44** (2002), 290–315.

[41] R. Curto and P. Muhly, C^*-algebras of multiplication operators on Bergman spaces, *J. Funct. Anal.* **64** (1985), 315–329.

[42] R. Curto, P. Muhly and J. Xia, Hyponormal pairs of commuting operators, *Operator Theory: Adv. Appl.* **35** (1988), 1–22.

[43] R. Curto, Y.T. Poon and J. Yoon, Subnormality of Bergman-like weighted shifts, *J. Math. Anal. Appl.* **308** (2005), 334–342.

[44] R. Curto and M. Putinar, Nearly subnormal operators and moment problems, *J. Funct. Anal.* **115** (1993), 480–497.

[45] R. Curto and M. Putinar, Polynomially hyponormal operators, in *A Glimpse at Hilbert Space Operators: Paul R. Halmos in Memoriam*, Oper. Theory Adv. Appl. **207** (2010), 195–207.

[46] R. Curto and K. Yan, Spectral theory of Reinhardt measures, *Bull. Amer. Math. Soc.* (N.S.) **24** (1991), 379–385.

[47] R. Curto and K. Yan, The spectral picture of Reinhardt measures, *J. Funct. Anal.* **131** (1995), 279–301.

[48] R. Curto and J. Yoon, Jointly hyponormal pairs of subnormal operators need not be subnormal, *Trans. Amer. Math. Soc.* **358** (2006), 5139–5159.

[49] R. Curto and J. Yoon, Disintegration-of-measure techniques for commuting multivariable weighted shifts, *Proc. London Math. Soc.* **92** (2006), 381–402.

[50] R. Curto and J. Yoon, Propagation phenomena for hyponormal 2-variable weighted shifts, *J. Operator Th.* **58** (2007), 175–203.

[51] R. Curto and J. Yoon, Spectral pictures of 2-variable weighted shifts, *C. R. Math. Acad. Sci. Paris, Ser. I* **343** (2006), 579–584.

[52] R. Curto and J. Yoon, When is hyponormality for 2-variable weighted shifts invariant under powers?, *Indiana Univ. Math. J.* **60** (2011), 997–1032.

[53] R. Curto and J. Yoon, Toral and spherical Aluthge transforms for 2-variable weighted shifts, *C. R. Acad. Sci. Paris* **354** (2016), 1200–1204.

[54] R. Curto and J. Yoon, Aluthge transforms of 2-variable weighted shifts, *Integral Equations Operator Theory* (2018) 90:52; 33 pp.

[55] R. Curto and J. Yoon, Spherically quasinormal pairs of commuting operators, *Trends in Math.*, to appear.

[56] M. Dritschel and S. McCullough, private communication.

[57] K. Dykema and H. Schultz, Brown measure and iterates of the Aluthge transform for some operators arising from measurable actions, *Trans. Amer. Math. Soc.* **361** (2009), 6583–6593.

[58] J. Eschmeier and M. Putinar, Some remarks on spherical isometries, *Operator Theory: Adv. Appl.* **129** (2001), 271–291.

[59] G. Exner, Subnormality of transformations of Bergman-like weighted shifts, IWOTA 2006 lecture, Seoul, August 2006.

[60] G. Exner, Aluthge transforms and n-contractivity of weighted shifts, *J. Operator Theory* **61** (2009), 419–438.

[61] E. Franks, Polynomially subnormal operator tuples, *J. Operator Theory* **31** (1994), 219–228.

[62] R. Gellar and L. J. Wallen, Subnormal weighted shifts and the Halmos-Bram criterion, *Proc. Japan Acad.* **46**(1970), 375–378.

[63] J. Gleason, Quasinormality of Toeplitz tuples with analytic symbols, *Houston J. Math.* **32** (2006) 293–298.

[64] P.R. Halmos, Ten problems in Hilbert space, *Bull. Amer. Math. Soc.* **76** (1970), 887–933.

[65] P.R. Halmos, *A Hilbert Space Problem Book*, Second Edition, Graduate Texts in Mathematics, Springer-Verlag, New York, 1982.

[66] Z.J. Jablonski, I.B. Jung and J. Stochel, Weighted shifts on directed trees, *Mem. Amer. Math. Soc.* **216** (2012), no. 1017, viii+106 pp.

[67] N.P. Jewell and A.R. Lubin, Commuting weighted shifts and analytic function theory in several variables, *J. Operator Theory* **1** (1979), 207–223.

[68] I.B. Jung, E. Ko and C. Pearcy, Aluthge transform of operators, *Integral Equations Operator Theory* **37** (2000), 437–448.

[69] I.B. Jung, E. Ko and C. Pearcy, Spectral pictures of Aluthge transforms of operators, *Integral Equations Operator Theory* **40** (2001), 52–60.

[70] I.B. Jung, E. Ko and C. Pearcy, The iterated Aluthge transform of an operator, *Integral Equations Operator Theory* **45** (2003), 375–387.

[71] J. Kim and J. Yoon, Aluthge transforms and common invariant subspaces for a commuting n-tuple of operators, *Integral Equations Operator Theory* **87** (2017) 245–262.

[72] J. Kim and J. Yoon, Taylor spectra and common invariant subspaces through the Duggal and generalized Aluthge transforms for commuting n-tuples of operators (preprint 2018).

[73] M.-K. Kim and E. Ko, Some connections between an operator and its Aluthge transform, *Glasg. Math. J.* **47** (2005), 167–175.

[74] D. Kimsey, The cubic complex moment problem, *Integral Equations Operator Theory* **80** (2014), 353–378.

[75] , D. Kimsey, The subnormal completion problem in several variables, *J. Math. Anal. Appl.* **434** (2016), 1504–1532.

[76] F. Kimura, Analysis of non-normal operators via Aluthge transformation, *Integral Equations Operator Theory* **50** (2004), 375–384.

[77] S. H. Lee, W. Y. Lee and J. Yoon, Subnormality of Aluthge transform of weighted shifts, *Integral Equations Operator Theory* **72** (2012), 241–251.

[78] P. Muhly and J. N. Renault, C*-algebras of multivariable Wiener-Hopf operators, *Trans. Amer. Math. Soc.* **274** (1982), 1–44.

[79] J. N. Renault, A groupoid approach to C*-algebras, in *Lecture Notes in Mathematics*, no. 793, Springer-Verlag, Berlin, 1980.

[80] A. Shields, Weighted shift operators and analytic function theory, *Math. Surveys*, no. 13, Amer. Math. Soc., Providence, 1974, pp. 49–128.

[81] J.L. Smul'jan, An operator Hellinger integral (Russian), *Mat. Sb.* **91** (1959), 381–430.

[82] J. Stampfli, Which weighted shifts are subnormal?, *Pacific J. Math.* **17** (1966), 367–379.

[83] J. L. Taylor, A joint spectrum for several commuting operators, *J. Funct. Anal.* **6** (1970), 172–191.

[84] J. L. Taylor, The analytic functional calculus for several commuting operators, *Acta Math.* **125** (1970), 1–48.

[85] Wolfram Research, Inc., *Mathematica*, Version 8.0, Wolfram Research Inc., Champaign, IL, 2011.

[86] T. Yamazaki, An expression of spectral radius via Aluthge transformation, *Proc. Amer. Math. Soc.* **130** (2002), 1131–1137.

[87] J. Yoon, Schur product techniques for commuting multivariable weighted shifts, *J. Math. Anal. Appl.* **333** (2007), 626–641.

Chapter 3

Commutants, Reducing Subspaces and von Neumann Algebras Associated with Multiplication Operators

Kunyu Guo
School of Mathematical Sciences, Fudan University, Shanghai, 200433, China
kyguo@fudan.edu.cn

Hansong Huang
Department of Mathematics, East China University of Science and Technology, Shanghai, 200237, China
hshuang@ecust.edu.cn

CONTENTS

3.1 Introduction

Let Ω be a bounded domain of $\mathbb{C}^n$ and let $\mathscr{H}$ be a Hilbert space consisting of holomorphic functions on Ω. If for every $\lambda \in \Omega$, the map

$$E_\lambda : f \to f(\lambda), f \in \mathscr{H}$$

defines a bounded evaluation functional, then $\mathscr{H}$ is called *a reproducing kernel Hilbert space* of holomorphic functions on Ω. For example, let dV denote the normalized Lebesgue measure over Ω and $L_a^2(\Omega)$ be the Bergman space of all holomorphic functions f over Ω satisfying

$$\|f\|^2 = \int_\Omega |f(z)|^2 dV(z) < \infty.$$

Then $L_a^2(\Omega)$ is a reproducing kernel Hilbert space. With $\Omega = \mathbb{D}$, the unit disk in $\mathbb{C}$, we obtain the Bergman space $L_a^2(\mathbb{D})$.

Another classic reproducing kernel Hilbert space is the Hardy space $H^2(\mathbb{D})$, which can be regarded as the completion of analytic polynomials p in $\|\cdot\|_2$-norm defined by

$$\|p\|_2 = \sqrt{\int_0^{2\pi} |p(e^{i\theta})|^2 \frac{d\theta}{2\pi}}.$$

An alternative way to define $H^2(\mathbb{D})$ is to consider it as the Hilbert space of all holomorphic functions f on $\mathbb{D}$ satisfying

$$\|f\|_2^2 = \sup_{0<r<1} \int_0^{2\pi} |f(re^{i\theta})|^2 \frac{d\theta}{2\pi} < \infty.$$

The Banach algebra of all bounded holomorphic functions over $\mathbb{D}$ is denoted by $H^\infty(\mathbb{D})$.

If T is a bounded linear operator on a Hilbert space $\mathscr{H}$, and if M is a closed subspace of $\mathscr{H}$ such that $TM \subseteq M$, then M is called an invariant subspace of T. If, in addition, M is invariant under T^*, then M is called a *reducing subspace* of T. A nonzero reducing subspace M is called minimal if it contains no reducing subspace other than $\{0\}$ and itself. We keep the same terminology if T is replaced with a tuple $\mathbf{T}$ of bounded operators on $\mathscr{H}$, and T^* with $\{R^* : R \in \mathbf{T}\}$.

Throughout this paper, $\mathscr{H}$ will denote the Hardy space $H^2(\mathbb{D})$ or the Bergman space $L_a^2(\Omega)$ over a bounded domain Ω in $\mathbb{C}^n$. For a bounded holomorphic function φ on Ω, the map $f \mapsto \varphi f$, $f \in \mathscr{H}$ always defines a bounded operator, and we call it *a multiplication operator* on $\mathscr{H}$ and denote it by M_φ. The function φ is then called the *symbol* for M_φ.

Recall that a von Neumann algebra is a unital C^*-algebra on a Hilbert space that is closed in the weak operator topology. We use $\mathscr{W}^*(\varphi)$ to denote the von Neumann algebra generated by M_φ and put $\mathscr{V}^*(\varphi) \triangleq \mathscr{W}^*(\varphi)'$, the commutant algebra of $\mathscr{W}^*(\varphi)$. It is clear that both $\mathscr{W}^*(\varphi)$ and $\mathscr{V}^*(\varphi)$ are von Neumann algebras, and that $\mathscr{V}^*(\varphi) = \mathbb{C}I$ if and only if M_φ has only trivial reducing subspaces.

In some cases, we will write $\mathscr{W}^*(\varphi,\Omega)$ for $\mathscr{W}^*(\varphi)$ and $\mathscr{V}^*(\varphi,\Omega)$ for $\mathscr{V}^*(\varphi)$ in order to emphasize that the underlying space is the Bergman space $L_a^2(\Omega)$ of the domain Ω. In the same manner, if we consider a tuple $(M_{\varphi_1},\cdots,M_{\varphi_n})$ (in short, M_Φ) of multiplication operators instead of a single operator, we will write $\mathscr{W}^*(\Phi)$ and $\mathscr{V}^*(\Phi)$ for $\mathscr{W}^*(\varphi)$ and $\mathscr{V}^*(\varphi)$, respectively.

Recall that each von Neumann algebra is generated by the projections in it. The following description will be of help in the study of reducing subspaces for M_φ from the point of view of von Neumann algebras. To be precise, for each reducing subspace M, let P_M denote the orthogonal projection onto M. Then P_M commutes with both M_φ and M_φ^*; that is, P_M lies in $\mathscr{V}^*(\varphi)$. Conversely, if P is a projection in $\mathscr{V}^*(\varphi)$, then the range of P is necessarily a reducing subspace for M_φ. Thus $M \mapsto P_M$ defines a bijective correspondence between all reducing subspaces of M_φ and the projections in $\mathscr{V}^*(\varphi)$. This indicates that the study of the lattice of reducing subspaces for M_φ is, in some sense, equivalent to that of projections in $\mathscr{V}^*(\varphi)$. Furthermore, this helps to understand the structures of both $\mathscr{V}^*(\varphi)$ and $\mathscr{W}^*(\varphi)$ in view of the von Neumann bicommutant theorem $\mathscr{V}^*(\varphi)' = \mathscr{W}^*(\varphi)$.

It is worthy pointing out that the unitary equivalence between reducing subspaces of M_φ can be identified with the equivalence between projections in $\mathscr{V}^*(\varphi)$. More precisely, for two reducing subspaces M and N of M_φ, if there is a unitary operator $U : M \to N$ such that $UM_\varphi = M_\varphi U$, then M is said to be unitarily equivalent to N [GSZZ, GH1]. It can be shown that M and N are unitarily equivalent if and only if there is a partial isometry W in $\mathscr{V}^*(\varphi)$ such that $W^*W = P_M$ and $WW^* = P_N$. This means that P_M and P_N, as projections, are equivalent in $\mathscr{V}^*(\varphi)$. For details, we call the readers' attention to [GH4, p. 47].

This survey is on recent developments concerning commutants, reducing subspaces and von Neumann algebras associated with multiplication operators that are defined on both the Hardy space and the Bergman spaces. Section 3.2 presents a historical outline of motivations and developments on commutants and reducing subspaces of multiplication operators, initiated on the Hardy space $H^2(\mathbb{D})$ in the 1970's. Some classic problems are posed along this line as well as some recent developments in this area. For example, some close connections are exhibited between the totally Abelian property of a large class of multiplication operators (defined on both $H^2(\mathbb{D})$ and $L_a^2(\mathbb{D})$) and geometric properties of their symbol curves [DGH]. In Section 3.3, we move from the Hardy space to the Bergman space $L_a^2(\mathbb{D})$, which diverges a lot. There, our main attention is on three classes of multiplication operators, namely, multiplication operators whose symbols are finite Blaschke products, thin Blaschke products, and covering Blaschke products (those Blaschke products that are covering maps). Considering the smoothness of the symbols of multiplication operators, in Section 3.4 we compare such operators defined on the space $L_a^2(\mathbb{D})$ and the Bergman spaces over polygons. Geometry of the polygon turns out to play an indispensable role in the study of reducing subspaces of these multiplication operators. Section 3.5 elaborates on multiplication operators on Bergman spaces of higher dimensional domains, where an interplay of geometry, function theory, and operator algebras will be seen. Finally, in Section 3.6 some open problems and conjectures are presented.

3.2 Commutants and reducing subspaces for multiplication operators on the Hardy space $H^2(\mathbb{D})$

In this section we will give a historical outline of motivations and developments on commutants and reducing subspaces of multiplication operators on the Hardy space $H^2(\mathbb{D})$ in the 1970's. Several considerable advances in this period were achieved, mainly by Abrahamse and Douglas [AD], Cowen [Cow1, Cow2, Cow3], Baker, Deddens and Ullman [BDU], Deddens and Wong [DW], Nordgren [Nor], and Thomson [T1, T2, T3, T4]. Another important reference is [RR].

In general, it is difficult to determine all reducing subspaces of a concrete operator. Nordgren gave a sufficient condition for multiplication operators M_φ on $H^2(\mathbb{D})$ to have no nontrivial reducing subspace [Nor]. To be precise, his theorem says that if M_φ is a multiplication operator, and if there is a Borel subset E of $\partial\mathbb{D}$ with positive Lebesgue measure such that the restriction $\varphi|_E$ is one-to-one and that $\varphi(E)\cap\varphi(\partial\mathbb{D}-E)=\emptyset$, then M_φ has no nontrivial reducing subspace. Besides, a direct observation shows that if ψ is an inner function, then M_ψ defines an isometric operator on the Hardy space $H^2(\mathbb{D})$, so M_ψ has a nontrivial reducing subspace if and only if $\dim H^2(\mathbb{D})\ominus\psi H^2(\mathbb{D})>1$, that is, ψ is not a Möbius transformation. A slightly more generalized result is that if $\varphi=h\circ\psi$, where $h\in H^\infty$ and ψ is an inner function different from a Möbius transformation, then M_φ has nontrivial reducing subspaces [Nor]. To show that its converse is not true, Abrahamse [A1] constructed a holomorphic covering map φ from the unit disk $\mathbb{D}$ onto the annulus $\{z\in\mathbb{C}:\frac{1}{2}<|z|<1\}$, namely,

$$\varphi(z)=\exp(\frac{i}{\pi}\ln\frac{1}{2}\ln\frac{1+z}{1-z}+\frac{1}{2}\ln\frac{1}{2}),\quad z\in\mathbb{D},$$

and proved that the multiplication operator M_φ on $H^2(\mathbb{D})$ has nontrivial reducing subspaces. Furthermore, if ω is an inner function satisfying $\varphi(z)=h(\omega(z))$ for all z in $\mathbb{D}$ with $h\in H^\infty(\mathbb{D})$, then ω is a Möbius transformation. Thus φ provides a desired counterexample.

As was mentioned in the introduction, each reducing subspace of M_φ is exactly the range of a projection in $\mathscr{V}^*(\varphi)$ and $\mathscr{V}^*(\varphi)=\{M_\varphi,M_\varphi^*\}'\subseteq\{M_\varphi\}'$. Thus the problem of characterizing reducing subspaces is generalized to that of determining the commutant $\{M_\varphi\}'$ of M_φ. It was pointed out in [Cow1] that, to understand the structure of an operator T on a Hilbert space, it will be of help and of interest to consider those operators commuting with T.

The first attack on the commutants of subnormal operators was probably made by Shields and Wallen [SWa]. Their method can be used to show that, for a univalent function φ on $\mathbb{D}$, we have $\{M_\varphi\}'=\{M_z\}'=\{M_f:f\in H^\infty(\mathbb{D})\}$; see [DW]. Also, in their paper [DW], Deddens and Wong studied the commutant of a multiplication operator on $H^2(\mathbb{D})$ and raised six interesting questions. These questions had stimulated much further work. See [AB, AC, ACR, AD, Cl, CDG, CGW, Cow1, Cow2, Cow3,

Cu, GL, GLX, Gu2, Gu1, GW1, GW2, JL, Nor, SZ1, SZ2, SZZ1, SZZ2, T1, T2, T3, T4, Zhu1, Zhu2].

Two of the aforementioned questions read as follows.

Question 1. For a function φ in $H^\infty(\mathbb{D})$, let $\varphi = \chi F$ be the inner-outer factorization of φ. Does it hold that $\{M_\varphi\}' = \{M_\chi\}' \cap \{M_F\}'$?

Question 2. Suppose that φ is a nonconstant function in $H^\infty(\mathbb{D})$. Is the zero operator the only compact operator in the commutant of M_φ?

As was mentioned in [DW], the answer to Question 2 is affirmative in the case of $\varphi(z) = z$. Note that the multiplication operator of an inner function is unitarily equivalent to the direct sum of some copies of M_z. Then an affirmative answer to Question 1 implies an affirmative answer to Question 2. Unfortunately, Question 1 is negatively answered by Abrahamse [A1]. Later, Thomson established a remarkable result on commutants of multiplication operators on $H^2(\mathbb{D})$ [T1], and thus provided an affirmative answer to Question 1 under a mild condition. Precisely, if φ is a nonconstant function holomorphic on $\overline{\mathbb{D}}$ and has the inner-outer factorization $\varphi = \chi F$, then $\{M_\varphi\}' = \{M_\chi\}' \cap \{M_F\}'$. Compared with Abrahamse's example [A1], it follows that the condition of φ being holomorphic on $\overline{\mathbb{D}}$ is sharp.

Cowen [Cow3] constructed a function φ in $H^\infty(\mathbb{D})$ such that M_φ commutes with a nonzero compact operator, thus answering Question 2 negatively. Besides, in [Cow1] two theorems are established to give sufficient conditions on h such that $\{M_h\}'$ contains no nonzero compact operator.

Another interesting question from [DW] is the following. *Suppose that $\mathscr{F}$ is a family of inner functions. Is it true that*

$$\{M_f : f \in \mathscr{F}\}' = \{M_B\}'$$

where B is a Blaschke product such that each function f in $\mathscr{F}$ has the form $f = g \circ B$, with $g \in H^\infty(\mathbb{D})$?

Work on the question above can be found in Baker, Deddens and Ullman [BDU], and Thomson [T1, T2]. Eventually, a satisfactory answer is given by Cowen [Cow1].

Theorem 2.1 (Cowen) *Let $\mathscr{F}$ be a family of $H^\infty(\mathbb{D})$-functions. If for some point $a \in \mathbb{D}$ the greatest common divisor of the inner parts of $\{h - h(a) : h \in \mathscr{F}\}$ is a finite Blaschke product, then there is a finite Blaschke product B such that $\cap_{h\in\mathscr{F}}\{M_h\}' = \{M_B\}'$ holds on the Hardy space $H^2(\mathbb{D})$, and for each function h in $\mathscr{F}$, there is an $H^\infty(\mathbb{D})$ function ψ such that $h = \psi(B)$.*

We call the readers' attention to the special situation where $\mathscr{F}$ contains exactly one member. Also note that, before Cowen [Cow1] and inspired by Deddens and Wong's work, Baker, Deddens and Ullman [BDU] proved that for each entire function f there is a positive integer k such that $\{M_f\}' = \{M_{z^k}\}'$ holds on $H^2(\mathbb{D})$. This can be regarded as a consequence of Theorem 2.1.

Thomson [T1, T2, T3, T4] worked on the problem of when the intersection of commutants or a single commutant equals $\{M_B\}'$ for some finite Blaschke product B. As a special case of Theorem 2.1, an easier version of Thomson's result reads as follows [T1].

Theorem 2.2 (Thomson) *If h is holomorphic on a neighborhood of $\overline{\mathbb{D}}$, then there is a finite Blaschke product B and a function g holomorphic on a neighborhood of $\overline{\mathbb{D}}$ such that $h = g \circ B$ and $\{M_h\}' = \{M_B\}'$.*

Theorem 2.2 holds not only on the Hardy space $H^2(\mathbb{D})$ but also on the Bergman space $L_a^2(\mathbb{D})$ and on the Dirichlet space $\mathscr{D}$. See [GH4, Chapter 4] for example. However, for a finite Blaschke product B, little has been done for $\mathscr{V}^*(B)$ defined on the Dirichlet space $\mathscr{D}$. See [CLY, Luo, Zhao].

Recall that an operator T on a Hilbert space H is called totally Abelian if $\{T\}'$ is Abelian, or equivalently, $\{T\}'$ is a maximal Abelian subalgebra of $B(H)$. Berkson and Rubel [BR] completely characterized totally Abelian operators on finite dimensional Hilbert spaces. More specifically, they proved in this case that T is totally Abelian if and only if T has a cyclic vector. For a separable and infinitely dimensional Hilbert space, they also characterized when normal operators (including unitary operators) and non-unitary isometric operators are totally Abelian. In particular, for holomorphic multiplication operators on $H^2(\mathbb{D})$, it is shown that if φ is an inner function, then M_φ is totally Abelian on $H^2(\mathbb{D})$ if and only if there exist a unimodular constant c and a point $\lambda \in \mathbb{D}$ such that $\varphi(z) = c\frac{\lambda - z}{1-\overline{\lambda}z}$ [BR]. Recall that $\{M_z\}' = \{M_h : h \in H^\infty(\mathbb{D})\}$ is maximal Abelian. It follows that a holomorphic multiplication operator M_φ is totally Abelian if and only if $\{M_\varphi\}' = \{M_z\}'$. But in general, it is hard to judge which holomorphic function φ satisfies $\{M_\varphi\}' = \{M_z\}'$.

In the context of holomorphic multiplication operators on both the Hardy and Bergman spaces, when the symbol is a meromorphic function on $\mathbb{C}$, it is shown that there exist some fascinating connections between the totally Abelian property of these operators and geometric properties of their symbol curves. It is found that winding numbers and multiplicities of self-intersection of symbol curves play an important role on this topic [DGH]. To be precise, let $\mathfrak{M}(\overline{\mathbb{D}})$ consist of all meromorphic functions over $\mathbb{C}$ which have no pole on the closed unit disk $\overline{\mathbb{D}}$. For a function φ in $\mathfrak{M}(\overline{\mathbb{D}})$, let $\mathbb{T}$ be the unit circle, and let $N(\varphi - \varphi(\eta), \mathbb{T})$ denote the cardinality of the set $\{w \in \mathbb{T} : \varphi(w) - \varphi(\eta) = 0\}$. We set

$$N(\varphi) = \min\{N(\varphi - \varphi(\eta), \mathbb{T}) : \eta \in \mathbb{T}\}$$

and call it *the multiplicity of self-intersection of the curve* $\varphi(z)\,(z \in \mathbb{T})$[DGH]. It is shown in [DGH] that for a nonconstant function φ in $\mathfrak{M}(\overline{\mathbb{D}})$, the multiplication operator M_φ is totally Abelian if and only if $N(\varphi) = 1$ if and only if there is a finite subset E of $\mathbb{T}$ such that each point $\xi \in \mathbb{T} \backslash E$ satisfies $N(\varphi - \varphi(\xi), \mathbb{T}) = 1$.

Four integer quantities are introduced in [DGH] and they turn out to be equal under a mild condition. More specifically, the finite Blaschke product B in Theorem 2.2 is unique modulo a composition of a holomorphic automorphism of $\mathbb{D}$, and the map φ itself completely determines the order of B, called the Cowen-Thomson order $b(\varphi)$ of φ [T1, DGH]. For $c \notin \varphi(\mathbb{T})$, let wind$(\varphi, c)$ denote the winding number of the curve $\varphi(z)\,(z \in \mathbb{T})$ around the point c and write

$$n(\varphi) = \min\{\mathrm{wind}(\varphi, \varphi(a)) : a \in \mathbb{D}, \varphi(a) \notin \varphi(\mathbb{T})\}.$$

For each continuous function φ on $\mathbb{T}$ define $G(\varphi)$ to be the set of all continuous maps ρ from $\mathbb{T}$ to $\mathbb{T}$ satisfying $\varphi \circ \rho = \varphi$ and put $o(\varphi) = \sharp G(\varphi)$. For a nonconstant function $\varphi \in \mathfrak{M}(\overline{\mathbb{D}})$ the multiplication operator M_φ is totally Abelian if and only if $o(\varphi) = 1$ [DGH]. More generally, Dan, Guo and Huang [DGH] showed that if φ is a nonconstant function in $\mathfrak{M}(\overline{\mathbb{D}})$, then

$$n(\varphi) = b(\varphi) = o(\varphi) = N(\varphi).$$

3.3 The case of the Bergman space $L_a^2(\mathbb{D})$

In the last section we reviewed the commutants and reducing subspaces of multiplication operators on the Hardy space $H^2(\mathbb{D})$ and stated some new results concerning the totally Abelian property in this area. As one moves from $H^2(\mathbb{D})$ to the Bergman space $L_a^2(\mathbb{D})$, the situation becomes much different and much more complicated. Thus in this section we elaborate on reducing subspaces of multiplication operators on the Bergman space $L_a^2(\mathbb{D})$. We will focus our attention on three classes of multiplication operators on $L_a^2(\mathbb{D})$, namely, operators whose symbols are finite Blaschke products, thin Blaschke products, and covering Blaschke products (those Blaschke products that are covering maps). Each of the three cases has its own story.

We first consider multiplication operators induced by finite Blaschke products. To begin with, some observations are in order. For one thing, multiplication operators induced by inner functions, including Blaschke products, play a significant role in the study of the commutant problem under certain conditions [BDU, DW, T1]. For another thing, an analogue of Thomson's theorem on commutants holds on the Bergman space as well. By the remark following Theorem 2.2, if h is holomorphic on a neighborhood of $\overline{\mathbb{D}}$, then there is a finite Blaschke product B and a function $g \in H^\infty(\mathbb{D})$ such that $h = g \circ B$ and $\{M_h\}' = \{M_B\}'$ on the Bergman space. In particular,

$$\mathscr{V}^*(h) = \mathscr{V}^*(B).$$

Thus, our attention turns to the study of M_B, where B is a finite Blaschke product.

On the Bergman space, the topic of commutants and reducing subspaces was initiated by Zhu's conjecture concerning the number of minimal reducing subspaces of the multiplication operator induced by a finite Blaschke product B [Zhu1]. For over a decade, steady progress had been made, and many interesting results mushroomed during this period to exhibit an interplay of analytical, geometrical, operator- and group- theoretical approaches [DSZ, DPW, GH1, GH2, GH3, GH4, GH5, GSZZ, SZZ1, SZZ2].

When a finite Blaschke product B is of order $n = 2$, it was shown that M_B on the Bergman space has exactly two distinct minimal reducing subspaces in [SW] and [Zhu1] independently. Motivated by this fact Zhu conjectured that for a finite Blaschke product B of order n, there are exactly n distinct minimal reducing subspaces [Zhu1]. To attack Zhu's conjecture, one faces a basic problem: *if B is a finite*

Blaschke product and B is not a Möbius map, is there a nontrivial reducing subspace for the multiplication operator M_B on $L_a^2(\mathbb{D})$?

An affirmative answer to the question above was provided by Stessin and Zhu [SZ2], and by Hu, Sun, Xu and Yu [HSXY]. In fact, it was shown in [HSXY] that there is a minimal reducing subspace M_0 such that $M_B|_{M_0}$ is unitarily equivalent to the Bergman shift $M_z : L_a^2(\mathbb{D}) \to L_a^2(\mathbb{D})$. This reducing subspace M_0, called the *distinguished reducing subspace*, turns out to be unique in an appropriate sense [GSZZ].

It is interesting to compare with the case of the Hardy space $H^2(\mathbb{D})$, where the multiplication operator induced by a Blaschke product (more generally, any inner function) is an isometry. It follows from this observation that, for an inner function η different from a Möbius map, the multiplication operator M_η on $H^2(\mathbb{D})$ always has infinitely many minimal reducing subspaces. On the other hand, the multiplication operator M_{z^n} acting on the Bergman space has exactly n distinct minimal reducing subspaces [SZ1], which are given by

$$\overline{\operatorname{span}\{z^{i+kn} : k = 0, 1, \cdots\}}, \quad i = 0, 1, \cdots, n-1.$$

Therefore, the situation of the Bergman space is quite distinct from the Hardy space.

We come back to Zhu's conjecture when the order n of B is greater than 2. In light of [SZZ2, Theorem 3.1], Zhu's conjecture holds for $n > 2$ only if $B(z) = \varphi(z)^n$ for a Möbius transformation φ. Consequently, the conjecture was later modified as follows: *the number of distinct minimal reducing subspaces of M_B on $L_a^2(\mathbb{D})$ is less than or equal to the order of B.* See [GH1].

After a careful study in [GSZZ] of the case when the order of B is equal to 3, Guo, Sun, Zheng and Zhong formulated a more delicate conjecture: *the number of minimal reducing subspaces of M_B on $L_a^2(\mathbb{D})$ equals the number of connected components of the Riemann surface $\mathscr{S}_B$ of $B^{-1} \circ B$ on the unit disk [GSZZ, DSZ].* Here by a Riemann surface we mean a complex manifold of complex dimension 1, not necessarily connected. The amazing part of this modified conjecture is that the operator-theoretic quantity (the number of minimal reducing subspaces of M_B) is accurately linked to a geometric quantity.

An elementary argument shows that the number of minimal reducing subspaces of M_B on $L_a^2(\mathbb{D})$ is less than or equal to the order of B if and only if $\mathscr{V}^*(B)$ is Abelian [GH4, Proposition 2.6.8]. In the cases when the order of B is $3, 4, 5, 6$, this was verified in [GSZZ, SZZ1, GH1], and different techniques involving function theory and operator theory were used at each step forward. Also, by using the techniques of local inverse and group-theoretic methods, Douglas, Sun and Zheng proved that if B is a finite Blaschke product of order 7 or 8, then $\mathscr{V}^*(B)$ is Abelian [DSZ].

The problem was finally solved by Douglas, Putinar and Wang [DPW] in full generality when they provided a conclusive answer to the modified Zhu conjecture. The following is their result.

Theorem 3.1 *For each finite Blaschke product B the von Neumann algebra $\mathscr{V}^*(B)$ is Abelian. Moreover, the following three integers are equal:*

(a) the number of minimal reducing subspaces of M_B on $L_a^2(\mathbb{D})$,

(b) *the dimension of* $\mathscr{V}^*(B)$,

(c) *the number of connected components of the Riemann surface* $\mathscr{S}_B$ *of* $B^{-1} \circ B$ *on the unit disk.*

Combining the theorem above with the second paragraph of this section, we see that a satisfactory conclusion has been reached on the study of reducing subspaces of multiplication operators induced by symbols that are holomorphic over $\overline{\mathbb{D}}$. Moreover, from Theorem 3.1 [GH3], in an appropriate sense, for "most" finite Blaschke products B, M_B has only two distinct minimal reducing subspaces.

However, there is a lot of mystery when it comes to the case of an infinite Blaschke product B. Not much is known about the structure of $\mathscr{V}^*(B)$ in this case. Instead of addressing the problem for a general Blaschke product, we focus on two special classes of infinite Blaschke products: thin Blaschke products and covering Blaschke products. Examples in both classes exist abundantly and they share nice properties.

Roughly speaking, a thin Blaschke product is one that possesses sparse zeros on $\mathbb{D}$ so that its zero sequence tends rapidly to the boundary of $\mathbb{D}$. For example, the zero sequence $\{1-\frac{1}{n!}\}_{n\geq 2}$ gives a thin Blaschke sequence.

Each discrete subset E in $\mathbb{D}$ corresponds to a holomorphic covering map f from $\mathbb{D}$ onto $\mathbb{D}\backslash E$. Furthermore, f is always a Blaschke product if, in addition, $0 \notin E$. See [GH4, Example 6.3.9]. The next question is quite natural.

If B is an infinite Blaschke product, does M_B have a nontrivial reducing subspace?

As was mentioned before, for a finite Blaschke product B of order greater than 1, there always exists a nontrivial reducing subspace for M_B. One may guess that a similar assertion holds for infinite Blaschke products B. However, a negative answer to this question was presented in [GH3] where the counterexample comes in the form of a certain thin Blaschke product. The construction of this thin Blaschke product uses the techniques of local inverse and analytic continuation, and is pretty technical. More specifically, in [GH3], the authors investigate the geometry of the Riemann surface $\mathscr{S}_B$ related to B and show that, for each thin Blaschke product B, $\mathscr{V}^*(B)$ is Abelian. Furthermore, under a mild condition, it holds that $\mathscr{V}^*(B) = \mathbb{C}I$. Consequently, the structure of $\mathscr{V}^*(B)$ is very simple for most thin Blaschke products.

Unexpectedly, things become very different and very complicated if we replace thin Blaschke products with covering Blaschke products. In fact, if φ is a covering map, not only does M_φ have infinitely many reducing subspaces, but also the lattice of these reducing subspaces has a complicated structure, which implies the complexity of the structure of $\mathscr{V}^*(\varphi)$. To make the statement clearer, we need the notation of group von Neumann algebras.

Given a group G, define the left regular representation of G on $l^2(G)$ as follows: for $a \in G$, $L_a f(x) = f(ax)$, where $x \in G, f \in l^2(G)$. Then the map $a \mapsto L_a$ is a unitary representation of G on $l^2(G)$, called the left regular representation of G. Let $\mathscr{L}(G)$ be the closure of the span of all L_a in the weak operator topology. Then $\mathscr{L}(G)$ is a von Neumann algebra, called the group von Neumann algebra associated with G. In general, we have the following result [GH2].

Theorem 3.2 *If φ is a holomorphic covering map from $\mathbb{D}$ onto a bounded planar domain Ω, then $\mathcal{V}^*(\varphi)$ is $*$-isomorphic to the group von Neumann algebra $\mathcal{L}(\pi_1(\Omega))$, where $\pi_1(\Omega)$ is the fundamental group of Ω. Consequently, $\mathcal{V}^*(\varphi)$ is Abelian if and only if the fundamental group $\pi_1(\Omega)$ of Ω is Abelian, or equivalently, Ω is conformally isomorphic to one of the following: a disk, an annulus, or a punctured disk.*

Given a bounded planar domain Ω, $\pi_1(\Omega)$ is either trivial or isomorphic to the free group $F_n(1 \leq n \leq \infty)$ on n generators. Let E be a discrete subset of $\mathbb{D}$ of cardinality n, and let $\Omega = \mathbb{D} - E$. Then $\pi_1(\Omega)$ is isomorphic to F_n, and thus $\mathcal{L}(\pi_1(\Omega)) = \mathcal{L}(F_n)$. By Theorem 3.2, the structure of $\mathcal{V}^*(\varphi)$ has a natural and fascinating connection to one of the long-standing problems in free group factors: that is, for $2 \leq n < m \leq \infty$, does it hold that

$$\mathcal{L}(F_n) \overset{*}{\cong} \mathcal{L}(F_m)?$$

In the language of function theory, this open problem can be reformulated as follows [GH4, Problem 6.5.10].

For $2 \leq n < m \leq \infty$, let Ω_1 and Ω_2 be the unit disk minus n and m discrete points in $\mathbb{D}$, respectively, and let $\varphi_i : \mathbb{D} \to \Omega_i$ be holomorphic covering maps for $i = 1, 2$. Is there a unitary operator U on $L^2_a(\mathbb{D})$ such that $U\mathcal{V}^*(\varphi_1)U^* = \mathcal{V}^*(\varphi_2)$?

Theorem 3.2 can also be generalized to regular holomorphic branched covering maps, which has a close connection with orbifold manifolds. More details are included in [GH2]. For more considerations, see [Huang] and [GH4, Chapter 6].

Abrahamse and Douglas [AD] constructed a class of subnormal operators related to multiply-connected domains, and they considered the von Neumann algebra generated by a single multiplication operator induced by a bounded holomorphic covering map acting on vector-valued Hardy spaces. However, the techniques developed in [GH2] are completely different from those in [AD].

3.4 The case of Bergman space over a polygon

This section concerns our recent study of multiplication operators over the Bergman space of a polygon and the von Neumann algebras induced by these operators. Such operator-theoretic problems turn out to have deep connections with the geometry of the underlying polygon.

A natural question is whether an analogue to Theorem 2.2 can be established on planar domains Ω different from $\mathbb{D}$. Let Ω be a simply connected domain in $\mathbb{C}$ and $\Omega \neq \mathbb{C}$. The well-known Riemann mapping theorem says that there is a biholomorphic map σ from Ω to $\mathbb{D}$. It is natural to make a conjecture analogous to Theorem 2.2 on Ω. Not surprisingly, it holds on those domains Ω with analytic boundaries. In fact, if the boundary of Ω is an analytic closed curve, the Riemann map σ and its inverse map are analytic on $\overline{\Omega}$ and $\overline{\mathbb{D}}$, respectively [BeK, CP]. This leads to

$$H^\infty(\overline{\Omega}) \circ \sigma^{-1} = H^\infty(\overline{\mathbb{D}}).$$

However, the above identity may fail if $\partial\Omega$ has a cusp point. For example, if Ω is a polygon, some new phenomena emerge as we will demonstrate now.

Recall that an n-gon Σ is a polygon with n sides. If h is holomorphic on some neighborhood of $\overline{\Sigma}$ and $h'(a) = 0$ for some point a on $\overline{\Sigma}$, then h is said to have a critical point at a. Huang and Zheng [HZ] prove the following theorem, which implies that for a non-equilateral triangle Σ, the von Neumann algebra $\mathscr{V}^*(h,\Sigma)$ is trivial for "most" of the functions h.

Theorem 4.1 *Suppose Σ is a q-gon for a prime q and h is a holomorphic function on some neighborhood of $\overline{\Sigma}$ with no critical point at the vertices of $\overline{\Sigma}$. If the q-gon is not equiangular, then the von Neumann algebra $\mathscr{V}^*(h,\Sigma)$ is trivial.*

Theorem 4.1 reveals some geometric properties of polygons in terms of the von Neumann algebra $\mathscr{V}^*(h,\Sigma)$, where h is holomorphic on some neighborhood of $\overline{\Sigma}$. However, things are quite different on the unit disk $\mathbb{D}$. For example, if B is a finite Blaschke product with order at least 2, then $\mathscr{V}^*(B)$ is nontrivial. This shows that there exist a lot of non-constant functions $h \in H^\infty(\overline{\mathbb{D}})$ such that $\mathscr{V}^*(h,\mathbb{D})$ is nontrivial. Theorem 4.1 simply states that this fails for most q-gons.

The condition of q being prime is sharp in Theorem 4.1. For example, let Σ be diamond-shaped centered at 0. Define the unitary operator U by

$$Uf(z) = f(-z), \qquad f \in L^2_a(\Sigma).$$

It is easy to see that U is a nontrivial element in $\mathscr{V}^*(z^2,\Sigma)$, so $\mathscr{V}^*(z^2,\Sigma)$ is non-trivial.

In the case when Σ is a quadrilateral, Huang and Zheng [HZ] proved the following result.

Theorem 4.2 *Suppose Σ is a quadrilateral and h is a holomorphic function on some neighborhood of $\overline{\Sigma}$ with no critical point at the vertices of $\overline{\Sigma}$. If Σ is not a parallelogram, then the von Neumann algebra $\mathscr{V}^*(h,\Sigma)$ is trivial.*

For two domains Ω_1 and Ω_2 in $\mathbb{C}$, let $\psi : \Omega_1 \to \Omega_2$ be a holomorphic proper map and let E denote the set of critical points of ψ. It is known that for each point $w \in \Omega_2 \setminus \psi(E)$, the cardinality of $\psi^{-1}(w)$ is a constant integer, which is called the multiplicity of ψ [Ru]. For example, if we write $\Omega_1 = \{z \in \mathbb{C} : \frac{1}{2} < |z| < 2\}$, $\psi(z) = z + \frac{1}{z}$, and $\Omega_2 = \psi(\Omega_1)$, then we can verify that $\psi : \Omega_1 \to \Omega_2$ is a proper map and its multiplicity equals 2. The following is a strong contrast with Theorem 2.2; see [HZ].

Theorem 4.3 *Let Σ be a regular n-gon and let h be holomorphic on some neighborhood of $\overline{\Sigma}$ such that h has no critical point at the vertices of $\overline{\Sigma}$. If $\mathscr{V}^*(h,\Sigma)$ is non-trivial, then there is a bounded holomorphic function $\widetilde{h}$ on Σ and a proper holomorphic map $\psi : \Sigma \to \Sigma$ such that*

$$h = \widetilde{h} \circ \psi,$$

$\mathscr{V}^*(h,\Sigma) = \mathscr{V}^*(\psi,\Sigma)$, and the multiplicity of ψ is a factor of n.

With some extra effort, we can prove that $\dim \mathscr{V}^*(h,\Sigma) \leq n$. However, when the polygon Σ is replaced with the unit disk $\mathbb{D}$, we note that in Theorem 2.2 there is no restriction on the dimension of $\mathscr{V}^*(h,\mathbb{D})$ if h is merely holomorphic on $\overline{\mathbb{D}}$. For example, for each $k \geq 1$, $\mathscr{V}^*(z^k,\mathbb{D}) = k$.

The following example provides more evidence that multiplication operators on the Bergman space of a polygon are quite different from multiplication operators on the Bergman space of the unit disk.

Example 4.4 *Let Σ be an equilateral triangle centered at zero with side length* 1. *In view of Theorem 4.3, it is shown in [HZ] that*

$$\mathscr{V}^*(z^3,\Sigma) = \mathscr{V}^*(z^{3k},\Sigma), \quad k > 1.$$

However, this does not hold on the unit disk, that is,

$$\mathscr{V}^*(z^3,\mathbb{D}) \neq \mathscr{V}^*(z^{3k},\mathbb{D}), \quad k > 1.$$

For the study of multiplication operators on the Bergman spaces over other planar domains, we call the readers' attention to [DK], where a special class of multiplication operators is studied on the Bergman spaces over an annulus.

3.5 The case of the Bergman space over high dimensional domains

In this section we consider multiplication operators on Bergman spaces of high dimensional domains and the associated von Neumann algebras. Again, we will see an interesting interplay of geometry, function theory, and operator algebras.

We begin with the definition of local inverse and analytic continuation. Let Ω_0 be a planar domain and f be a holomorphic function on Ω_0. If ρ is a map defined on a sub-domain V of Ω_0 such that $\rho(V) \subseteq \Omega_0$ and $f(\rho(z)) = f(z)$ for $z \in V$, then ρ is called a local inverse of f on V [T1]. If a local inverse ρ of f admits unrestricted continuation in a open subset of Ω_0, then ρ is called admissible.

For example, a finite Blaschke product B is considered as a holomorphic function on $\mathbb{D}$, and each local inverse of B defined on a sub-domain of $\mathbb{D}$ is admissible [T1].

Before we consider the multi-variable case, we will illustrate how $\mathscr{V}^*(B,\mathbb{D})$ is connected with local inverses of a finite Blaschke product B. Recall from Theorem 2.2 that if φ is a nonconstant holomorphic function over $\overline{\mathbb{D}}$, then there exist a finite Blaschke product B and an $H^\infty(\mathbb{D})$-function ψ such that $\varphi = \psi(B)$ and

$$\{M_\varphi\}' = \{M_B\}'.$$

This immediately gives

$$\mathscr{V}^*(\varphi,\mathbb{D}) = \mathscr{V}^*(B,\mathbb{D}).$$

Since $\dim \mathscr{V}^*(B,\mathbb{D}) < \infty$ for each finite Blaschke product B [DSZ, GH1], we obtain $\dim \mathscr{V}^*(\varphi,\mathbb{D}) < \infty$.

In the study of operators in $\mathscr{V}^*(\varphi,\mathbb{D})$, admissible local inverses prove to be an effective strategy [T1, DSZ]. Both φ and B share the same family of admissible local inverses and each local inverse of B is necessarily admissible.

For a local inverse ρ, let $[\rho]$ denote the equivalence class of ρ, which consists of all analytic continuations of ρ. We define an operator $\mathscr{E}_{[\rho]}$ on the Bergman space $L_a^2(\mathbb{D})$ to be

$$\mathscr{E}_{[\rho]}h = \sum_{\sigma\in[\rho]} h\circ\sigma J\sigma, h \in L_a^2(\mathbb{D}), \tag{3.1}$$

where the sum involves only finitely many terms and $J\sigma$ denotes the determinant of the Jacobian of σ (here, the derivative of σ) for a holomorphic map σ. The local inverses can be defined for a holomorphic map $F : \Omega \to \mathbb{C}^d$, so the definition of the operator $\mathscr{E}_{[\rho]}$ in (3.1) extends naturally to the multi-variable case.

Note that $\mathscr{V}^*(\varphi,\mathbb{D})$ equals the linear span of $\mathscr{E}_{[\rho]}$, where ρ runs over all admissible local inverses of φ. For each finite Blaschke product B, the dimension $\dim \mathscr{V}^*(B,\mathbb{D})$ equals the number of components of the Riemann surface $\mathscr{S}_B$ defined by B [DSZ]. A nontrivial analogue of these proves valid in the multi-variable case.

Theorem 5.1 *Suppose that Ω is a bounded domain in $\mathbb{C}^d$. If the interior of $\overline{\Omega}$ equals Ω, $\Phi : \Omega \to \mathbb{C}^d$ is holomorphic on $\overline{\Omega}$, and the image of Φ has an interior point, then $\mathscr{V}^*(\Phi,\Omega)$ is a finite dimensional von Neumann algebra. Moreover, $\mathscr{V}^*(\Phi,\Omega)$ is generated by $\mathscr{E}_{[\rho]}$, where ρ runs over admissible local inverses of Φ.*

It is worth mentioning that the term "admissible" in the multi-variable case diverges a lot from the case of one variable. The reason is that a local inverse for a holomporphic map may encounter a block of singular points, and yet in the one variable case such points constitute a discrete subset. To conquer this obstacle, Huang and Zheng [HZ2] generalized the notion of admissible local inverses to the multi-variable case. To be precise, for a holomorphic map $\Phi : \Omega \to \mathbb{C}^d$, we assume the image of Φ contains an interior point. Let

$$A = \overline{\Phi^{-1}(\Phi(Z))},$$

where Z denotes the zero variety of the Jacobian $J\Phi$. A local inverse ρ of $\Phi : \Omega \to \mathbb{C}^d$ is called *admissible* if for each curve γ in $\Omega \setminus A$, ρ admits an analytic continuation along γ with values in Ω. Note that ρ may initially be only defined on a smaller sub-domain of Ω.

There are many domains satisfying the following condition on Ω: *the interior of $\overline{\Omega}$ equals* Ω. These include domains with C^1-boundary, star-shaped domains, circled bounded domains, convex domains, bounded symmetric domains, and strictly pseudoconvex domains [HZ2]. Theorem 5.1 was obtained for bounded smooth pseudoconvex domains with some additional assumptions [Ti].

Theorem 5.1 implies the following criterion of when $\mathscr{V}^*(\Phi,\Omega)$ is nontrivial.

Corollary 5.2 *Suppose that Ω is a bounded domain in $\mathbb{C}^d$. If the interior of $\overline{\Omega}$ equals Ω, $\Phi : \Omega \to \mathbb{C}^d$ is holomorphic on $\overline{\Omega}$, and the image of Φ has an interior point, then $\mathscr{V}^*(\Phi,\Omega)$ is nontrivial if and only if there exists an admissible local inverse of Φ distinct from the identity map.*

On the complex plane $\mathbb{C}$, the assumption in Corollary 5.2 reads that Φ is non-constant and holomorphic over $\overline{\Omega}$. For the case of $\Omega = \mathbb{D}$, Corollary 5.2 is implied in [T1]. On a polygon Ω, it was illustrated in Section 3.4 that $\mathscr{V}^*(\Phi,\Omega)$ is often trivial because the nontrivialness of $\mathscr{V}^*(\Phi,\Omega)$ requires some geometric conditions of the polygon Ω [HZ].

On the other hand, it is meaningful to study the structure of $\mathscr{V}^*(\Phi,\Omega)$ for a holomorphic proper map Φ from Ω to $\Phi(\Omega)$, as holomorphic proper maps play the same role on the bounded domain Ω as finite Blaschke products do on the unit disk. The following result shows that in most cases $\mathscr{V}^*(\Phi,\Omega)$ is nontrivial for a holomorphic proper map Φ on Ω [HZ2].

Theorem 5.3 *Suppose $\Phi : \Omega \to \Omega'$ is a holomorphic proper map. Then $\mathscr{V}^*(\Phi,\Omega)$ is nontrivial if and only if Φ is not biholomorphic.*

Since a holomorphic proper map is onto [Ru, Proposition 15.1.5], it is biholomorphic if and only if it is univalent. In higher dimensional complex spaces, "nontrivial" holomorphic proper maps can arise from polynomials. For example, both (z_1+z_2, z_1z_2) and $(z_1^2z_2^5, z_1^2+z_2^5)$ are holomorphic proper maps on the bidisk $\mathbb{D}^2$. On the other hand, (z_1+z_2, z_1z_2) is also a holomorphic proper map on the unit ball $\mathbb{B}_2$ of $\mathbb{C}^2$, but the map $(z_1^2z_2^5, z_1^2+z_2^5)$ on $\mathbb{B}_2$ is not. For more examples, the reader can refer to [HZ2, Section 6].

As was done in [DSZ], for a holomorphic map $\Phi : \Omega \to \mathbb{C}^d$ we define

$$\mathscr{S}_\Phi = \{(z,w) \in \Omega^2 : \Phi(z) = \Phi(w), z \notin \Phi^{-1}(\Phi(\overline{Z}))\},$$

where Z denotes the zero variety of the Jacobian $J\Phi$ of Φ. Then $\mathscr{S}_\Phi$ is a complex manifold in $\mathbb{C}^{2d}$. The following is an analogue to Theorem 7.6 in [DSZ]; see [HZ2].

Theorem 5.4 *Suppose Φ is a holomorphic proper map on Ω. Then $\mathscr{V}^*(\Phi,\Omega)$ is generated by $\mathscr{E}_{[\rho]}$, where ρ are local inverses of Φ. In particular, the dimension of $\mathscr{V}^*(\Phi,\Omega)$ equals the number of components of $\mathscr{S}_\Phi$, and it also equals the number of equivalence classes of local inverses of Φ.*

However, von Neumann algebras induced by holomorphic proper maps of a single variable are considerably different from the multi-variable case. For example, on the unit disk $\mathbb{D}$, $\mathscr{V}^*(\Phi,\mathbb{D})$ is Abelian and the deck transformation group $G(\Phi)$ is cyclic for a holomorphic proper map Φ. In the multi-variable situation, things become complicated. On the one hand, $\mathscr{V}^*(\Phi,\Omega)$ can be non-Abelian for certain maps Φ (such as $(z_1^2z_2^4, z_1^2+z_2^4)$ defined on $\mathbb{D}^2$). On the other hand, for some maps Ψ the deck transformation group $G(\Psi)$ can be non-cyclic and Abelian (for example, let $\Psi = (z_1z_2^2, z_1+z_2^2)$, $(z_1,z_2) \in \mathbb{D}^2$) [HZ2]. Under a mild condition, it was shown in [HZ2] that $\mathscr{V}^*(\Phi,\Omega)$ is $*$-isomorphic to the group von Neumann algebra $\mathscr{L}(G(\Phi))$.

Here is a nontrivial example. For an index $\beta = (\beta_1, \cdots, \beta_d)$ in $\mathbb{Z}_+^d$, define

$$z^\beta = \prod_{j=1}^d z_j^{\beta_j}, z = (z_1, \cdots, z_d) \in \mathbb{C}^d.$$

Let A denote a $d \times d$ $\mathbb{Z}_+$-entry matrix and let $\alpha^1, \cdots, \alpha^d$ be d row vectors from A. Define

$$\Phi_A(z) = (z^{\alpha^1}, \cdots, z^{\alpha^d}), \quad z \in \mathbb{D}^d.$$

Then by the discussions in [GH2, Section 2], Φ_A is a holomorphic proper map from $\mathbb{D}^d$ to its image if and only if $\det A \neq 0$. It follows from further computations that $\mathscr{V}^*(\Phi_A, \mathbb{D}^d)$ is trivial if and only if A satisfies $\det A = \pm 1$ [DH2, Theorem 1.1]. The following is a more concrete example.

Example 5.5 *For $k \in \mathbb{Z}_+$, let*

$$A_1 = \begin{pmatrix} 17 & 2 & 0 \\ 8 & 1 & k \\ 0 & 0 & 1 \end{pmatrix}, A_2 = \begin{pmatrix} 0 & 1 & 0 \\ 0 & 1 & 1 \\ 1 & 0 & 1 \end{pmatrix}, A_3 = \begin{pmatrix} 1 & 1 & 0 \\ 0 & 1 & 1 \\ 1 & 0 & 1 \end{pmatrix}.$$

We have $\det A_1 = \det A_2 = 1$. *Note that* $\Phi_{A_1} = (z_1^{17} z_2^2, z_1^8 z_2 z_3^k, z_3)$, $\Phi_{A_2} = (z_2, z_2 z_3, z_1 z_3)$, *and by [DH2, Theorem 1.1] both* $\mathscr{V}^*(\Phi_{A_1}, \mathbb{D}^3)$ *and* $\mathscr{V}^*(\Phi_{A_2}, \mathbb{D}^3)$ *are trivial. But* $\det A_3 = 2 \neq \pm 1$ *and* $\mathscr{V}^*(\Phi_{A_3}, \mathbb{D}^3) \neq \mathbb{C}I$, *where*

$$\Phi_{A_3} = (z_1 z_2, z_2 z_3, z_1 z_3).$$

Note that A_3 and A_2 have close forms.

For the study of special classes of a single multiplication operator, we call the reader's attention to some recent work in [DH1, GW1, LZ, SL, WDH]. Lu and Zhou [LZ] and Shi and Lu [SL] have investigated reducing subspaces of a single multiplication operator defined by a monomial on $L_{a,\alpha}^2(\mathbb{D}) \otimes L_{a,\alpha}^2(\mathbb{D})$, $\alpha > -1$, where $L_{a,\alpha}^2(\mathbb{D})$ stands for the weighted Bergman space with the weighted measure $(1-|z|^2)^\alpha dA(z)$. All minimal reducing subspaces are characterized in [SL]. Dan and Huang completely characterized all minimal reducing subspaces of $M_{z_1^k + z_2^l}$ $(k \geq 1, l \geq 1)$ on the unweighted Bergman space $L_a^2(\mathbb{D}^2)$ and they obtained that $\mathscr{V}^*(z_1^k + z_2^l, \mathbb{D}^2)$ is C*-isomorphic to the direct sum of finitely many full matrix algebras $M_2(\mathbb{C})$ and $\mathbb{C}$. This result was then generalized to weighted Bergman spaces in [WDH]. Later, Guo and Wang generalized the above results in an operator-theoretic setting, and as an application, they studied the reducing subspaces of multiplication operators $M_{z_1^k + \alpha z_2^l}$ $(\alpha \neq 0)$ and some related problems on general function spaces [GW1, GW2].

In addition to multiplication operators on analytic function spaces, non-analytic Toeplitz operators also began to receive attention in recent years. Albaseer, Lu, and Shi [ALU] completely characterized the non-trivial reducing subspaces for the Toeplitz operators $T_{z_1^n \bar{z}_2^m}$ on the Bergman space $L_a^2(\mathbb{D}^2)$, where n and m are positive integers. Very recently, Gu [Gu2] characterized reducing subspaces for a class of non-analytic Toeplitz operators on Hilbert spaces of holomorphic functions over the bidisk, including the Hardy space and the Dirichlet space.

3.6 Further questions

This section contains some further conjectures and questions, which partially come from [GH4, Chapter 7].

Recall that Section 3.3 mainly deals with the structure of $\mathcal{V}^*(\varphi)$, where φ is from one of the three classes of holomorphic functions: finite Blaschke products, thin Blaschke products, and covering Blaschke products. Thus only a very "tiny" part of all Blaschke products has been investigated. Inspired by the results in Section 3.3, it is natural to raise the following question [GH4].

Question 6.1 *For an infinite Blaschke product B, is there a connection between the commutativity of $\mathcal{V}^*(B)$ on $L_a^2(\mathbb{D})$ and the density of the zero set $Z(B)$ of B in $\mathbb{D}$? If so, how do we describe it?*

Inspired by Theorems 3.1 and 3.2, we make the following conjecture [GH4].

Conjecture 6.2 *For each Blaschke product B, $\mathcal{V}^*(B)$ defined on $L_a^2(\mathbb{D})$ is a finite von Neumann algebra.*

Concerning the commutativity of $\mathcal{V}^*(\varphi)$ defined on $L_a^2(\mathbb{D})$, two conjectures are included in [GH4, Appendix C].

Conjecture 6.3 *Suppose that $\varphi \in H^\infty(\mathbb{D})$ satisfies $Z(\varphi) \cap \mathbb{D} \neq \emptyset$. If M is a reducing subspace of M_φ such that $\dim M \ominus \varphi M < \infty$, then P_M lies in the center of $\mathcal{V}^*(\varphi)$.*

Note that if Conjecture 6.3 holds, then under the same assumption the von Neumann algebra $P_M \mathcal{V}^*(\varphi)|_M$ is Abelian. This follows from an easy observation: any orthogonal projection in $P_M \mathcal{V}^*(\varphi)|_M$ has the form P_N, where N is a reducing subspace satisfying $N \subseteq M$, and thus

$$\dim N \ominus \varphi N \leq \dim M \ominus \varphi M < \infty.$$

By Conjecture 6.3 all projections P_N lie in the center of $\mathcal{V}^*(\varphi)$ and hence $P_M \mathcal{V}^*(\varphi)|_M$ is Abelian.

Conjecture 6.4 *Under the assumption of Conjecture 6.3, the von Neumann algebra $P_M \mathcal{V}^*(\varphi)|_M$ is Abelian.*

In Section 3.4 a tuple of multiplication operators are discussed. It is of interest to ask whether an analogue of Theorem 2.2 holds in the multi-variable case. Note that a finite Blaschke product is a holomorphic proper map from the unit disk $\mathbb{D}$ to $\mathbb{D}$ itself. Thus we raise the following open problem.

Let $\Omega = \mathbb{B}_d$ or $\Omega = \mathbb{D}^d$. If $\Phi : \Omega \to \mathbb{C}^d$ is holomorphic over $\overline{\Omega}$, then is there a holomorphic proper map $\Psi : \Omega \to \Psi(\Omega)$ and a bounded holomorphic map F on $\Psi(\Omega)$ such that $\Phi = F \circ \Psi$ and $\{M_\Phi\}' = \{M_\Psi\}'$ holds on $L_a^2(\Omega)$? Recall that for $\Phi = (\varphi_1, \cdots, \varphi_d)$, $\{M_\Phi\}' = \{M_{\varphi_1}, \cdots, M_{\varphi_d}\}'$.

The above problem remains open even in the case of Φ being a tuple of polynomials. It is also natural to raise an analogue of the above problem in the case of multiply connected planar domains. Specifically, for $0 < r < 1$ put

$$A_r = \{z \in \mathbb{C} : r < |z| < \frac{1}{r}\}.$$

Noting that $z \mapsto z^n + \frac{1}{z^n}$ defines a holomorphic proper map from A_r onto its image, we ask the following question.

Suppose $\varphi : A_r \to \mathbb{C}$ is holomorphic over $\overline{A_r}$. Is there a positive integer n such that φ can be written as a function of $\psi(z) = z^n + \frac{1}{z^n}$ or $\psi(z) = z^n$ and $\{M_\varphi\}' = \{M_\psi\}'$ holds on $L_a^2(A_r)$?

It turns out that a simple application of the methods in [T1] and [T2] is not enough to answer the above question. A partial reason lies in that the product of all local inverses of the map $\psi(z) = z^n + \frac{1}{z^n}$,

$$z, \frac{1}{z}; \omega z, \frac{1}{\omega z}; \cdots, \omega^{n-1} z, \frac{1}{\omega^{n-1} z}, \quad (\omega = e^{\frac{2\pi i}{n}})$$

equals 1, although the product of all local inverses of a finite Blaschke product equals itself modulo the composition of a holomorphic automorphism of $\mathbb{D}$.

Acknowledgment

This work is supported by the National Natural Science Foundation of China.

References

[A1] M. Abrahamse, Analytic Toeplitz operators with automorphic symbol, *Proc. Amer. Math. Soc.* **52** (1975), 297-302.

[AB] M. Abrahamse and J. Ball, Analytic Toeplitz operators with automorphic symbol, II, *Proc. Amer. Math. Soc.* **59** (1976), 323-328.

[AC] S. Axler and Z. Cuckovic, Commuting Toeplitz operators with harmonic symbols, *Integr. Equ. Oper. Theory* **14** (1991), 1-12.

[ACR] S. Axler, Z. Cuckovic and N. Rao, Commutants of analytic Toeplitz operators on the Bergman space, *Proc. Amer. Math. Soc.* **128** (2000), 1951-1953.

[AD] M. Abrahamse and R. Douglas, A class of subnormal operators related to multiply-connected domains, *Adv. Math.* **19** (1976), 106-148.

[ALU] M. Albaseer, Y. Lu, and Y. Shi, Reducing subspaces for a class of Toeplitz operators on the Bergman space of the bidisk, *Bull. Korean Math. Soc.* **52** (2015), 1649-1660.

[BDU] I. Baker, J. Deddens and J. Ullman, A theorem on entire functions with applications to Toeplitz operators, *Duke Math. J.* **41** (1974), 739-745.

[BeK] S. Bell and S. Krantz, Smoothness to the boundary of conformal maps, *Rocky Mountain J. Math.* **17** (1987), 23-40.

[BR] E. Berkson and L. Rubel, Totally Abelian operators and analytic functions, *Math. Ann.* **204** (1973), 57-63.

[CDG] L. Chen, R. Douglas and K. Guo, On the double commutant of Cowen-Douglas operators, *J. Funct. Anal.* **260** (2011), 1925-1943.

[CGW] G. Cheng, K. Guo and K. Wang, Transitive algebras and reductive algebras on reproducing analytic Hilbert spaces, *J. Funct. Anal.* **258** (2010), 4229-4250.

[Cl] B. Cload, Toeplitz operators in the commutant of a composition operator, *Studia Math.* **133** (1999), 187-196.

[CLY] Y. Chen, Y. J. Lee and T. Yu, Reducibility and unitary equivalence for a class of multiplication operators on the Dirichlet space, *Studia Math.* **220** (2014), 141-156.

[Cow1] C. Cowen, The commutant of an analytic Toeplitz operator, *Trans. Amer. Math. Soc.* **239** (1978), 1-31.

[Cow2] C. Cowen, The commutant of an analytic Toeplitz operator, II, *Indiana Univ. Math. J.* **29** (1980), 1-12.

[Cow3] C. Cowen, An analytic Toeplitz operator that commutes with a compact operator and a related class of Toeplitz operators, *J. Funct. Anal.* **36** (1980), 169-184.

[CP] S. Choy and S. Pai, On the regularity of the Riemann mapping function in the plane, http://maths.sogang.ac.kr/shcho/pdf/P18.pdf

[Cu] Z. Cuckovic, Commutants of Toeplitz operators on the Bergman space, *Pacific J. Math.* **162** (1994), 277-285.

[DGH] H. Dan, K. Guo and H. Huang, Totally Abelian Toeplitz operators and geometric invariants associated with their symbol curves, *J. Funct. Anal.* **273** (2017), 559-597.

[DH1] H. Dan and H. Huang, Multiplication operators defined by a class of polynomials on $L_a^2(\mathbb{D}^2)$, *Integr. Equ. Oper. Theory* **80** (2014), 581-601.

[DH2] H. Dan and H. Huang, Multiplication operators on Bergman spaces over polydisks associated with integer matrix, *Bull. Kor. Math. Soc.* **55** (2018), 41-50.

[DK] R. Douglas and Y. Kim, Reducing subspaces on the annulus, *Integr. Equ. Oper. Theory* **70** (2011), 1-15.

[DPW] R. Douglas, M. Putinar and K. Wang, Reducing subspaces for analytic multipliers of the Bergman space, *J. Funct. Anal.* **263** (2012), 1744-1765.

[DW] J. Deddens and T. Wong, The commutant of analytic Toeplitz operators, *Trans. Amer. Math. Soc.* **184** (1973), 261-273.

[DSZ] R. Douglas, S. Sun and D. Zheng, Multiplication operators on the Bergman space via analytic continuation, *Adv. Math.* **226** (2011), 541-583.

[GH1] K. Guo and H. Huang, On multiplication operators of the Bergman space: Similarity, unitary equivalence and reducing subspaces, *J. Operator Theory* **65** (2011), 355-378.

[GH2] K. Guo and H. Huang, Multiplication operators defined by covering maps on the Bergman space: the connection between operator theory and von Neumann algebras, *J. Funct. Anal.* **260** (2011), 1219-1255.

[GH3] K. Guo and H. Huang, Geometric constructions of thin Blaschke products and reducing subspace problem, *Proc. London Math. Soc.* **109** (2014), 1050-1091.

[GH4] K. Guo and H. Huang, *Multiplication operators on the Bergman space*, Lecture Notes in Mathematics **2145**, Springer, Heidelberg, 2015.

[GH5] K. Guo and H. Huang, Reducing subspaces of multiplication operators on function spaces: Dedicated to the memory of Chen Kien-Kwong on the 120th anniversary of his birth, *Appl. Math. J. Chinese Univ.* **28** (2013), 395-404.

[Gu1] C. Gu, Reducing subspaces of weighted shifts with operator weights, *Bull. Kor. Math. Soc.* **53** (2016), 1471-1481.

[Gu2] C. Gu, Reducing subspaces of non-analytic Toeplitz operators on weighted Hardy and Dirichlet spaces of the bidisk, *J. Math. Anal. Appl.* **459** (2018), 980-996.

[GL] C. Gu and S. Luo, Composition and multiplication operators on the derivative Hardy space $S^2(\mathbb{D})$, *Complex Var. Elliptic Equ.* **63** (2018), 599-624.

[GLX] C. Gu, S. Luo and J. Xiao, Reducing subspaces for multiplication operators on the Dirichlet space through local inverses and Riemann surfaces, *Complex Manifolds* **4** (2017), 84-119.

[GW1] K. Guo and X. Wang, Reducing subspaces of tensor products of weighted shifts, *Sci. China Ser. A.* **59** (2016), 715-730.

[GW2] K. Guo and X. Wang, The graded structure induced by operators on a Hilbert space, *J. Math. Soc. Japan* **70** (2018), 853-875.

[GSZZ] K. Guo, S. Sun, D. Zheng and C. Zhong, Multiplication operators on the Bergman space via the Hardy space of the bidisk, *J. Reine Angew. Math.* **629** (2009), 129-168.

[Huang] H. Huang, von Neumann algebras generated by multiplication operators on the weighted Bergman space: a function-theory view into operator theory, *Sci. China Ser. A.* **56** (2013), 811-822.

[HSXY] J. Hu, S. Sun, X. Xu and D. Yu, Reducing subspace of analytic Toeplitz operators on the Bergman space, *Integr. Equ. Oper. Theory* **49** (2004), 387-395.

[HZ] H. Huang and D. Zheng, Operators on the Bergman spaces on polygons vs geometry of polygons, *J. Math. Anal. Appl.* **456** (2017), 1049-1061.

[HZ2] H. Huang and D. Zheng, Multiplication operators on the Bergman space of bounded domains in $\mathbb{C}^d$, preprint.

[JL] C. Jiang and Y. Li, The commutant and similarity invariant of analytic Toeplitz operators on Bergman space, *Sci. China Ser. A.* **5** (2007), 651-664.

[Luo] S. Luo, Reducing subspaces of multiplication operators on the Dirichlet space, *Integr. Equ. Oper. Theory* **85** (2016), 539-554.

[LZ] Y. Lu and X. Zhou, Invariant subspaces and reducing subspaces of weighted Bergman space over bidisk, *J. Math. Soc. Japan* **62** (2010), 745-765.

[Nor] E. Nordgren, Reducing subspaces of analytic Toeplitz operators, *Duke Math. J.* **34** (1967), 175-181.

[RR] H. Radjavi and P. Rosenthal, *Invariant Subspaces*, Springer-Verlag, New York, 1973.

[Ru] W. Rudin, *Function Theory in the Unit Ball of* $\mathbb{C}^n$, Grundlehren der Math. **241**, Springer, New York, 1980.

[SL] Y. Shi and Y. Lu, Reducing subspaces for Toeplitz operators on the polydisk, *Bull. Kor. Math. Soc.* **50** (2013), 687-696.

[SW] S. L. Sun and Y. Wang, Reducing subspaces of certain analytic Toeplitz operators on the Bergman space, *Northeastern Math. J.* **14** (1998), 147-158.

[SWa] A. Shields and L. Wallen, The commutants of certain Hilbert space operators, *Indiana Univ. Math. J.* **20** (1970/71), 777-788.

[SZ1] M. Stessin and K. Zhu, Reducing subspaces of weighted shift operators, *Proc. Amer. Math. Soc.* **130** (2002), 2631-2639.

[SZ2] M. Stessin and K. Zhu, Generalized factorization in Hardy spaces and the commutant of Toeplitz operators, *Canad. J. Math.* **55** (2003), 379-400.

[SZZ1] S. Sun, D. Zheng and C. Zhong, Classification of reducing subspaces of a class of multiplication operators via the Hardy space of the bidisk, *Canad. J. Math.* **62** (2010), 415-438.

[SZZ2] S. Sun, D. Zheng and C. Zhong, Multiplication operators on the Bergman space and weighted shifts, *J. Operator Theory* **59** (2008), 435-452.

[T1] J. Thomson, The commutant of a class of analytic Toeplitz operators, *Amer. J. Math.* **99** (1977), 522-529.

[T2] J. Thomson, The commutant of a class of analytic Toeplitz operators, II, *Indiana Univ. Math. J.* **25** (1976), 793-800.

[T3] J. Thomson, The commutant of certain analytic Toeplitz operators, *Proc. Amer. Math. Soc.* **54** (1976), 165-169.

[T4] J. Thomson, Intersections of commutants of analytic Toeplitz operators, *Proc. Amer. Math. Soc.* **52** (1975), 305-310.

[Ti] A. Tikaradze, Multiplication operators on the Bergman spaces of pseudo-convex domains, *New York J. Math.* **21** (2015), 1327-1345.

[WDH] X. Wang, H. Dan and H. Huang, Reducing subspaces of multiplication operators with the symbol $\alpha z^k + \beta w^l$ on $L_a^2(\mathbb{D}^2)$, *Sci. China Ser. A.* **58** (2015), 2167-2180.

[Zhao] L. Zhao, Reducing subspaces for a class of multiplication operators on the Dirichlet space, *Proc. Amer. Math. Soc.* **139** (2009), 3091-3097.

[Zhu1] K. Zhu, Reducing subspaces for a class of multiplication operators, *J. London Math. Soc.* **62** (2000), 553-568.

[Zhu2] K. Zhu, Irreducible multiplication operators on spaces of analytic functions, *J. Operator Theory* **51** (2004), 377-385.

Chapter 4

Operators in the Cowen-Douglas Class and Related Topics

Gadadhar Misra
Department of Mathematics, Indian Institute of Science, Bangalore 560012, India
gm@iisc.ac.in

CONTENTS

Linear spaces with an Euclidean metric are ubiquitous in mathematics, arising both from quadratic forms and inner products. Operators on such spaces also occur naturally. In recent years, the study of multivariate operator theory has made substantial progress. Although, the study of self adjoint operators goes back a few decades, the non-self adjoint theory has developed at a slower pace. While several approaches to this topic have been developed, the one that has been most fruitful is clearly the study of Hilbert spaces that are modules over natural function algebras like $\mathscr{A}(\Omega)$, where $\Omega \subseteq \mathbb{C}^m$ is a bounded domain, consisting of complex valued functions which are holomorphic on some open set U containing $\overline{\Omega}$, the closure of Ω. The book [29] showed how to recast many of the familiar theorems of operator theory in the language of Hilbert modules. The books [31] and [14] provide an account of the achievements from the recent past. The impetus for much of what is described below comes from the interplay of operator theory with other areas of mathematics like complex geometry and representation theory of locally compact groups.

4.1 Introduction

The first half of this expository article describes several elementary properties of the operators in the Cowen-Douglas class. This is divided into five separate themes. In the second half of the article, we elaborate a little more on each of these themes.

4.1.1 Operators in the Cowen-Douglas class

In the paper [17], Cowen and Douglas initiated a systematic study of a class of bounded linear operators on a complex separable Hilbert space possessing an open set of eigenvalues of constant (and finite) multiplicity. Let Ω be the set of eigenvalues of the operator $T : \mathscr{H} \to \mathscr{H}$ in this class. Assuming that $\operatorname{ran}(T - wI) = \mathscr{H}$, using elementary Fredholm theory, they prove: For a fixed but arbitrary $w_0 \in \Omega$, there is an open neighbourhood U of w_0 and holomorphic functions

$$\gamma_i : U \to \mathscr{H},\ (T - w)\gamma_i(w) = 0,\ 1 \leq i \leq n,$$

such that the vectors $\{\gamma_1(w), \dots, \gamma_n(w)\}$ are linearly independent, $w \in U$. They also show that such an operator T defines a holomorphic Hermitian vector bundle E_T:

$$t : \Omega \to \operatorname{Gr}(n, \mathscr{H}),\ t(w) = \ker(T - w) \subseteq \mathscr{H}.$$

This means, for any fixed but arbitrary point $w_0 \in \Omega$, there exists a holomorphic map γ_T of the form $\gamma_T(w) = (\gamma_1(w), \dots, \gamma_n(w))$, $(T - w)\gamma_i(w) = 0$ in some open neighbourhood U of w_0. It is called a holomorphic frame for the operator T. Finally, Cowen and Douglas also assume that the linear span of $\{\gamma_1(w), \dots, \gamma_n(w) : w \in \Omega\}$ is dense in $\mathscr{H}$. Let $B_n(\Omega)$ denote this class of operators.

One of the striking results of Cowen and Douglas says that there is a one to one correspondence between the unitary equivalence class of the operators T and the (local) equivalence classes of the holomorphic Hermitian vector bundles E_T determined by them. As a result of this correspondence set up by the Cowen-Douglas theorem, the invariants of the vector bundle E_T like the curvature, the second fundamental form, etc. now serve as unitary invariants for the operator T, although finding a complete set of tractable invariants, not surprisingly, is much more challenging. Examples were given in [51, Example 2.1] to show that the class of the curvature alone does not determine the class of the vector bundle except in the case of a line bundle. Before we consider this case in some detail, let us recall the interesting notion of a spanning section. A holomorphic function $s : \Omega \to \mathscr{H}$ is called a spanning section for an operator T in the Cowen-Douglas class if $\ker(T - w)s(w) = 0$ and the closed linear span of $\{s(w) : w \in \Omega\}$ is $\mathscr{H}$. Kehe Zhu in [62] proved the existence of a spanning section for an operator T in $B_n(\Omega)$ and showed that it can be used to characterize Cowen-Douglas operators of rank n up to unitary equivalence and similarity. Unfortunately, the existential nature of the spanning section makes it difficult to apply this result in concrete examples.

First note that the holomorphic frame γ_T is *not* uniquely determined even if the rank $n = 1$. If γ_T is any given holomorphic frame for the operator T defined on an open set $\Omega \subseteq \mathbb{C}$ and $\varphi : \Omega \to \mathbb{C}$ is a non-vanishing holomorphic function, then $\varphi\gamma_T$ is also a holomorphic frame for the line bundle E_T. Therefore, a holomorphic frame can't possibly determine the unitary equivalence class of the operator T. How does one get rid of this ambiguity in the holomorphic frame to obtain an invariant? It is evident that

$$\mathscr{K}_T(w) = -\frac{\partial^2}{\partial w \partial \bar{w}} \log \|\gamma_T(w)\|^2,\ w \in \Omega_0, \tag{4.1}$$

is the same for all holomorphic frames of the form $\varphi\gamma_T$, where φ is any non-vanishing holomorphic function on some open set $\Omega_0 \subseteq \Omega$. Since any two holomorphic frames of the operator T must differ by such a holomorphic change of frame, we conclude that $\mathscr{K}_T$ is a unitary invariant for the operator T. The converse is also valid and is well-known: The curvature $\mathscr{K}_T$ of the line bundle E_T is defined by the formula (4.1) and is a complete invariant for the line bundle E_T.

To see the usefulness of this invariant, consider the weighted unilateral shift W_λ determined by the weight sequence $\{\sqrt{\frac{n+1}{n+\lambda}}\}$, $\lambda > 0$, acting on the Hilbert space ℓ^2 of square summable sequences. Clearly, the adjoint W_λ^* admits a holomorphic frame. For instance, one may choose $\gamma_{W_\lambda^*}(w) = (1, c_1 w, c_2 w^2, \ldots)$, $w \in \mathbb{D}$, where c_k^2 is the co-efficient of x^k in the binomial expansion of the function $(1-x)^{-\lambda}$. It then follows that $\|\gamma_{W_\lambda^*}(w)\|^2 = (1-|w|^2)^{-\lambda}$ and that $\mathscr{K}_{W_\lambda^*}(w) = -\lambda(1-|w|^2)^{-2}$, $w \in \mathbb{D}$. Consequently, using the Cowen-Douglas theorem, we conclude that none of the operators W_λ are unitarily equivalent among themselves.

Finding similarity invariants for operators in the class $B_n(\Omega)$ has been somewhat difficult from the beginning. The conjecture made by Cowen and Douglas in [17] was shown to be false [15, 16]. However, significant progress on the question of similarity has been made recently (cf. [36, 37, 23, 40]).

After a model and a complete set of unitary invariants were provided for the operators in the class $B_n(\Omega)$ in [17], it was only natural to follow it up with the study of a commuting tuple of operators with similar properties. This was started in [18] and followed up in the papers [20] and [21]. The approaches in the papers [20] and [21] are quite different. We recall below the definition of the class $B_n(\Omega)$, $\Omega \subset \mathbb{C}^m$, from the paper [21]. This definition also appears in [18] and is implicit in [20].

Let $T = (T_1, \dots, T_m)$ be an m-tuple of commuting bounded linear operators on a separable complex Hilbert space $\mathscr{H}$. For $w = (w_1, \dots, w_m)$ in Ω, let $T - w$ denote the m-tuple $(T_1 - w_1, \dots, T_m - w_m)$. Define the operator $D_T : \mathscr{H} \to \mathscr{H} \oplus \cdots \oplus \mathscr{H}$ by

$$D_T(x) = (T_1 x, \dots, T_m x),\ x \in \mathscr{H}.$$

Definition 1.1 *For a connected, bounded and open subset Ω of $\mathbb{C}^m$, a m-tuple T is said to be in the Cowen-Douglas class $B_n(\Omega)$ of rank n, $n \in \mathbb{N}$, if*

(i) ran D_{T-w} *is closed for all* $w \in \Omega$

(ii) span $\{\ker D_{T-w} : w \in \Omega\}$ *is dense in* $\mathscr{H}$

(iii) dim ker $D_{T-w} = n$ *for all* $w \in \Omega$.

For $m = 1$, it is shown in [17, Proposition 1.12] that if T is in $B_n(\Omega)$, then there exists a choice of n eigenvectors in $\ker(T - w)$, which are holomorphic as functions of $w \in \Omega$ making

$$E_T := \{(w,x) : w \in \Omega, x \in \ker(T - w)\} \subseteq \Omega \times \mathscr{H},$$

$\pi : E_T \to \Omega$, $\pi(w,x) = w$, a rank n holomorphic Hermitian vector bundle over Ω. Here is one of the main results from [17].

Theorem (Cowen and Douglas). *The operators T and $\hat{T}$ in $B_n(\Omega)$ are unitarily equivalent if and only if the corresponding holomorphic Hermitian vector bundles E_T and $E_{\hat{T}}$ are equivalent on some open subset Ω_0 of Ω.*

The existence of the vector bundle E_T follows from [21, Theorem 2], while [21, Theorem 3.7] provides the analogue of the Cowen-Douglas Theorem for an arbitrary m. Finally, a complete set of invariants in this case is given in [20].

Crucial in any study of such a class is the problem of finding a canonical model and a set of invariants. For normal operators, the spectral theorem provides a model in the form of a multiplication operator and a complete set of invariants is given by the spectrum, the spectral measure and the multiplicity function. Similarly, the Sz.-Nagy – Foias theory provides a model for a pure completely nonunitary contraction and the characteristic function serves as a complete invariant. Now, we describe a model for the operators, resp. commuting tuples, in the Cowen-Douglas class.

Let V be a n-dimensional Hilbert space and $\mathscr{L}(V)$ denote the vector space of all linear transformations on V. A function $K : \Omega \times \Omega \to \mathscr{L}(V)$, satisfying

$$\sum_{i,j=1}^{N} \langle K(w_i, w_j)\zeta_j, \zeta_i \rangle_V \geq 0,\ w_1, \dots, w_N \in \Omega,\ \zeta_1, \dots, \zeta_N \in V, N \geq 1 \qquad (4.2)$$

is said to be a *non negative definite (nnd) kernel* on Ω. Given such an nnd kernel K on Ω, it is easy to construct a Hilbert space $\mathscr{H}$ of functions on Ω taking values in V with the property

$$\langle f(w), \zeta\rangle_V = \langle f, K(\cdot,w)\zeta\rangle_{\mathscr{H}},\ w \in \Omega,\ \zeta \in V,\ f \in \mathscr{H}. \tag{4.3}$$

The Hilbert space $\mathscr{H}$ is simply the completion of the linear span of all vectors of the form $K(\cdot,w)\zeta$, $w \in \Omega$, $\zeta \in V$, with inner product defined by (4.3).

Conversely, let $\mathscr{H}$ be any Hilbert space of functions on Ω taking values in V. Let $e_w : \mathscr{H} \to V$ be the evaluation functional defined by $e_w(f) = f(w)$, $w \in \Omega$, $f \in \mathscr{H}$. If e_w is bounded for each $w \in \Omega$ then it is easy to verify that the Hilbert space $\mathscr{H}$ possesses a reproducing kernel $K(z,w) = e_z e_w^*$, that is, $K(\cdot,w)\zeta \in \mathscr{H}$ for each $w \in \Omega$ and K has the reproducing property (4.3). Finally, the reproducing property (4.3) determines the kernel K uniquely. We let $(\mathscr{H},K)$ be the Hilbert space $\mathscr{H}$ equipped with the reproducing kernel K.

Remark 1.2 *Let $K : \Omega \times \Omega \to \mathscr{M}_k(\mathbb{C})$ be a non-negative definite kernel. For every $i \in \mathbb{Z}_+^m$, $\eta \in \mathbb{C}^k$ and $w \in \Omega$, we have*

(i) $\bar{\partial}^i K(\cdot,w)\eta$ *is in* $(\mathscr{H},K)$,

(ii) $\langle f, \bar{\partial}^i K(\cdot,w)\eta\rangle = \langle (\partial^i f)(w), \eta\rangle_{\mathbb{C}^k}, f \in (\mathscr{H},K)$.

The proof follows from the uniform boundedness principle [33, Proposition 2.1.3].

Given any m-tuple of operators T in $B_n(\Omega)$, there exists an open subset U of Ω and n linearly independent vectors $\gamma_1(w),\ldots,\gamma_n(w)$ in $\ker D_{T-w}$, $w \in U$, such that each of the maps $w \mapsto \gamma_i(w)$ is holomorphic on U, see [17, Proposition 1.11] and [21, Theorem 2.2]. Define $\hat{\Gamma} : U \to \mathscr{L}(\mathbb{C}^n,\mathscr{H})$ by setting

$$\hat{\Gamma}(w)\zeta = \sum_{i=1}^n \zeta_i \gamma_i(w),\ \zeta = (\zeta_1,\ldots,\zeta_n) \in \mathbb{C}^n.$$

Let $\mathscr{O}(U,\mathbb{C}^n)$ denote the linear space of holomorphic functions on U taking values in $\mathbb{C}^n$. Set $U^* := \{w : \overline{w} \in U\}$. Define $\Gamma : \mathscr{H} \to \mathscr{O}(U^*,\mathbb{C}^n)$ by

$$(\Gamma x)(w) = \hat{\Gamma}(\overline{w})^* x,\ x \in \mathscr{H},\ w \in U^*. \tag{4.4}$$

Define a sesqui-linear form on $\mathscr{H}_\Gamma = \operatorname{ran}\Gamma$ by $\langle \Gamma f, \Gamma g\rangle_\Gamma = \langle f, g\rangle$, $f,g \in \mathscr{H}$. The map Γ is linear and injective. Hence $\mathscr{H}_\Gamma$ is a Hilbert space of $\mathbb{C}^n$-valued holomorphic functions on U^* with inner product $\langle\cdot,\cdot\rangle_\Gamma$ and Γ is unitary. Then it is easy to verify the following (cf. [17, pp. 194] and [21, Remarks 2.6]).

a) $K(z,w) = \hat{\Gamma}(\bar{z})^*\hat{\Gamma}(\overline{w})$, $z,w \in U^*$ is the reproducing kernel for the Hilbert space $\mathscr{H}_\Gamma$.

b) $M_i^*\Gamma = \Gamma T_i$, where $(M_i f)(z) = z_i f(z)$, $z \in U^*$.

Thus any commuting m-tuple T of operators in the class $B_n(\Omega)$ may be realized as the adjoint of the m-tuple $M := (M_1, \ldots, M_m)$ of multiplication by the coordinate functions on some Hilbert space $\mathscr{H}$ of holomorphic functions defined on U^* possessing a reproducing kernel K. In this representation, clearly, $\Gamma(\gamma_i(w)) = K(\cdot, \bar{w})\varepsilon_i$, $1 \leq i \leq n$ is a holomorphic frame.

We give this correspondence for any commuting tuple of operators in $B_1(\Omega)$ adding that except for a slight increase in the notational complexity, the same proof works in general.

Let γ be a non-zero holomorphic section defined on some open subset U of Ω for the operator T acting on the Hilbert space $\mathscr{H}$. Consider the map $\Gamma : \mathscr{H} \to \mathscr{O}(U^*)$ defined by $\Gamma(x)(z) = \langle x, \gamma(\bar{z})\rangle$, $z \in U^*$. Transplant the inner product from $\mathscr{H}$ on the range of Γ. The map Γ is now unitary from $\mathscr{H}$ onto $\operatorname{ran}\Gamma$. Define K to be the function $K(z,w) = \Gamma\big(\gamma(\bar{w})\big)(z) = \langle \gamma(\bar{w}), \gamma(\bar{z})\rangle$, $z, w \in U^*$. Set $K_w(\cdot) := K(\cdot, w)$. Thus K_w is the function $\Gamma\big(\gamma(\bar{w})\big)$. It is then easily verified that K has the reproducing property, that is,

$$\begin{aligned}
\langle \Gamma(x)(z), K(z,w)\rangle_{\operatorname{ran}\Gamma} &= \langle (\langle x, \gamma(\bar{z})\rangle), (\langle \gamma(\bar{w}), \gamma(\bar{z})\rangle)\rangle_{\operatorname{ran}\Gamma} \\
&= \langle \Gamma x, \Gamma(\gamma(\bar{w}))\rangle_{\operatorname{ran}\Gamma} = \langle x, \gamma(\bar{w})\rangle_{\mathscr{H}} \\
&= \Gamma(x)(w),\ x \in \mathscr{H},\ w \in U^*.
\end{aligned}$$

It follows that $\|K_w(\cdot)\|^2 = K(w,w)$, $w \in U^*$. Also, $K_w(\cdot)$ is an eigenvector for the operator $\Gamma T_i \Gamma^*$, $1 \leq i \leq m$, with eigenvalue $\bar{w}_i$:

$$\begin{aligned}
\Gamma T_i \Gamma^*(K_w(\cdot)) &= \Gamma T_i \Gamma^*\big(\Gamma(\gamma(\bar{w}))\big) \\
&= \Gamma T_i \gamma(\bar{w}) \\
&= \Gamma \bar{w}_i \gamma(\bar{w}) \\
&= \bar{w}_i K_w(\cdot),\ w \in U^*.
\end{aligned}$$

Since the linear span of the vectors $\{K_w : w \in U^*\}$ is dense in $(\mathscr{H}, K)$ (see [17, Corollary 1.13]), it follows that $\Gamma T_i \Gamma^*$ is the adjoint M_i^* of the multiplication operator M_i acting on $(\mathscr{H}, K)$. We therefore assume, without loss of generality, that an operator T in $B_1(\Omega)$ has been realized as the adjoint M^* of the multiplication operator M on some Hilbert space $(\mathscr{H}, K)$ of holomorphic functions on U^* possessing a reproducing kernel K.

Moreover, starting from any nnd kernel K defined on Ω taking values in $\mathscr{M}_n(\mathbb{C})$ and fixing a w_0 in Ω, we note that the function

$$K_0(z,w) = K(w_0,w_0)^{\frac{1}{2}} \varphi(z)^{-1} K(z,w) \overline{\varphi(w)^{-1}} K(w_0,w_0)^{\frac{1}{2}}$$

is defined on some open neighbourhood U of w_0 on which $\varphi(z) = K(z, w_0)$ is holomorphic and non-zero. Thus, the m - tuple M defined on $(\mathscr{H}, K)$ is unitarily equivalent to the the m - tuple M on $(\mathscr{H}_0, K_0)$, see [17, 21].

The kernel K_0 is said to be normalized at w_0 in the sense that $K_0(z, w_0) = I_n$ for each $z \in U$.

The commuting m-tuple of multiplication operators acting on the Hilbert space $\mathscr{H}_\Gamma$ is called the canonical model. This terminology is justified by [21, Theorem 4.12(a)]; it says, "the canonical models associated with two generalized Bergman kernels are unitarily equivalent if and only if the normalized forms of the kernels are unitarily equivalent via a unitary that does not depend on points of Ω."

It is possible to impose conditions on a kernel function $K : \Omega \times \Omega \to \mathbb{C}$ so that each of the multiplication operators $M_1, \ldots, M_m$ are bounded on the Hilbert space $(\mathscr{H}, K)$. Additional conditions, explicitly given in [21], on K ensure that $M^* := (M_1^*, \ldots, M_m^*)$ is in $B_1(\Omega^*)$. If we set the curvature $\mathscr{K}$ of the m-tuple M^* to be the $(1,1)$ - form

$$\mathscr{K}(w) := -\sum_{i,j=1}^{m} \mathscr{K}_{i,j}(w) dw_i \wedge d\bar{w}_j,$$

where $\mathscr{K}_{i,j}(\bar{w}) = \left(\frac{\partial^2}{\partial w_i \partial \bar{w}_j} \log K\right)(w,w)$, then the unitary equivalence class of operators T in $B_1(\Omega^*)$, which we assume is of the form M^* on some reproducing kernel Hilbert space $(\mathscr{H}, K)$, is determined by the curvature $(1,1)$ form.

In the case of a commuting m-tuple of operators T in the Cowen-Douglas class $B_n(\Omega)$, the existence of a spanning section was proved in [32]. Some examples of spanning sections are given in [8].

4.1.2 Curvature inequalities

We may assume, without loss of generality, that an operator T in $B_1(\Omega)$ has been realized as the adjoint M^* of the multiplication operator M on some Hilbert space $(\mathscr{H}, K)$ of holomorphic functions on Ω^* possessing a reproducing kernel $K : \Omega^* \times \Omega^* \to \mathbb{C}$. For the unit disc $\mathbb{D}$, the distinction between $\mathbb{D}$ and $\mathbb{D}^*$ disappears and we write $K(z,w)$, when strictly speaking, we should be writing $K(\bar{z}, \bar{w})$, $z, w \in \mathbb{D}$. The curvature of the operator M^* may be also written in the form

$$\mathscr{K}(w) = -\frac{\|\gamma(w)\|^2 \|\gamma'(w)\|^2 - |\langle \gamma'(w), \gamma(w)\rangle|^2}{\|\gamma(w)\|^4}$$

for some holomorphic frame γ. In particular, choosing $\gamma(\bar{w}) = K(\cdot, w)$, $w \in \mathbb{D}$, we also have

$$\mathscr{K}(\bar{w}) = -\frac{\partial^2}{\partial w \partial \bar{w}} \log K(w,w) = -\frac{K(w,w)(\partial\bar{\partial}K)(w,w) - |(\partial K)(w,w)|^2}{K(w,w)^2}.$$

In either case, since K is nnd, the Cauchy - Schwarz inequality applies, and we see that the numerator is non-negative. Therefore, $\frac{\partial^2}{\partial w \partial \bar{w}} \log K(w,w)$ must be a non-negative function.

The contractivity of the adjoint M^* of the multiplication operator M on some reproducing kernel Hilbert space $(\mathscr{H}, K)$ is equivalent to the requirement that $K^{\ddagger}(z,w) := (1 - z\bar{w})K(z,w)$ is nnd on $\mathbb{D}$. This is easy to prove as long as K is positive definite. However, with a little more care, one can show this assuming only that K is nnd; see [33, Lemma 2.1.10].

Now, let T be any contraction in $B_1(\mathbb{D})$ realized in the form of the adjoint M^* of the multiplication operator M on some reproducing kernel Hilbert space $(\mathscr{H}, K)$. Then we have

$$\frac{\partial^2}{\partial w \partial \bar{w}} \log K(w,w) = \frac{\partial^2}{\partial w \partial \bar{w}} \log \frac{1}{(1-|w|^2)} + \frac{\partial^2}{\partial w \partial \bar{w}} \log K^{\ddagger}(w,w),\ w \in \mathbb{D}.$$

Let S be the unilateral shift acting on ℓ^2. Choosing a holomorphic frame γ_{S^*}, say $\gamma_{S^*}(w) = (1, w, w^2, \ldots)$, it follows that $\|\gamma_{S^*}(w)\|^2 = (1-|w|^2)^{-1}$ and that $\mathscr{K}_{S^*}(w) = -(1-|w|^2)^{-2}$, $w \in \mathbb{D}$. We can, therefore, rewrite the previous equality in the form

$$\mathscr{K}_{M^*}(w) - \mathscr{K}_{S^*}(w) = -\frac{\partial^2}{\partial w \partial \bar{w}} \log K^{\ddagger}(w,w) \leq 0,\ w \in \mathbb{D}.$$

In consequence, we have

$$\mathscr{K}_{M^*}(w) \leq \mathscr{K}_{S^*}(w),\ w \in \mathbb{D}.$$

Thus the the operator S^* is an extremal operator in the class of all contractive Cowen-Douglas operators in $B_1(\mathbb{D})$. The extremal property of the operator S^* prompts the following question due to R. G. Douglas.

A question of R. G. Douglas. For a contraction T in $B_1(\mathbb{D})$, if $\mathscr{K}_T(w_0) = -(1-|w_0|^2)^{-2}$ for some fixed w_0 in $\mathbb{D}$, then does it follow that T must be unitarily equivalent to the operator S^*?

It is known that the answer is negative in general. However, it has an affirmative answer if, for instance, T is a homogeneous contraction in $B_1(\mathbb{D})$; see [47]. From the simple observation that $\mathscr{K}_T(\bar{\zeta}) = -(1-|\zeta|^2)^{-2}$ for some $\zeta \in \mathbb{D}$ if and only if the two vectors $K^{\ddagger}_{\zeta}$ and $\bar{\partial} K^{\ddagger}_{\zeta}$ are linearly dependent, it follows that the question of Douglas has an affirmative answer in the class of contractive, co-hyponormal backward weighted shifts. The Question of Douglas for all those operators T in $B_1(\mathbb{D})$ possessing two additional properties, namely, T^* is 2 hyper-contractive and $(\varphi(T))^*$ has the wandering subspace property for any bi-holomorphic automorphism φ of $\mathbb{D}$ mapping ζ to 0. This is Theorem 3.6 of the paper [50].

Now suppose that the domain Ω is not simply connected. In this case, replacing the contractivity of the operator T by the contractivity of the homomorphism ρ_T induced by an operator T, namely, $\rho_T(r) = r(T)$, $r \in \mathrm{Rat}(\Omega^*)$, the algebra of rational functions with poles off $\overline{\Omega}^*$, we assume that $\|r(T)\| \leq \|r\|_{\Omega^*,\infty}$. For such operators T, the curvature inequality

$$\mathscr{K}_T(\bar{w}) \leq -4\pi^2 (S_{\Omega^*}(\bar{w},\bar{w}))^2, \quad \bar{w} \in \Omega^*,$$

where S_{Ω^*} is the Szëgo kernel of the domain Ω^*, was established in [48]. Equivalently, since $S_{\Omega}(z,w) = S_{\Omega^*}(\bar{w},\bar{z}),\ z,w \in \Omega$, the curvature inequality takes the form

$$\frac{\partial^2}{\partial w \partial \bar{w}} \log K(w,w) \geq 4\pi^2 (S_{\Omega}(w,w))^2, \quad w \in \Omega. \tag{4.5}$$

The curvature inequality in (4.5) is for operators T in $B_1(\Omega^*)$ for which $\overline{\Omega}^*$ is a spectral set. It is not known if there exists an extremal operator T in $B_1(\Omega^*)$, that is, if

$\mathscr{K}_T(w) = -4\pi^2(S_{\Omega^*}(w,w))^2$, $w \in \Omega^*$, for some operator T in $B_1(\Omega^*)$. Indeed, from a result of Suita (cf. [58]), it follows that the adjoint of the multiplication operator on the Hardy space $(H^2(\Omega), ds)$ is not extremal. It was shown in [48] that for any fixed but arbitrary $w_0 \in \Omega$, there exists an operator T in $B_1(\Omega^*)$ for which equality is achieved, at $w = w_0$, in the inequality (4.5). The question of Douglas is the question of uniqueness of such an operator. It was partially answered recently in [55]. The precise result is that these "point-wise" extremal operators are determined uniquely within the class of the adjoint of the bundle shifts introduced in [1]. It was also shown in the same paper that each of these bundle shifts can be realized as a multiplication operator on a Hilbert space of weighted Hardy space and conversely. Some very interesting inequalities involving what the authors call "higher order curvature" are given in [59].

4.1.3 Homogeneous operators

The question of Douglas discussed before has an affirmative answer in the class of homogeneous operators in the Cowen-Douglas class. An operator T with its spectrum $\sigma(T)$ contained in the closed unit disc $\bar{\mathbb{D}}$ is said to be homogeneous if $U_\varphi^* T U_\varphi = \varphi(T)$ for each bi-holomorphic automorphism φ of the unit disc and some unitary U_φ. It is then natural to ask what are all the homogeneous operators. Let us describe (see [47, 60, 7]) the homogeneous operators in $B_1(\mathbb{D})$. We first show that the equivalence class of a holomorphic Hermitian line bundle L defined on a bounded planar domain Ω is determined by its curvature $\mathscr{K}_L$.

Proposition 1.3 *Suppose that E and F are two holomorphic Hermitian line bundles defined on some bounded domain $\Omega \subseteq \mathbb{C}^m$. Then they are locally equivalent as holomorphic Hermitian bundles if and only if $\mathscr{K}_E = \mathscr{K}_F$.*

Proof For simplicity, first consider the case of $m = 1$. Suppose that E is a holomorphic line bundle over the domain $\Omega \subseteq \mathbb{C}$ with a hermitian metric $G(w) = \langle \gamma_w, \gamma_w \rangle$, where γ is a holomorphic frame. The curvature $\mathscr{K}_E$ is given by the formula $\mathscr{K}_E(w) = -(\frac{\partial^2}{\partial w \partial \bar{w}} \log G)(w)$, for $w \in \Omega$. Clearly, in this case, $\mathscr{K}(w) \equiv 0$ on Ω is the same as saying that $\log G$ is harmonic on Ω. Let F be a second line bundle over the same domain Ω with the metric H with respect to a holomorphic frame η. Suppose that the two curvatures $\mathscr{K}_E$ and $\mathscr{K}_F$ are equal. It then follows that $u = \log(G/H)$ is harmonic on Ω and thus there exists a harmonic conjugate v of u on any simply connected open subset Ω_0 of Ω. For $w \in \Omega_0$, define $\tilde{\eta}_w = e^{(u(w)+iv(w))/2}\eta_w$. Then clearly, $\tilde{\eta}_w$ is a new holomorphic frame for F, which we can use without loss of generality. Consequently, we have the metric $H(w) = \langle \tilde{\eta}_w, \tilde{\eta}_w \rangle$ relative to the frame $\tilde{\eta}$ for the vector bundle F. We have that

$$
\begin{aligned}
H(w) &= \langle \tilde{\eta}_w, \tilde{\eta}_w \rangle \\
&= \langle e^{(u(w)+iv(w))/2}, \eta_w e^{(u(w)+iv(w))/2} \eta_w \rangle
\end{aligned}
$$

$$\begin{aligned} &= e^{u(w)}\langle \eta_w, \eta_w\rangle \\ &= G(w). \end{aligned}$$

This calculation shows that the map $\tilde{\eta}_w \mapsto \gamma_w$ defines an isometric holomorphic bundle map between the vector bundles E and F.

To complete the proof in the general case, recall that

$$\sum_{i,j=1}^{m} \big(\frac{\partial^2}{\partial w_i \partial \bar{w}_j} \log G/H\big)(w) dw_i \wedge d\bar{w}_j = 0$$

means that the function is $u := H/G$ is pluriharmonic. The proof then follows exactly the same way as in the case of $m = 1$. Indeed, as in [18, Theorem 1], the map

$$U\Big(\sum_{|I|\leq n} \alpha_I (\bar{\partial}^I \tilde{\eta})(w_0)\Big) = \sum_{|I|\leq n} \alpha_I (\bar{\partial}^I \gamma)(w_0),\ \alpha_I \in \mathbb{C}, \tag{4.6}$$

where w_0 is a fixed point in Ω and I is a multi-index of length n, is well-defined, extends to a unitary operator on the Hilbert space spanned by the vectors $(\bar{\partial}^I \tilde{\eta})(w_0)$ and intertwines the two m-tuples of operators in $\mathrm{B}_1(\Omega)$ corresponding to the vector bundles E and F.

As shown in [17], it now follows that the curvature $\mathscr{K}_T$ of an operator T in $B_1(\Omega)$ determines the unitary equivalence class of T and conversely.

Theorem (Cowen-Douglas) *Two operators T and $\tilde{T}$ belonging to the class $B_1(\Omega)$ are unitarily equivalent if and only if $\mathscr{K}_T = \mathscr{K}_{\tilde{T}}$.*

Proof Let $\gamma_T(w)$ be a holomorphic frame for the line bundle E_T over Ω corresponding to an operator T in $\mathrm{B}_1(\Omega)$. Thus the real analytic function $G_T(w) := \langle \gamma_T(w), \gamma_T(w)\rangle$ is the Hermitian metric for the bundle E_T. Similarly, let $\gamma_{\tilde{T}}$ and $G_{\tilde{T}}$ be the holomorphic frame and the Hermitian metric corresponding to the operator $\tilde{T}$. If T and $\tilde{T}$ are unitarily equivalent, then the eigenvector $\gamma_{\tilde{T}}(w)$ must be a multiple, say c depending on w, of the eigenvector $\gamma_T(w)$. However, since both $\gamma_{\tilde{T}}$ and γ_T are holomorphic, it follows that c must be holomorphic. Hence $G_{\tilde{T}}(w) = |c(w)|^2 G_T(w)$ and we see that $\mathscr{K}_T = \mathscr{K}_{\tilde{T}}$.

Conversely, if the two curvatures are equal, from Proposition 1.3, we find that we may choose, without loss of generality, a holomorphic frame $\gamma_{\tilde{T}}$ for the operator $\tilde{T}$ such that $G_{\tilde{T}} = G_T$. Since the linear span of the vectors $\gamma_T(w)$ and $\gamma_{\tilde{T}}(w)$ are dense, it follows that the map U taking $\gamma_T(w)$ to $\gamma_{\tilde{T}}(w)$ is isometric. Extending it linearly, we obtain a unitary operator that intertwines T and $\tilde{T}$.

We now explain how the curvature can be extracted directly from an operator $T : \mathscr{H} \to \mathscr{H}$ which is in the class $B_1(\Omega)$. Let γ be a holomorphic frame for the operator T. Recall that $\gamma'(w)$, $w \in \Omega$, is also in the Hilbert space $\mathscr{H}$. The restriction $N(w)$ of the operator $T - wI$ to the two dimensional subspaces $\{\gamma(w), \gamma'(w)\}$, $w \in \Omega$ is nilpotent and encodes important information about the operator T.

With a little more effort, one may work with commuting tuples of bounded operators on a Hilbert space possessing an open set $\Omega \subseteq \mathbb{C}^m$ of joint eigenvalues. We postpone the details to Section 4.2.1.1.

Let $\mathscr{N}(w) \subseteq \mathscr{H}$, $w \in \Omega$, be the subspace consisting of the two linearly independent vectors $\gamma(w)$ and $\gamma'(w)$. There is a natural nilpotent action $N(w) := (T - wI)_{|\mathscr{N}(w)}$ on the space $\mathscr{N}(w)$ determined by the rule

$$\gamma'(w) \xrightarrow{N(w)} \gamma(w) \xrightarrow{N(w)} 0.$$

Let $e_0(w), e_1(w)$ be the orthonormal basis for $\mathscr{N}(w)$ obtained from $\gamma(w), \gamma'(w)$ by the Gram-Schmidt orthonormalization. The matrix representation of $N(w)$ with respect to this orthonormal basis is of the form $\begin{pmatrix} 0 & h(w) \\ 0 & 0 \end{pmatrix}$. It is easy to compute $h(w)$. Indeed, we have

$$h(w) = \frac{\|\gamma(w)\|^2}{(\|\gamma'(w)\|^2\|\gamma(w)\|^2 - |\langle\gamma'(w), \gamma(w)\rangle|^2)^{\frac{1}{2}}}.$$

We observe that $\mathscr{K}_T(w) = -h(w)^{-2}$.

Let φ be a bi-holomorphic automorphism of the unit disc $\mathbb{D}$. Thus $\varphi(z)$ is of the form $e^{i\theta}\frac{z-\alpha}{1-\bar{\alpha}z}$ for some θ, α, $0 \le \theta < 2\pi$, $\alpha \in \mathbb{D}$. Since the spectrum of T is contained in $\bar{\mathbb{D}}$ and $q(z) = 1 - \bar{\alpha}z$ does not vanish on it, we can define $\varphi(T)$ to be the operator $p(T)q(T)^{-1}$, where $p(z) = z - \alpha$. This definition coincides with the usual holomorphic functional calculus. It is not hard to prove that $\varphi(T)$ is in $B_1(\mathbb{D})$, whenever T is in $B_1(\mathbb{D})$, see [50].

Theorem 1.4 *Let T be an operator in* $\mathrm{B}_1(\mathbb{D})$. *Suppose that $\varphi(T)$ unitarily equivalent to T for each bi-holomorphic automorphism of φ of $\mathbb{D}$. Then*

$$\mathscr{K}_T(\alpha) = (1 - |\alpha|^2)^{-2}\mathscr{K}_T(0),$$

where $-\lambda = \mathscr{K}_T(0) < 0$ is arbitrary.

Proof For each fixed but arbitrary $w \in \mathbb{D}$, we have

$$\varphi(T)_{|\ker(\varphi(T)-\varphi(w))^2} = \varphi\big(T_{|\ker(T-w)^2}\big).$$

Since $T_{|\ker(T-w)^2}$ is of the form $\begin{pmatrix} w & h_T(w) \\ 0 & w \end{pmatrix}$ and

$$\varphi\begin{pmatrix} w & h_T(w) \\ 0 & w \end{pmatrix} = \begin{pmatrix} \varphi(w) & \varphi'(w)h_T(w) \\ 0 & \varphi(w) \end{pmatrix},$$

it follows that

$$\varphi(T)_{|\ker(\varphi(T)-\varphi(w))^2} = \begin{pmatrix} \varphi(w) & \varphi'(w)h_T(w) \\ 0 & \varphi(w) \end{pmatrix} \cong \begin{pmatrix} \varphi(w) & |\varphi'(w)|h_T(w) \\ 0 & \varphi(w) \end{pmatrix},$$

where we have used the symbol $\cong$ for unitary equivalence. Finally, we have

$$\begin{aligned}\left(-\mathscr{K}_{\varphi(T)}(\varphi(w))\right)^{-1/2} &= h_{\varphi(T)}(\varphi(w)) \\ &= |\varphi'(w)| h_T(w) \\ &= |\varphi'(w)| \left(-\mathscr{K}_T(w)\right)^{-1/2}.\end{aligned}$$

This is really a "change of variable formula for the curvature", which can be obtained directly using the chain rule.

Put $w = 0$, choose $\varphi = \varphi_\alpha$ such that $\varphi(0) = \alpha$. In particualr, take $\varphi_\alpha(z) = \frac{\alpha - z}{1 - \bar{\alpha} z}$. Then

$$\begin{aligned}\mathscr{K}_{\varphi_\alpha(T)}(\alpha) &= \mathscr{K}_{\varphi_\alpha(T)}(\varphi_\alpha(0)) = |\varphi'(0)|^{-2} \mathscr{K}_T(0) \\ &= (1 - |\alpha|^2)^{-2} \mathscr{K}_T(0), \quad \alpha \in \mathbb{D}. \end{aligned} \tag{4.7}$$

Now, suppose that $\varphi_\alpha(T)$ is unitarily equivalent to T for all φ_α, $\alpha \in \mathbb{D}$. Then $K_{\varphi_\alpha(T)}(w) = \mathscr{K}_T(w)$ for all $w \in \mathbb{D}$. Hence

$$(1 - |\alpha|^2)^{-2} \mathscr{K}_T(0) = K_{\varphi_\alpha(T)}(\varphi_\alpha(0)) = K_T(\alpha), \quad \alpha \in \mathbb{D}. \tag{4.8}$$

(Here the first equality is the change of variable formula given in (4.7) and the second equality follows from equality of the curvature of two unitarily equivalent operators.)

Corollary 1.5 *If T is a homogeneous operator in $B_1(\mathbb{D})$, then it must be the adjoint of the multiplication operator on the reproducing kernel Hilbert space $\mathscr{H}^{(\lambda)}$ determined by the reproducing kernel $K^{(\lambda)}(z,w) := (1 - z\bar{w})^{-2\lambda}$.*

Proof It follows from the Theorem that if the operator T is homogeneous, then the corresponding metric G for the bundle E_T, which is determined up to the square of the absolute value of a holomorphic function, is of the form:

$$G(w) = (1 - |w|^2)^{-2\lambda}, \quad w \in \mathbb{D}, \lambda > 0.$$

This corresponds to the reproducing kernel K (obtained via polarization of the real analytic function G) of the form:

$$K^{(\lambda)}(z,w) = (1 - z\bar{w})^{-2\lambda}, \quad \lambda > 0;\ z, w \in \mathbb{D}.$$

The kernel $B(z,w) = (1 - z\bar{w})^{-2}$ is the reproducing kernel of the Hilbert space of square integrable (with respect to area measure) holomorphic functions defined on the unit disc $\mathbb{D}$ and is known as the Bergman kernel. The kernel $K^{(\lambda)}$ is therefore a power of the Bergman kernel and the Hilbert space $\mathscr{H}^{(\lambda)}$ determined by $K^{(\lambda)}$ is known as the weighted Bergman space. The adjoint of the multiplication operator M on the Hilbert space $\mathscr{H}^{(\lambda)}$ corresponding to the reproducing kernel $K^{(\lambda)}$ is in $B_1(\mathbb{D})$.

If $\lambda > \frac{1}{2}$, then the multiplication operator $M^{(\lambda)}$ is subnormal and the inner product in the Hilbert space $\mathscr{H}^{(\lambda)}$ is induced by the measure

$$d\mu(z) = \tfrac{2\lambda - 1}{\pi}(1 - |z|^2)^{2\lambda - 2} dz d\bar{z}.$$

For $\lambda = \frac{1}{2}$, this operator is the usual shift on the Hardy space. However, for $\lambda < \frac{1}{2}$, there is no such measure and the corresponding operator $M^{(\lambda)}$ is not a contraction, not even power bounded and therefore not subnormal.

4.1.4 Quotient and submodules

The interaction of one-variable function theory and functional analysis with operator theory over the past half century has been extremely fruitful. Much of the progress in multivariable spectral theory during the last two decades was made possible by the use of methods from several complex variables, complex analytic and algebraic geometry. A unifying approach to many of these problems is possible in the language of Hilbert modules.

For any ring R and an ideal $I \subseteq R$, the study of the pair I and R/I as modules over the ring R is natural in algebra. However, if one were to assume that the ring R has more structure, for instance, if R is taken to be the polynomial ring $\mathbb{C}[z]$, in m - variables, equipped with the supremum norm over some bounded domain Ω in $\mathbb{C}^m$, then the study of a pair analogous to I and R/I, as above, still makes sense and is important. Fix an inner product on the algebra $\mathbb{C}[z]$. The completion of $\mathbb{C}[z]$ with respect to this inner product is a Hilbert space, say $\mathscr{M}$. It is natural to assume that the natural action of point-wise multiplication module action $\mathbb{C}[z] \times \mathbb{C}[z] \to \mathbb{C}[z]$ extends continuously to $\mathbb{C}[z] \times \mathscr{M} \to \mathscr{M}$ making $\mathscr{M}$ a module over $\mathbb{C}[z]$. Natural examples are the Hardy and Bergman spaces on some bounded domain $\Omega \subseteq \mathbb{C}^m$. Here the module map is induced by point-wise multiplication, namely, $\mathbf{m}_p(f) = pf$, $p \in \mathbb{C}[z]$, $f \in \mathscr{M}$. A Hilbert module need not be obtained as the completion of the polynomial ring. More generally, a Hilbert module is simply a Hilbert space equipped with an action of a ring R. When this action is continuous in both the variables, the Hilbert space $\mathscr{M}$ is said to be a Hilbert module over the polynomial ring $\mathbb{C}[z]$.

A closed subspace $\mathscr{S}$ of $\mathscr{M}$ is said to be a submodule of $\mathscr{M}$ if $\mathbf{m}_p h \in \mathscr{S}$ for all $h \in \mathscr{S}$ and $p \in \mathbb{C}[z]$. The quotient module $\mathscr{Q} := \mathscr{M}/\mathscr{S}$ is the Hilbert space $\mathscr{S}^\perp$, where the module multiplication is defined to be the compression of the module multiplication on $\mathscr{M}$ to the subspace $\mathscr{S}^\perp$; that is, the module action on $\mathscr{Q}$ is given by $\mathbf{m}_p(h) = P_{\mathscr{S}^\perp}(\mathbf{m}_p h)$, $h \in \mathscr{S}^\perp$. Two Hilbert modules $\mathscr{M}_1$ and $\mathscr{M}_2$ over $\mathbb{C}[z]$ are said to be isomorphic if there exists a unitary operator $U : \mathscr{M}_1 \to \mathscr{M}_2$ such that $U(p \cdot h) = p \cdot Uh$, $p \in \mathbb{C}[z]$, $h \in \mathscr{M}_1$. A Hilbert module $\mathscr{M}$ over the polynomial ring $\mathbb{C}[\mathbf{z}]$ is said to be in $B_n(\Omega)$ if $\dim \mathscr{H}/\mathfrak{m}_w\mathscr{M} = n < \infty$ for all $w \in \Omega$, and $\cap_{w \in \Omega}\mathfrak{m}_w\mathscr{M} = \{0\}$, where $\mathfrak{m}_w$ is the maximal ideal in $\mathbb{C}[\mathbf{z}]$ at w. In practice, it is enough to work with analytic Hilbert modules defined below.

Definition 1.6 *A Hilbert module $\mathscr{M}$ over $\mathbb{C}[z]$ is called an analytic Hilbert module if the Hilbert space $\mathscr{M}$ consists of holomorphic functions on some bounded domain $\Omega \subseteq \mathbb{C}^m$, $\mathbb{C}[z] \subseteq \mathscr{M}$ is dense in $\mathscr{M}$, and $\mathscr{H}$ possesses a reproducing kernel $K : \Omega \times \Omega \to \mathscr{L}(V)$, where V is some finite dimensional linear space.*

The module action in an analytic Hilbert module $\mathscr{M}$ is given by point-wise multiplication, that is, $\mathfrak{m}_p(h)(z) = p(z)h(z)$, $z \in \Omega$.

There are many closely related notions like the locally free module and the quasi-free module (cf. [13, 26]). No matter which notion one adopts, the goal is to ensure the existence of a holomorphic Hermitian vector bundle such that the equivalence class of the module and that of the vector bundle are in one to one correspondence. The generalized Bergman kernel and the sharp kernel appearing in [21] and [2] achieve a similar objective. The polynomial density in an analytic Hilbert module ensures that the joint kernel

$$\cap_{i=1}^{m} \ker(M_i - w_i)^*,\ w \in \Omega^*,$$

is of constant dimension, see [25, Remark, pp. 5]. Let K be the reproducing kernel of the analytic Hilbert module $\mathscr{M}$ and $\{\varepsilon_i\}_{i=1}^{n}$, be a basis of the linear space V. Evidently, the map

$$\gamma_i : w \to K(\cdot, \bar{w})\varepsilon_i,\ w \in \Omega^*, i = 1, \ldots, n,$$

serves as a holomorphic frame for the Hilbert module $\mathscr{M}$. This way, we obtain a holomorphic Hermitian vector bundle $E_{\mathscr{M}}$.

Let $\mathscr{Z} \subseteq \Omega$ be an analytic submanifold and $T\Omega = T\mathscr{Z} \dot{+} N\mathscr{Z}$ be the decomposition of the tangent bundle of Ω. Pick a basis for the normal bundle $N\mathscr{Z}$, say ∂_i, $1 \leq i \leq p$. Fix the submodule $\mathscr{M}_0 \subseteq \mathscr{M}$ of all functions f in $\mathscr{M}$ vanishing on $\mathscr{Z}$ to total order k, that is, $\partial^\alpha f_{|\mathrm{res}\ \mathscr{Z}} = 0$ for all multi index α of length less or equal to k. We now have a a short exact sequence

$$0 \longleftarrow \mathscr{Q} \longleftarrow \mathscr{M} \overset{X}{\longleftarrow} \mathscr{M}_0 \longleftarrow 0,$$

where $\mathscr{Q} = \mathscr{M} \ominus \mathscr{M}_0$ is the quotient module and X is the inclusion map.

One of the fundamental problems is to find a canonical model and obtain a (complete) set of unitary invariants for the quotient module $\mathscr{Q}$.

If the submodule is taken to be the maximal set of functions vanishing on an analytic hypersurface $\mathscr{Z}$ in Ω, then appealing to an earlier result of Aronszajn [4] the following theorem was proved in [24] to analyze the quotient module $\mathscr{Q}$. Set

$$\mathscr{M}_{\mathrm{res}} = \{f_{|\mathscr{Z}} \mid f \in \mathscr{M}\}$$

and

$$\|f_{|\mathscr{Z}}\|_{\mathrm{res}} = \inf\{\|g\|_{\mathscr{M}} \mid g \in \mathscr{M}, g_{|\mathscr{Z}} = f_{|\mathscr{Z}}\}.$$

Theorem (Aronszajn). *The restriction $\mathscr{M}_{res}$ is a Hilbert module over $\mathbb{C}[z]$ possessing a reproducing kernel K_{res}, which is the restriction of K to $\mathscr{Z}$, that is, $K_{res}(\cdot, w) = K(\cdot, w)|_{\mathscr{Z}}$ for w in $\mathscr{Z}$. Furthermore, $\mathscr{Q}$ is isometrically isomorphic to $\mathscr{M}_{res}$.*

As an example, consider the Hardy module $H^2(\mathbb{D}^2)$. Since the Sžego kernel

$$\mathbb{S}_{\mathbb{D}^2}(\mathbf{z}, \mathbf{w}) = \frac{1}{(1 - z_1\bar{w}_1)} \frac{1}{(1 - z_2\bar{w}_2)}$$

is the reproducing kernel of $H^2(\mathbb{D}^2)$, restricting it to the hyper-surface $z_1 - z_2 = 0$ and using new coordinates $u_1 = \frac{z_1 - z_2}{2}$, $u_2 = \frac{z_1 + z_2}{2}$, we see that $K_{\mathscr{Q}}(u, v) = \frac{1}{(1 - u\bar{v})^2}$ for

u, v in $\{z_1 - z_2 = 0\}$. This is a multiple of the kernel function for the Bergman space $L^2_{\rm hol}(\mathbb{D})$. Hence the quotient module is isometrically isomorphism to the Bergman module since multiplication by a constant doesn't change the isomorphism class of a Hilbert module.

Thus the extension of Aronszajn's result provides a model for the quotient module. However, Hilbert modules determined by different kernel functions may be equivalent. To obtain invariants one approach is to appeal to the inherent complex geometry. Assume that the m - tuple of multiplication operators by the coordinate functions on the Hilbert module $\mathscr{M}$ belongs to $B_1(\Omega)$. Then the results from [17] apply and show that the curvature is a complete unitary invariant. Therefore, if the quotient module belongs to $B_1(\mathscr{Z})$, we need to compute its curvature. It is shown in [24] that the curvature of $\mathscr{Q}$ is the restriction of the curvature $(1,1)$ form of $\mathscr{M}$ to the hyper-surface $\mathscr{Z}$ followed by a projection to the $(1,1)$ forms on $\mathscr{Z}$.

The submodule in [28] is taken to be the (maximal) set of functions which vanish to some given order k on the hypersurface $\mathscr{Z}$. As in the previous case, two descriptions are provided for the quotient module. The first one, produces a Hilbert space of holomorphic functions taking values in $\mathbb{C}^k$ via what is known as the jet construction. The kernel function now takes values in $k \times k$ matrices. The second one provides a rank k holomorphic Hermitian vector bundle. Although, describing a complete set of invariants is much more complicated in this case (cf. [27]).

In the paper [19], it was observed that all the submodules of the Hardy module are isomorphic to the Hardy module; that is, there exists an intertwining unitary module map between them. Applying the von Neumann-Wold decomposition, it is not difficult to obtain a description of all the submodules of the Hardy module. Following this new proof of the Beurling's theorem, it was natural to determine isomorphism classes of submodules of other Hilbert modules. For instance, the situation is much more complicated for the Hardy module in m - variables. Indeed, the submodule

$$H^2_0(\mathbb{D}^2) := \{f \in H^2(\mathbb{D}^2) \mid f(0,0) = 0\}$$

is not equivalent to the Hardy module. This is easy to see: the dimension of the joint kernel $\cap_{i=1,2} \ker M_i^*$ of the two multiplication operators on $H^2(\mathbb{D}^2)$ is 1, while it is 2 on $H^2_0(\mathbb{D}^2)$.

The fundamental question of which submodules of a Hilbert module are equivalent was answered in [30] after making reasonable hypothesis on the nature of the submodule. A different approach to the same problem is outlined in [10]. A sheaf model to study submodules like the Hardy module $H^2_0(\mathbb{D}^2)$ of functions vanishing at $(0,0)$ is given in [11].

4.1.5 Flag structure

Fix a bounded planar domain Ω. Let E be a holomorphic Hermitian vector bundle of rank n in $\Omega \times \mathscr{H}$. By a well-known theorem of Grauert, every holomorphic vector bundle over a plane domain is trivial. So, there exists a holomorphic frame $\gamma_1, \ldots, \gamma_n$:

$\Omega \to \mathscr{H}$. A Hermitian metric on E relative to this frame is given by the formula $G_\gamma(w) = (\langle \gamma_i(w), \gamma_j(w) \rangle)$. The curvature of the vector bundle E is a complex $(1,1)$ form which is given by the formula

$$\mathscr{K}_E(w) = \sum_{i,j=1}^{m} \mathscr{K}_{i,j}(w)\, dw_i \wedge d\bar{w}_j,$$

where

$$\mathscr{K}_{i,j}(w) = -\frac{\partial}{\partial \bar{w}_j}\big(G_\gamma^{-1}(w)\big(\frac{\partial}{\partial w_i} G_\gamma(w)\big)\big).$$

It clearly depends on the choice of the frame γ except when $n = 1$. A complete set of invariants is given in [17, 18]. However, these invariants are not easy to compute. So, finding a tractable set of invariants for a smaller class of vector bundles which is complete would be worthwhile. For instance, in the paper [38], irreducible holomorphic Hermitian vector bundles, possessing a flag structure have been isolated. For these, the curvature together with the second fundamental form (relative to the flag) is a complete set of invariants. As an application, at least for $n = 2$, it is shown that the homogeneous holomorphic Hermitian vector bundles are in this class. A complete description of these is then given. This is very similar to the case of $n = 1$ except that now the second fundamental form associated with the flag has to be also considered along with the curvature. All the vector bundles in this class and the operators corresponding to them are irreducible. The flag structure they possess by definition is rigid which aids in the construction of a canonical model and in finding a complete set of unitary invariants. The study of commuting tuples of operators in the Cowen-Douglas class possessing a flag structure is under way.

The definition of the smaller class $\mathscr{F}B_2(\Omega)$ of operators in $B_2(\Omega)$ below is from [38]. We will discuss the class $\mathscr{F}B_n(\Omega)$, $n > 1$ separately at the end.

Definition 1.7 *We let $\mathscr{F}B_2(\Omega)$ denote the set of bounded linear operators T for which we can find operators T_0, T_1 in $\mathscr{B}_1(\Omega)$ and a non-zero intertwiner S between T_0 and T_1, that is, $T_0 S = S T_1$ so that*

$$T = \begin{pmatrix} T_0 & S \\ 0 & T_1 \end{pmatrix}.$$

An operator T in $B_2(\Omega)$ admits a decomposition of the form $\left(\begin{smallmatrix} T_0 & S \\ 0 & T_1 \end{smallmatrix}\right)$ for some pair of operators T_0 and T_1 in $B_1(\Omega)$ (cf. [40, Theorem 1.49]) . Conversely, an operator T, which admits a decomposition of this form for some choice of T_0, T_1 in $B_1(\Omega)$ can be shown to be in $B_2(\Omega)$. In defining the new class $\mathscr{F}B_2(\Omega)$, we are merely imposing one additional condition, namely that $T_0 S = S T_1$.

An operator T is in the class $\mathscr{F}B_2(\Omega)$ if and only if there exists a frame $\{\gamma_0, \gamma_1\}$ of the vector bundle E_T such that $\gamma_0(w)$ and $t_1(w) := \frac{\partial}{\partial w}\gamma_0(w) - \gamma_1(w)$ are orthogonal for all w in Ω. This is also equivalent to the existence of a frame $\{\gamma_0, \gamma_1\}$ of the vector bundle E_T such that

$$\tfrac{\partial}{\partial w}\|\gamma_0(w)\|^2 = \langle \gamma_1(w), \gamma_0(w) \rangle, \quad w \in \Omega.$$

Our first main theorem on unitary classification is given below.

Theorem 1.8 *Let*

$$T = \begin{pmatrix} T_0 & S \\ 0 & T_1 \end{pmatrix}, \qquad \tilde{T} = \begin{pmatrix} \tilde{T}_0 & \tilde{S} \\ 0 & \tilde{T}_1 \end{pmatrix}$$

be two operators in $\mathscr{F}B_2(\Omega)$. Also let t_1 and $\tilde{t}_1$ be non-zero sections of the holomorphic Hermitian vector bundles E_{T_1} and $E_{\tilde{T}_1}$ respectively. The operators T and $\tilde{T}$ are equivalent if and only if $\mathscr{K}_{T_0} = \mathscr{K}_{\tilde{T}_0}$ (or $\mathscr{K}_{T_1} = \mathscr{K}_{\tilde{T}_1}$) and

$$\frac{\|S(t_1)\|^2}{\|t_1\|^2} = \frac{\|\tilde{S}(\tilde{t}_1)\|^2}{\|\tilde{t}_1\|^2}.$$

Cowen and Douglas point out in [18] that an operator in $B_1(\Omega)$ must be irreducible. However, determining which operators in $B_n(\Omega)$ are irreducible is a formidable task. It turns out that the operators in $\mathscr{F}B_2(\Omega)$ are always irreducible. Indeed, if we assume S is invertible, then T is strongly irreducible.

Recall that an operator T in the Cowen-Douglas class $B_n(\Omega)$, up to unitary equivalence, is the adjoint of the multiplication operator M on a Hilbert space $\mathscr{H}$ consisting of holomorphic functions on $\Omega^* := \{\bar{w} : w \in \Omega\}$ possessing a reproducing kernel K. What about operators in $\mathscr{F}B_n(\Omega)$? For $n = 2$, a model for these operators is described below.

For an operator $T \in \mathscr{F}B_2(\Omega)$, there exists a holomorphic frame $\gamma = (\gamma_0, \gamma_1)$ with the property $\gamma_1(w) := \frac{\partial}{\partial w}\gamma_0(w) - t_1(w)$ and that $t_1(w)$ is orthogonal to $\gamma_0(w)$, $w \in \Omega$, for some holomorphic map $t_1 : \Omega \to \mathscr{H}$. In what follows, we fix a holomorphic frame with this property. Then the operator T is unitarily equivalent to the adjoint of the multiplication operator M on a Hilbert space $\mathscr{H}_\Gamma \subseteq \mathrm{Hol}(\Omega^*, \mathbb{C}^2)$ possessing a reproducing kernel $K_\Gamma : \Omega^* \times \Omega^* \to \mathscr{M}_2(\mathbb{C})$. The details are in [38]. It is easy to write down the kernel K_Γ explictly: For $z, w \in \Omega^*$, we have

$$\begin{aligned} K_\Gamma(z,w) &= \begin{pmatrix} \langle \gamma_0(\bar{w}), \gamma_0(\bar{z})\rangle & \langle \gamma_1(\bar{w}), \gamma_0(\bar{z})\rangle \\ \langle \gamma_0(\bar{w}), \gamma_1(\bar{z})\rangle & \langle \gamma_1(\bar{w}), \gamma_1(\bar{z})\rangle \end{pmatrix} \\ &= \begin{pmatrix} \langle \gamma_0(\bar{w}), \gamma_0(\bar{z})\rangle & \frac{\partial}{\partial \bar{w}}\langle \gamma_0(\bar{w}), \gamma_0(\bar{z})\rangle \\ \frac{\partial}{\partial z}\langle \gamma_0(\bar{w}), \gamma_0(\bar{z})\rangle & \frac{\partial^2}{\partial z \partial \bar{w}}\langle \gamma_0(\bar{w}), \gamma_0(\bar{z})\rangle + \langle t_1(\bar{w}), t_1(\bar{z})\rangle \end{pmatrix}, w \in \Omega. \end{aligned}$$

Setting $K_0(z,w) = \langle \gamma_0(\bar{w}), \gamma_0(\bar{z})\rangle$ and $K_1(z,w) = \langle t_1(\bar{w}), t_1(\bar{z})\rangle$, we see that the reproducing kernel K_Γ has the form:

$$K_\Gamma(z,w) = \begin{pmatrix} K_0(z,w) & \frac{\partial}{\partial \bar{w}} K_0(z,w) \\ \frac{\partial}{\partial z} K_0(z,w) & \frac{\partial^2}{\partial z \partial \bar{w}} K_0(z,w) + K_1(z,w) \end{pmatrix}. \tag{4.9}$$

All the irreducible homogeneous operators in $B_2(\mathbb{D})$ belong to the class $\mathscr{F}B_2(\mathbb{D})$. An application of Theorem 1.8 determines the curvature and the second fundamental form of these operators. It is then not hard to identify these operators (up to unitary equivalence) as shown in [38].

4.2 Some future directions and further thoughts

In this second half of the paper, we discuss some of the topics mentioned only briefly in the first half. Along the way, we also mention some of the open problems in these topics.

4.2.1 Operators in the Cowen-Douglas class

In this section, we give a description of the local operators $T_{|\mathcal{N}(w)}$, where

$$\mathcal{N}(w) := \cap_{i,j=1}^{m} \ker(T_i - w_i I)(T_j - w_j I).$$

The matrix representation of these local operators contains one of the most important geometric invariants, namely, the curvature. Over the past three decades, this has been used to obtain various curvature inequalities. We also discuss a class of pure subnormal operators T studied in detail by Abrahamse and Douglas, see [1]. We describe briefly, a new realization of such operators of rank 1 from the paper [55], as multiplication operators on ordinary weighted Hardy spaces. For operators of rank > 1, such a description perhaps exists but has not been found yet.

4.2.1.1 Local operators

Fix a m - tuple of operators T in $\mathrm{B}_1(\Omega)$ and let $N_T(w)$ be the m-tuple of operators $(N_1(w), \ldots, N_m(w))$, where $N_i(w) = (T_i - w_i I)|_{\mathcal{N}(w)}$, $i = 1, \ldots, m$. Clearly, $N_i(w)N_j(w) = 0$ for all $1 \leq i, j \leq m$. Since $(T_i - w_i I)\gamma(w) = 0$ and $(T_i - w_i I)(\partial_j \gamma)(w) = \delta_{i,j}\gamma(w)$ for $1 \leq i, j \leq m$, we have the matrix representation $N_k(w) = \left(\begin{smallmatrix} 0 & e_k \\ 0 & 0 \end{smallmatrix}\right)$, where e_k is the vector $(0, \ldots, 1, \ldots, 0)$ with the 1 in the $k\mathtt{th}$ slot, $k = 1, \ldots, m$. Representing $N_k(w)$ with respect to an orthonormal basis in $\mathcal{N}(w)$, it is possible to read off the curvature of T at w using the relationship:

$$\left(-\mathcal{K}_T(w)^{\mathrm{t}}\right)^{-1} = \left(\mathrm{tr}\left(N_k(w)\overline{N_j(w)}^{\mathrm{t}}\right)\right)_{k,j=1}^{m} = A(w)^{\mathrm{t}}\overline{A(w)}, \tag{4.10}$$

where the $k\mathtt{th}$-column of $A(w)$ is the vector α_k (depending on w) which appears in the matrix representation of $N_k(w)$ with respect to an appropriate choice of an orthonormal basis in $\mathcal{N}(w)$ which we describe below.

This formula is established for a pair of operators in $\mathrm{B}_1(\Omega)$ (cf. [18, Theorem 7]). However, let us verify it for an m-tuple T in $\mathrm{B}_1(\Omega)$ for any $m \geq 1$ following [49].

Fix w_0 in Ω. We may assume without loss of generality that $\|\gamma(w_0)\| = 1$. The function $\langle \gamma(w), \gamma(w_0) \rangle$ is invertible in some neighborhood of w_0. Then setting $\hat{\gamma}(w) := \langle \gamma(w), \gamma(w_0) \rangle^{-1}\gamma(w)$, we see that

$$\langle \partial_k \hat{\gamma}(w_0), \gamma(w_0) \rangle = 0, \; k = 1, 2, \ldots, m.$$

Thus $\hat{\gamma}$ is another holomorphic section of E. The norms of the two sections γ and $\hat{\gamma}$

differ by the absolute square of a holomorphic function, that is

$$\frac{\|\hat{\gamma}(w)\|}{\|\gamma(w)\|} = |\langle \gamma(w), \gamma(w_0)\rangle|.$$

Hence the curvature is independent of the choice of the holomorphic section. Therefore, without loss of generality, we will prove the claim assuming, for a fixed but arbitrary w_0 in Ω, that

(i) $\|\gamma(w_0)\| = 1$,

(ii) $\gamma(w_0)$ is orthogonal to $(\partial_k \gamma)(w_0)$, $k = 1, 2, \ldots, m$.

Let G be the Grammian corresponding to the $m+1$ dimensional space spanned by the vectors

$$\{\gamma(w_0), (\partial_1 \gamma)(w_0), \ldots, (\partial_m \gamma)(w_0)\}.$$

This is just the space $\mathcal{N}(w_0)$. Let v, w be any two vectors in $\mathcal{N}(w_0)$. Find $c = (c_0, \ldots, c_m), d = (d_0, \ldots, d_m)$ in $\mathbb{C}^{m+1}$ such that $v = \sum_{i=0}^m c_i \partial_i \gamma(w_0)$ and $w = \sum_{j=0}^m d_j \partial_j \gamma(w_0)$. Here $(\partial_0 \gamma)(w_0) = \gamma(w_0)$. We have

$$\begin{aligned}
\langle v, w\rangle &= \langle \sum_{i=0}^m c_i \partial_i \gamma(w_0), \sum_{j=0}^m d_j \partial_j \gamma(w_0)\rangle \\
&= \langle G^{\mathrm{t}}(w_0) c, d\rangle_{\mathbb{C}^{m+1}} \\
&= \langle (G^{\mathrm{t}})^{\frac{1}{2}}(w_0) c, (G^{\mathrm{t}})^{\frac{1}{2}}(w_0) d\rangle_{\mathbb{C}^{m+1}}.
\end{aligned}$$

Let $\{\varepsilon_i\}_{i=0}^m$ be the standard orthonormal basis for $\mathbb{C}^{m+1}$. Also, let $(G^{\mathrm{t}})^{-\frac{1}{2}}(w_0)\varepsilon_i := \alpha_i(w_0)$, where $\alpha_i(j)(w_0) = \alpha_{ji}(w_0)$, $i = 0, 1, \ldots, m$. We see that the vectors $\varepsilon_i := \sum_{j=0}^m \alpha_{ji} (\partial_j \gamma)(w_0)$, $i = 0, 1, \ldots, m$, form an orthonormal basis in $\mathcal{N}(w_0)$:

$$\begin{aligned}
\langle \varepsilon_i, \varepsilon_l\rangle &= \Big\langle \sum_{j=0}^m \alpha_{ji} \partial_j \gamma(w_0), \sum_{p=0}^m \alpha_{pl} \partial_p \gamma(w_0)\Big\rangle \\
&= \langle (G^{\mathrm{t}})^{-\frac{1}{2}}(w_0)\alpha_i, (G^{\mathrm{t}})^{-\frac{1}{2}}(w_0)\alpha_l\rangle \\
&= \delta_{il},
\end{aligned}$$

where δ_{il} is the Kronecker delta. Since $N_k\big((\partial_j \gamma)(w_0)\big) = \gamma(w_0)$ for $j = k$ and 0 otherwise, we have $N_k(\varepsilon_i) = \begin{pmatrix} 0 & \alpha_k^{\mathrm{t}} \\ 0 & 0 \end{pmatrix}$. Hence

$$\begin{aligned}
\operatorname{tr}\big(N_i(w_0) N_j^*(w_0)\big) &= \alpha_i(w_0)^{\mathrm{t}} \overline{\alpha}_j(w_0) \\
&= \big((G^{\mathrm{t}})^{-\frac{1}{2}}(w_0)\varepsilon_i\big)^{\mathrm{t}} \overline{\big((G^{\mathrm{t}})^{-\frac{1}{2}}(w_0)\varepsilon_j\big)} \\
&= \langle G^{-\frac{1}{2}}(w_0)\varepsilon_i, G^{-\frac{1}{2}}(w_0)\varepsilon_j\rangle = (G^{\mathrm{t}})^{-1}(w_0)_{i,j}.
\end{aligned}$$

Since the curvature, computed with respect to the holomorphic section γ satisfying the conditions (i) and (ii), is of the form

$$
\begin{aligned}
-\mathscr{K}_T(w_0)_{i,j} &= \frac{\partial^2}{\partial w_i \partial \bar{w}_j} \log \|\gamma(w)\|^2_{|w=w_0} \\
&= \Big(\frac{\|\gamma(w)\|^2 (\frac{\partial^2 \gamma}{\partial w_i \partial \bar{w}_j})(w) - (\frac{\partial \gamma}{\partial w_i})(w)(\frac{\partial \gamma}{\partial \bar{w}_j})(w)}{\|\gamma(w)\|^4} \Big)_{|w=w_0} \\
&= (\frac{\partial^2 \gamma}{\partial w_i \partial \bar{w}_j})(w_0) = G(w_0)_{i,j},
\end{aligned}
$$

we have verified the claim (4.10).

The local description of the m - tuple of operators T shows that the curvature is indeed obtained from the holomorphic frame and the first order derivatives using the Gram-Schmidt orthonormalization. The following theorem was proved for $m = 2$ in (cf. [18, Theorem 7]). However, for any natural number m, the proof is evident from the preceding discussion. The case of an m -tuple of operators in $B_n(\Omega)$ for an arbitrary $n \in \mathbb{N}$ is discussed in [50].

Theorem 2.1 *Two m-tuples of operators T and $\tilde{T}$ in $\mathrm{B}_1(\Omega)$ are unitarily equivalent if and only if $N_k(w)$ and $\tilde{N}_k(w)$, $1 \leq k \leq m$, are simultaneously unitarily equivalent for w in some open subset of Ω.*

Proof Let us fix an arbitrary point w in Ω. In what follows, the dependence on this w is implicit. Suppose that there exists a unitary operator $U : \mathscr{N} \to \widetilde{\mathscr{N}}$ such that $UN_i = \tilde{N}_i U$, $i = 1, \ldots, m$. For $1 \leq i, j \leq m$, we have

$$
\begin{aligned}
\mathrm{tr}\big(\tilde{N}_i \tilde{N}_j{}^*\big) &= \mathrm{tr}\big((UN_iU^*)(UN_jU^*)^*\big) \\
&= \mathrm{tr}\big(UN_iN_j^*U^*\big) \\
&= \mathrm{tr}\big(N_iN_j^*U^*U\big) \\
&= \mathrm{tr}\big(N_iN_j^*\big).
\end{aligned}
$$

Thus the curvature of the operators T and $\tilde{T}$ coincide making them unitarily equivalent proving the Theorem in one direction. In the other direction, observe that if the operators T and $\tilde{T}$ are unitarily equivalent then this unitary must map $\mathscr{N}$ to $\tilde{\mathscr{N}}$. Thus the restriction of U to the subspace $\mathscr{N}$ intertwines N_k and $\tilde{N}_k$ simultaneously for $k = 1, \cdots, m$.

4.2.1.2 Pure subnormal operators

The unilateral shift U_+ or the multiplication M by the coordinate function on the Hardy space $H^2(\mathbb{D})$ is a very special kind of subnormal operator in that it is a pure isometry. The spectrum of its minimal normal extension, namely, the bi-lateral shift U or the operator of multiplication M by the coordinate function on $L^2(\mathbb{T})$ is the unit circle $\mathbb{T} = \partial\overline{\mathbb{D}} = \partial\sigma(U_+)$. These two properties determine such an operator uniquely;

that is, if a pair consisting of a pure isometry S and its minimal normal extension N has the spectral inclusion property: $\sigma(N) \subseteq \partial\sigma(S)$, then S must be unitarily equivalent to the direct sum of a number of copies of the operator U_+. The situation for the annulus is more complicated and was investigated in [57]. Following this, Abrahamse and Douglas initiated the study of a class of pure subnormal operators S which share this property. Thus the spectrum $\sigma(S)$ of the operator S is assumed to be a subset of the closure $\overline{\Omega}$ of a bounded domain Ω and the spectrum $\sigma(N)$ of its normal extension N is contained in $\partial\sigma(S)$.

Let Ω be a bounded domain in $\mathbb{C}$ and let $\mathscr{O}(\Omega)$ be the space of holomorphic functions on Ω. For $f \in \mathscr{O}(\Omega)$, the real analytic function $|f|$ is subharmonic and hence admits a least harmonic majorant, which is the function

$$u_f(z) := \inf\{u(z) : |f| \leq u, u \text{ is harmonic on } \Omega\};$$

that is, u_f is either harmonic or infinity throughout Ω.

Fix a point $w \in \Omega$. The Hardy space on Ω is defined to be the Hilbert space

$$H^2(\Omega) := \{f \in \mathscr{O}(\Omega) : u_f(w) < \infty\}.$$

It is easily verified that $\|f\| := u_f(w)$ defines a norm and a different choice of $w \in \Omega$ induces an equivalent topology. Let $\mathscr{O}_{\mathfrak{h}}(\Omega)$ denote the space of holomorphic functions f defined on Ω taking values in some Hilbert space $\mathfrak{h}$. Again, $z \mapsto \|f(z)\|_{\mathfrak{h}}$ is a subharmoinic function and admits a least harmonic majorant u_f. We define the Hardy space

$$H^2_{\mathfrak{h}}(\Omega) := \{f \in \mathscr{O}_{\mathfrak{h}}(\Omega) : \|f\|_w := u_f(w) < \infty\}$$

as before. Let m be the harmonic measure relative to the point $w \in \Omega$ and let $L^2(\partial\Omega, m)$ denote the space of square integrable functions defined on $\partial\Omega$ with respect to m. The closed subspace

$$H^2(\partial\Omega) := \{f \in L^2(\partial\Omega, m) : \int_{\partial\Omega} fg\,dm = 0, g \in \mathscr{A}(\Omega)\},$$

where $\mathscr{A}(\Omega)$ is the space of all functions which are holomorphic in some open neighbourhood of the closed set $\overline{\Omega}$, is the Hardy space of $\partial\Omega$. Let $L^2_{\mathfrak{h}}(\partial\Omega)$ denote the space of square integrable functions, with respect to the measure m, taking values in the Hilbert space $\mathfrak{h}$. The Hardy space $H^2_{\mathfrak{h}}(\partial\Omega)$ is the closed subspace of $L^2_{\mathfrak{h}}(\partial\Omega)$ consisting of those functions in $L^2_{\mathfrak{h}}(\partial\Omega)$ for which $\int_{\partial\Omega} fg\,dm = 0$, $g \in \mathscr{A}(\Omega)$.

Boundary values. A function f in the Hardy space admits a boundary value $\hat{f}$. This means that the $\lim_{z\to\lambda} f(z)$ exists (almost everywhere relative to m) as z approaches $\lambda \in \partial\Omega$ through any non-tangential path in Ω. Define the function $\hat{f} : \partial\Omega \to \mathbb{C}$ by setting $\hat{f}(\lambda) = \lim_{z\to\lambda} f(z)$. It then follows that the map $f \mapsto \hat{f}$ is an isometric isomorphism between the two Hardy spaces $H^2(\Omega)$ and $H^2(\partial\Omega)$; see [56]. This correspondence works for the case of $H^2_{\mathfrak{h}}(\Omega)$ and $H^2_{\mathfrak{h}}(\partial\Omega)$ as well.

A topological space E is said to be a vector bundle of rank n over Ω if there is a continuous map $p : E \to \Omega$ such that $E_z := p^{-1}(z)$ is a linear space. A co-ordinate chart relative to an open cover $\{U_s\}$ of the domain Ω is a set of homeomorphisms

$\varphi_s : U_s \times \mathfrak{h} \to p^{-1}(U_s)$ such that $\varphi_s(z,k)$ is in E_z for any fixed $z \in U_s$ and that the restriction $\varphi_s(z) := \varphi_s|_{\{z\}\times\mathfrak{h}}$ is a continuous linear isomorphism. Therefore, $\varphi_{st}(z) := \varphi_s^{-1}(z)\varphi_t(z)$ is in $\mathrm{GL}(\mathfrak{h})$, $z \in U_s \cap U_t$. If the transition functions φ_{st} are holomorphic for some choice of co-ordinate functions φ_s, then the vector bundle E is said to be a holomorphic vector bundle. A section f of E is said to be holomorphic if $\varphi_s^{-1} f$ is a holomorphic function from U_s to $\mathfrak{h}$. Finally, if $\varphi_s^{-1}(z)\varphi_t(z)$ are chosen to be unitary, $z \in U_s \cap U_t$, then one says that E is a flat unitary vector bundle. Thus for a flat unitary vector bundle E, we have that

$$\|\varphi_s(z)^{-1}f(z)\|_{\mathfrak{h}} = \|\varphi_t(z)^{-1}f(z)\|_{\mathfrak{h}}, \quad z \in U_s \cap U_t.$$

This means that the functions

$$h_f^{(s)}(z) := \|\varphi_s(z)^{-1}f(z)\|_{\mathfrak{h}}, \quad z \in U_s,$$

agree on the overlaps and therefore define a function h_f^E on all of Ω. Since $z \mapsto \varphi_s(z)^{-1}f(z)$ is holomorphic and subharmonicity is a local property, it follows that h_f is subharmonic on Ω. As before, the function h_f^E admits a least harmonic majorant u_f^E and the Hardy space $H_E^2(\Omega)$ is the subspace of holomorphic sections f of E for which

$$\|f\|_w^E := u_f^E(w) < \infty$$

for some fixed but arbitrary point $w \in \Omega$.

A function f which is bounded and holomorphic on Ω defines a bounded linear operator M_f on the Hardy space $H_{\mathfrak{h}}^2(\Omega)$ via pointwise multiplication. Let $T_{\mathfrak{h}}$ denote the multiplication by the coordinate function z. It is then evident that $T_{\mathfrak{h}}$ is unitarily equivalent to the tensor product $T \otimes I_{\mathfrak{h}}$, where T is the multiplication induced by the coordinate function z on the Hardy space $H^2(\Omega)$. The Hardy space $H_E^2(\Omega)$ is also invariant under multiplication by any function which is holomorphic and bounded on Ω. In particular, let T_E be the operator of multiplication by z on the Hardy space $H_E^2(\Omega)$. All the three theorems listed below are from [1].

Theorem 2.2 *If $\mathfrak{h}$ is a Hilbert space and E is a flat unitary vector bundle over Ω of rank* $\dim \mathfrak{h}$, *then T_E is similar to $T_{\mathfrak{h}}$; that is, there is a bounded invertible operator $L : H_{\mathfrak{h}}^2(\Omega) \to H_E^2(\Omega)$ with the property $L^{-1}T_E L = T_{\mathfrak{h}}$.*

Theorem 2.3 *The operators T_E and T_F are unitarily equivalent if and only if E and F are equivalent as flat unitary vector bundles.*

Theorem 2.4 *If E is a flat unitary vector bundle of dimension n, then for $z \in \Omega$, the dimension of the kernel of $(T_E - z)^*$ is n.*

The operator T_E^* on $H_E^2(\Omega)$, where E is a flat unitary line bundle, is in $B_1(\Omega^*)$; see [48, Corollary 2.1]. Consequently, it defines a holomorphic Hermitian line bundle, say $\mathscr{E}$. One of the main questions that remains unanswered is the relationship between the flat unitary bundle E and the holomorphic Hermitian bundle $\mathscr{E}$. Thus we are asking which operators in $B_1(\Omega)$ are pure subnormal operators with the spectral inclusion property and conversely, how to find a model for such operators in $B_1(\Omega)$. Any answer involving the intrinsic complex geometry will be very interesting. In spite of the substantial work in [61] and [12], this question remains elusive. We discuss this a little more in Section 4.2.2.1. Also, the question of commuting tuples of subnormal operators has not been addressed yet. These are very interesting directions for future research.

There is yet another way, which exploits the covering map $\pi : \mathbb{D} \to \Omega$, to define the Hardy space on Ω. Let G be the group of deck transformations for the covering map π. Thus G consists of those bi-holomorphic automorphisms $\varphi : \mathbb{D} \to \mathbb{D}$ satisfying $\pi \circ \varphi = \pi$. A function f on $\mathbb{D}$ is said to be G-automorphic if $f \circ \varphi = f$ for $\varphi \in G$. A function f defined on $\mathbb{D}$ is G-automorphic if and only if it is of the form $g \circ \pi$ for some function g defined on Ω.

Let $w = \pi(0)$ and m be the harmonic measure on $\partial\Omega$ relative to w. This means that m is supported on $\partial\Omega$ and if u is a function which is continuous on $\overline{\Omega}$ and harmonic on Ω, then

$$u(w) = \int_{\partial\Omega} u\, dm.$$

For any $f \in L^1(\partial\Omega, m)$, it may be shown that

$$\int_{\partial\Omega} f\, dm = \int_{\partial\mathbb{D}} f \circ \pi\, d\mu,$$

where μ is the Lebesgue measure on $\partial\mathbb{D}$. Thus $m = \mu \circ \pi^{-1}$.

The Hilbert spaces $L^2_{\mathfrak{h}}(\partial\Omega)$, $H^2_{\mathfrak{h}}(\Omega)$ and $H^2_{\mathfrak{h}}(\partial\Omega)$ now lift to closed subspaces of the respective Hilbert spaces on the unit disc $\mathbb{D}$ via the map $f \mapsto f \circ \pi$.

The dual $\hat{G}$ of the group G is the group of homomorphisms from G into the circle group $\mathbb{T}$. For each character $\alpha \in \hat{G}$, define the Hardy space

$$H^2_\alpha(\mathbb{D}) := \{f \in H^2(\mathbb{D}) : f \circ \varphi = \alpha(\varphi) f\}.$$

This is a closed subspace of $H^2(\mathbb{D})$ which is invariant under any function which is G-automorphic, holomorphic and bounded on $\mathbb{D}$. The quotient map π is such a function and we set T_α to be the operator of multiplication by π on $H^2_\alpha(\mathbb{D})$.

We can define vector valued Hardy spaces $H^2_\alpha(\mathbb{D}, \mathfrak{h})$ in a similar manner except that α is in $\mathrm{Hom}(G, \mathscr{U}(\mathfrak{h}))$, the group of unitary representations of G on the Hilbert space $\mathfrak{h}$. The operator T_α on $H^2_\alpha(\mathbb{D}, \mathfrak{h})$ is multiplication by π as before. Since G, in this case, is isomorphic to the fundamental group of Ω, there is a bijective correspondence between equivalence classes of the unitary representations of G on $\mathfrak{h}$ and the equivalence classes of flat unitary vector bundles on Ω. Again, the two theorems stated below are from [1].

Theorem 2.5 *The operators T_α and T_β are unitarily equivalent if and only if α and β are equivalent representations.*

Theorem 2.6 *If E is the flat unitary vector bundle on Ω determined by the representation α, then the operator T_E is unitarily equivalent to the operator T_α.*

The choice of the harmonic measure in the definition of the Hardy space over the domain Ω may appear to be somewhat arbitrary but it has the advantage of being a conformal invariant and it is closely related to the Greens' function g of Ω with pole at t:

$$dm(z) = -\frac{1}{2\pi}\frac{\partial}{\partial \eta_z}\big(g(z,t)\big)ds(z), \;\; z \in \partial\Omega,$$

where $g(z,t)$ denotes the Green's function for the domain Ω at the point t and $\frac{\partial}{\partial \eta_z}$ is the directional derivative along the outward normal direction (with respect to the positively oriented $\partial\Omega$). Exploiting this, in the paper [50], a description of the bundle shift is given as multiplication operators on Hardy spaces defined on the domain Ω with respect to a weighted arc length measure. Briefly, starting with a positive continuous function λ on $\partial\Omega$, define the weighted measure λds on $\partial\Omega$. Since the harmonic measure m is boundedly mutually absolutely continuous with respect to the arc length measure ds, it follows that $\big(H^2(\partial\Omega), \lambda ds\big)$ acquires the structure of a Hilbert space and the operator M on it is a pure, rationally cyclic, subnormal operator with spectrum equal to $\overline{\Omega}$ and the spectrum of the minimal normal extension is $\partial\Omega$. Consequently, the operator M on $\big(H^2(\partial\Omega), \lambda ds\big)$ must be unitarily equivalent to the bundle shift T_α on $\big(H^2_\alpha(\Omega), dm\big)$ for some character α. In the paper [50], this character has been explicitly described in Equation (2.2). It is then shown that given any character α, there exists a function λ such that the operator M on $H^2(\partial\Omega, \lambda ds)$ is unitarily equivalent to the bundle shift T_α.

4.2.2 Curvature inequalities

The local description of the Cowen-Douglas operator T naturally leads to curvature inequalities related to many well known extremal problems assuming that the homomorphism induced by the operator T is contractive. Among other things, we discuss a) the question of uniqueness of the extremal operators and b) if the curvature inequalities with additional hypothesis implies contractivity.

4.2.2.1 The Douglas question

Let

$$\mathscr{B}[w] := \{f \in \mathscr{O}(\Omega) : \|f\|_\infty < 1, f(w) = 0\}.$$

It is well-known that the extremal problem

$$\sup\{|f'(w)| : f \in \mathscr{B}[w]\} \tag{4.11}$$

admits a solution, say, $F_w \in \mathscr{B}[w]$. The function F_w is called the Ahlfor's function and maps Ω onto $\mathbb{D}$ in a n to 1 fashion if the connectivity of the region Ω is n. Indeed, F_w is a branched covering map with branch point at w. Also, $F_w'(w)$ is a real analytic function and polarizing it, we get a new function, holomorphic in the first variable and anti-holomorphic in the second, which is the Sžego kernel of the domain Ω. It is also the reproducing kernel of the Hardy space $H^2(\Omega, ds)$, which we denote by the symbol S_Ω. The relationship of the Hardy space with the solution to the extremal problem (4.11) yields a curvature inequality for contractive homomorphisms ρ_T induced by operators T in $B_1(\Omega)$, where $\rho_T(r) = r(T)$, $r \in \mathrm{Rat}(\Omega)$. Recall that the local operator $N(w) + wI = T_{|\mathcal{N}(w)}$ is of the form $\begin{pmatrix} w & h(w) \\ 0 & w \end{pmatrix}$, where

$$h(w) = \big(-\mathcal{K}_T(w)\big)^{-\frac{1}{2}}.$$

Consequently,

$$r(T_{|\mathcal{N}(w)}) = \begin{pmatrix} r(w) & r'(w)h(w) \\ 0 & r(w) \end{pmatrix}.$$

We have the obvious inequalities

$$\begin{aligned} & h(w)\sup\{|r'(w)|; r \in \mathrm{Rat}(\Omega),\ r(w) = 0\} \\ &= \sup\{\| \begin{pmatrix} 0 & r'(w)h(w) \\ 0 & 0 \end{pmatrix} \| : r \in \mathrm{Rat}(\Omega), r(w) = 0\} \\ &\leq \sup\{\| \begin{pmatrix} r(w) & r'(w)h(w) \\ 0 & r(w) \end{pmatrix} \| : r \in \mathrm{Rat}(\Omega)\} \\ &\leq \sup\{\|\rho_T(r)\| : r \in \mathrm{Rat}(\Omega)\}. \end{aligned}$$

The contractivity of ρ_T then gives the curvature inequality

$$\begin{aligned} \mathcal{K}_T(w) &= -(\tfrac{1}{h(w)})^2 \\ &\leq -\big(\sup\{|r'(w)|; r \in \mathrm{Rat}(\Omega),\ r(w) = 0\}\big)^2 \\ &\leq -\sup\{|f'(w)| : f \in \mathcal{B}[w]\} \\ &\leq -S_\Omega(w,w)^2. \end{aligned} \tag{4.12}$$

If Ω is the unit disc, then from the Schwarz Lemma, it follows that $F'_w(w) = \frac{1}{(1-|w|^2)}$. For the unit disc, taking $T = M^*$, the adjoint of the multiplication operator on the Hardy space $H^2(\mathbb{D})$, we see that we have equality for all w in $\mathbb{D}$ in the string of inequalities (4.12). One can easily see that it makes no difference if we replace the unit disc by any simply connected domain. However, if Ω is not simply connected, it is shown in [58] that the inequality is strict if we take the operator T to be the adjoint of the multiplication operator on the Hardy space $H^2(\Omega, ds)$.

Let K be a positive definite kernel on Ω. Assume that the adjoint of the multiplication operator M is in $B_1(\Omega^*)$. What we have shown is that if ρ_M is contractive, then we have the inequality

$$\mathcal{K}(\bar{w}) := -\frac{\partial^2}{\partial w \partial \bar{w}} \log K(w,w) \leq -S_\Omega(w,w)^2,\ w \in \Omega,$$

where $\mathcal{K}$ denotes the curvature of the operator M^* on $(\mathcal{H}, K)$. Since the obvious candidate which might have served as an extremal operator, namely, the multiplication operator on the Hardy space $H^2(\Omega, ds)$ is not extremal, one may ask if for a fixed but arbitrary $w_0 \in \Omega$, there exists a kernel K_0, depending on w_0, for which

$\mathscr{K}(w_0) = -F_{w_0}(w_0)^2$. The existence of a kernel K_0 with this property was established in [48], see also [55]. In the paper [55], more is proved. For instance, it is established that the multiplication operator on $(\mathscr{H}, K_0)$ is uniquely determined answering the question of Douglas within the class of pure subnormal operators T with $\sigma(T) \subseteq \overline{\Omega}$ and $\sigma(N) \subseteq \partial\sigma(T)$, where N is the minimal normal extension of T. This is Theorem 2.6 of [55]. The question of Douglas has an affirmative answer in a much larger class of operators when considering contractions in $B_1(\mathbb{D})$. The main theorem in [50] is reproduced below.

Theorem 2.7 *Fix an arbitrary point $\zeta \in \mathbb{D}$. Let T be an operator in $B_1(\mathbb{D})$ such that T^* is a 2 hyper-contraction. Suppose that the operator $(\varphi_\zeta(T))^*$ has the wandering subspace property for an automorphism φ_ζ of the unit disc $\mathbb{D}$ mapping ζ to 0. If $\mathscr{K}_T(\zeta) = -(1-|\zeta|^2)^{-2}$, then T must be unitarily equivalent to U_+^*, the backward shift operator.*

We believe, the condition in the Theorem requiring the operator $(\varphi_\zeta(T))^*$ to have the wandering subspace property must follow from the assumption that T is in $B_1(\mathbb{D})$. We haven't been able to prove this yet.

Continuing our previous question of which holomorphic vector bundles come from flat unitary ones, one might have imagined that the operators in $B_1(\Omega)$ which are extremal would do the trick. However, it is shown in [55] that many of the bundle shifts cannot be extremal at any point in the domain. This is discussed at the end of Section 3 of the paper [55]. So, this question remains open.

4.2.2.2 Infinite divisibility

For a contraction in the Cowen-Douglas class $B_1(\mathbb{D})$, we have established that $\mathscr{K}_T(w) \leq \mathscr{K}_{S^*}(w)$, $w \in \mathbb{D}$, where S is the forward shift operator. Clearly, this is equivalent to saying that the restriction of T to the 2 - dimensional subspace $\mathscr{N}_T(w)$ spanned by the two vectors $\gamma_T(w), \gamma_T'(w)$ is contractive. Therefore, it is unreasonable to expect that the curvature inequality for an operator T would force it to be contractive. Examples are given in [9]. A natural question is to ask if the curvature inequality can be strengthened to obtain contractivity. Let K be a positive definite kernel and the adjoint of the multiplication operator M is in $B_1(\mathbb{D})$. Clearly,

$$K^{\ddagger}(z,w) := (1 - z\bar{w})K(z,\bar{w})$$

need not be nnd (unless M is contractive). However, it is Hermitian symmetric, i.e.

$$K^{\ddagger}(z,w) = \overline{K^{\ddagger}(w,z)}.$$

Now, observe that the curvature inequality is equivalent to the inequality

$$\frac{\partial^2}{\partial w \partial \bar{w}} \log K^{\ddagger}(w,w) \geq 0.$$

But the function

$$\frac{\partial^2}{\partial w \partial \bar{w}} \log K^{\ddagger}(w,w)$$

is real analytic; its polarization is a function of two complex variables and it is Hermitian symmetric in those variables. Thus we ask what if we make the stronger assumption that

$$\frac{\partial^2}{\partial w \partial \bar{w}} \log K^{\ddagger}(w,w)$$

is nnd. For a Hermitian symmetric function K, we write $K \succeq 0$ to mean that K is nnd. Similarly, this stronger form of the curvature inequality implies $K^{\ddagger}$ must be infinitely divisible; that is, not only $K^{\ddagger}$ is nnd but all its positive real powers $K^{\ddagger^t}$ are nnd as well. In particular, it follows that the operator M must be contractive. This is Corollary 4.2 of the paper [9] which is reproduced below. For two Hermitian symmetric functions K_1 and K_2, the inequality $K_1 \preceq K_2$ means $K_1 - K_2$ is nnd.

Theorem 2.8 *Let K be a positive definite kernel on the open unit disc $\mathbb{D}$. Assume that the adjoint M^* of the multiplication operator M on the reproducing kernel Hilbert space $\mathscr{H}_K$ belongs to $B_1(\mathbb{D})$. The function*

$$\frac{\partial^2}{\partial w \partial \bar{w}} \log\left((1-|w|^2)K(w,w)\right)$$

is positive definite, or equivalently

$$\mathscr{K}_{M^*}(w) \preceq \mathscr{K}_{S^*}(w),\ w \in \mathbb{D},$$

if and only if the multiplication operator M is an infinitely divisible contraction.

4.2.2.3 The multi-variable case

We say that a commuting tuple of multiplication operators M is an infinitely divisible row contraction if $(1-\langle z,w\rangle)K(z,w)$ is infinitely divisible, that is,

$$\left((1-\langle z,w\rangle)K(z,w)\right)^t$$

is positive definite for all $t > 0$.

Recall that R_m^* is the adjoint of the joint weighted shift operator on the Drury-Arveson space H_m^2. The following theorem is a characterization of infinitely divisible row contractions.

Proposition 2.9 *([9, Corollary 4.3]) Let K be a positive definite kernel on the Euclidean ball $\mathbb{B}_m$. Assume that the adjoint M^* of the multiplication operator M on the reproducing kernel Hilbert space $\mathscr{H}_K$ belongs to $B_1(\mathbb{B}_m)$. The function*

$$\left(\left(\frac{\partial^2}{\partial w_i \partial \bar{w}_j} \log\left((1-\langle w,w\rangle)K(w,w)\right)\right)\right)_{i,j=1}^m, \quad w \in \mathbb{B}_m,$$

is positive definite, or equivalently

$$\mathscr{K}_{M^*}(w) \preceq \mathscr{K}_{R_m^*}(w), \quad w \in \mathbb{B}_m,$$

if and only if the multiplication operator M is an infinitely divisible row contraction.

In the case of the polydisc, we say a commuting tuple M of multiplication by the co-ordinate functions acting on the Hilbert space $\mathscr{H}_K$ is infinitely divisible if

$$\big(S(z,w)^{-1}K(z,w)\big)^t,$$

where

$$S(z,w) := \prod_{i=1}^{m}(1 - z_i\bar{w}_i)^{-1}, \quad z,w \in \mathbb{D}^m,$$

is positive definite for all $t > 0$. Every commuting tuple of contractions M^* need not be infinitely divisible. Let S_m be the commuting m - tuple of the joint weighted shift defined on the Hardy space $H^2(\mathbb{D}^m)$.

Corollary 2.10 *([9, Corollary 4.4]) Let K be a positive definite kernel on the polydisc $\mathbb{D}^m$. Assume that the adjoint M^* of the multiplication operator M on the reproducing kernel Hilbert space $\mathscr{H}_K$ belongs to $B_1(\mathbb{D}^m)$. The function*

$$\Big(\Big(\frac{\partial^2}{\partial w_i \partial \bar{w}_j}\log\big(S(w,w)^{-1}K(w,w)\big)\Big)\Big)_{i,j=1}^m, \quad w \in \mathbb{D}^m,$$

is positive definite, or equivalently

$$\mathscr{K}_{M^*}(w) \preceq \mathscr{K}_{S_m^*}(w), \quad w \in \mathbb{D}^m,$$

if and only if the multiplication operator M is an infinitely divisible m-tuple of row contractions.

Exploiting the explicit description of the local operators $N_1(w),\dots,N_m(w)$, a very general curvature inequality for a commuting tuple of operators T in $B_n(\Omega)$ is given in [50, Theorem 2.4]. We reproduce below a simple instance of such inequalities taking Ω to be the unit ball $\mathbb{B}_m$ in $\mathbb{C}^m$ and setting $n = 1$.

Theorem 2.11 *([49, Theorem 4.2]) Let θ_w be a bi-holomorphic automorphism of $\mathbb{B}_m$ such that $\theta_w(w) = 0$. If ρ_T is a contractive homomorphism of $\mathscr{O}(\mathbb{B}_m)$ induced by the localization $N_{\mathbf{T}}(w)$, $T \in \mathrm{B}_1(\mathbb{B})$, then*

$$\mathscr{K}_T(w) \le -\overline{D\theta_w(w)}^t D\theta_w(w) = \tfrac{1}{m+1}\mathscr{K}_B(w), \; w \in \mathbb{B}_m,$$

where $\mathscr{K}_B$ is the curvature

$$-\Big(\Big(\frac{\partial^2}{\partial w_i \partial \bar{w}_j}\log B(w,w)\Big)\Big)_{i,j=1}^m$$

and

$$B(z,w) = \big(\tfrac{1}{1-\langle z,w\rangle}\big)^{m+1}$$

is the Bergman kernel of the ball $\mathbb{B}_m$.

4.2.3 Homogeneous operators

Let $\mathscr{D}$ be a bounded symmetric domain. The typical examples are the matrix unit ball $(\mathbb{C}^{p\times q})_1$ of size $p \times q$, which includes the case of the Euclidean ball, i.e., $q = 1$. Let $G := \mathrm{Aut}(\mathscr{D})$ be the bi-holomorphic automorphism group of $\mathscr{D}$.

For the matrix unit ball, $G := \mathrm{SU}(p,q)$, which consists of all linear automorphisms leaving the form $\begin{pmatrix} I_n & 0 \\ 0 & -I_m \end{pmatrix}$ on $\mathbb{C}^{p+q}$ invariant. Thus $g \in \mathrm{SU}(p,q)$ is of the form $\begin{pmatrix} a & b \\ c & d \end{pmatrix}$. The group $\mathrm{SU}(p,q)$ acts on $(\mathbb{C}^{p\times q})_1$ via the map

$$g = \begin{pmatrix} a & b \\ c & d \end{pmatrix} : z \mapsto (az + bz)(cz + dz)^{-1},\ z \in (\mathbb{C}^{p\times q})_1.$$

This action is transitive. Indeed $(\mathbb{C}^{p\times q})_1 \cong \mathrm{SU}(p,q)/\mathbf{K}$, where $\mathbf{K}$ is the stabilizer of $\mathbf{0}$ in $(\mathbb{C}^{p\times q})_1$.

When $\mathscr{D}$ is a bounded symmetric domain of dimension m and $\mathscr{H}$ is any Hilbert space, an m-tuple $T = (T_1, \ldots, T_m)$ of commuting bounded operators acting on $\mathscr{H}$ is said to be homogeneous (cf. [52, 5]) if their joint Taylor spectrum is contained in $\overline{\mathscr{D}}$ and for every holomorphic automorphism g of $\mathscr{D}$, there exists a unitary operator U_g such that

$$g(T_1, \ldots, T_m) = (U_g^{-1} T_1 U_g, \ldots, U_g^{-1} T_m U_g).$$

Imprimitivity. More generally, let G be a locally compact second countable (lcsc) topological group and $\mathscr{D}$ be an lcsc G-space. Suppose that $U : G \to \mathscr{U}(\mathscr{H})$ is a unitary representation of the group G on the Hilbert space $\mathscr{H}$ and that $\rho : \mathbf{C}(\mathscr{D}) \to \mathscr{L}(\mathscr{H})$ is a $*$ - homomorphism of the C^* - algebra of continuous functions $\mathbf{C}(\mathscr{D})$ on the algebra $\mathscr{L}(\mathscr{H})$ of all bounded operators acting on the Hilbert space $\mathscr{H}$. Then the pair (U, ρ) is said to be a representation of the G-space $\mathscr{D}$ if

$$\rho(g \cdot f) = U(g)^* \rho(f) U(g),\ f \in \mathbf{C}(\mathscr{D}),\ g \in G,$$

where $(g \cdot f)(w) = f(g^{-1} \cdot w)$, $w \in \mathscr{D}$. This is the generalization due to Mackey of the imprimitivity relation of Frobenius. These are exactly the homogenous commuting tuples of normal operators.

As before, let $\mathbf{K}$ be the stabilizer group of $\mathbf{0}$ in G. Thus $G/\mathbf{K} \cong \mathscr{D}$, where the identification is obtained via the map: $g\mathbf{K} \to g\mathbf{0}$. The action of G on $\mathscr{D}$ is evidently transitive. Given any unitary representation σ of $\mathbf{K}$, one may associate a representation (U^σ, ρ^σ) of the G -space $\mathscr{D}$. The correspondence

$$\sigma \to (U^\sigma, \rho^\sigma)$$

is an equivalence of categories. The representation U^σ is the representation of G induced by the representation σ of the group $\mathbf{K}$. For a semi-simple group G, induction from the parabolic subgroups is the key to producing irreducible representations. Along with holomorphic induction, this method gives almost all the irreducible unitary representations of the semi-simple group G.

We ask what happens if the algebra of continuous functions is replaced by the polynomial ring and the $*$ - homomorphism ρ is required to be merely a homomorphism of this ring. These are the commuting tuples of homogeneous operators.

4.2.3.1 Quasi-invariant kernels

Let $\mathscr{M} \subseteq \text{Hol}(\mathscr{D})$ be a Hilbert space possessing a reproducing kernel, say, K. These holomorphic imprimitivities are exactly homogeneous operators. Assume that $\mathscr{M}$ is a Hilbert module over the polynomial ring $\mathbb{C}[\mathbf{z}]$. Let $U : G \to \mathscr{U}(\mathscr{M})$ be a unitary representation. What are the pairs (U, ρ) that satisfy the imprimitivity relation, namely,

$$U_g^* \rho(p) U_g = \rho(p \circ g^{-1}),\ g \in G,\ p \in \mathbb{C}[\mathbf{z}]?$$

Suppose that the kernel function K transforms according to the rule

$$J_g(z) K(g(z), g(w)) J_g(w)^* = K(z, w),\ g \in G,\ z, w \in \mathscr{D},$$

for some holomorphic function $J_g : \mathscr{D} \to \mathbb{C}$. Then the kernel K is said to be quasi-invariant, which is equivalent to saying that the map $U_g : f \to J_g\,(f \circ g^{-1}),\ g \in G$, is unitary. If we further assume that the $J_g : \mathscr{D} \to \mathbb{C}$ is a cocycle, then U is a homomorphism. The pair (U, ρ) is a representation of the G-space $\mathscr{D}$ and conversely.

Therefore, our question becomes that of a characterization of all the quasi-invariant kernels defined on $\mathscr{D}$, or equivalently, finding all the holomorphic cocycles, which is also equivalent to finding all the holomorphic Hermitian homogeneous vector bundles over $\mathscr{D}$.

Let $K : \Omega \times \Omega \to \mathscr{M}_m(\mathbb{C})$ be a kernel function. We will assume the function K is holomorphic in the first variable and anti-holomorphic in the second. For two functions of the form $K(\cdot, w_i)\zeta_i$, ζ_i in $\mathbb{C}^m$ $(i = 1, 2)$ define their inner product by the reproducing property, that is,

$$\langle K(\cdot, w_1)\zeta_1, K(\cdot, w_2)\zeta_2 \rangle = \langle K(w_2, w_1)\zeta_1, \zeta_2 \rangle.$$

This extends to an inner product on the linear span if and only if K is positive definite in the sense that

$$\begin{aligned} \sum_{j,k=1}^{n} \langle K(z_j, z_k)\zeta_k, \zeta_j \rangle &= \Big\langle \sum_{k=1}^{n} K(\cdot, z_k)\zeta_k, \sum_{j=1}^{n} K(\cdot, z_j)\zeta_j \Big\rangle \\ &= \Big\| \sum_{k=1}^{n} \langle K(\cdot, z_k)\zeta_k \Big\|^2 \geq 0. \end{aligned}$$

Let $G : \Omega \times \Omega \to \mathscr{M}_m(\mathbb{C})$ be the Grammian $G(z, w) = (\langle e_p(w), e_q(z) \rangle)_{p,q}$ of a set of m antiholomorphic functions $e_p : \Omega \to \mathscr{H}$, $1 \leq p \leq m$, taking values in some Hilbert space $\mathscr{H}$. Then

$$\begin{aligned} \sum_{j,k=1}^{n} \langle G(z_j, z_k)\zeta_k, \zeta_j \rangle &= \sum_{j,k=1}^{n} \Big(\sum_{pq=1}^{m} \langle e_p(z_k), e_q(z_j) \rangle \zeta_k(p) \overline{\zeta_j(q)} \Big) \\ &= \Big\| \sum_{j,k} \zeta_k(p) e_p(z_k) \Big\|^2 > 0. \end{aligned}$$

We therefore conclude that $G(z, w)^{\text{tr}}$ defines a positive definite kernel on Ω.

For an anti-holomorphic function $s : \Omega \to \mathbb{C}^m$, let us define the norm, at w, $\|s(w)\|^2 := \|K(\cdot,w)s(w)\|^2$, where the norm on the right hand side is the norm of the Hilbert space $\mathscr{H}$ defined by the positive definite kernel K. Let ε_i, $1 \le i \le m$, be the standard basis vectors in $\mathbb{C}^m$. We have

$$\begin{aligned}\Big\|K(\cdot,w)s(w)\Big\|^2 &= \Big\|\sum_i s_i(w)K(\cdot,w)\varepsilon_i\Big\|^2\\ &= \sum_{i,j}\langle s_i(w)K(\cdot,w)\varepsilon_i, s_j(w)K(\cdot,w)\varepsilon_j\rangle\\ &= \sum_{i,j} s_i(w)\overline{s_j(w)}\langle K(\cdot,w)\varepsilon_i, K(\cdot,w)\varepsilon_j\rangle\\ &= \sum_{i,j} K(w,w)_{j,i}\, s_i(w)\overline{s_j(w)}\\ &= \overline{(s(w))}^{\mathrm{tr}} K(w,w)^{\mathrm{tr}} s(w)\end{aligned}$$

For w in Ω, and p in the set $\{1,\dots,m\}$, let $e_p : \Omega \to \mathscr{H}$ be the antiholomorphic function:

$$e_p(w) := K_w(\cdot)\otimes\frac{\partial}{\partial\bar{w}_p}K_w(\cdot) - \frac{\partial}{\partial\bar{w}_p}K_w(\cdot)\otimes K_w(\cdot).$$

Then

$$\frac{1}{2}\langle e_p(w), e_q(z)\rangle = K(z,w)\frac{\partial^2}{\partial z_q\partial\bar{w}_p}K(z,w) - \frac{\partial}{\partial\bar{w}_p}K(z,w)\frac{\partial}{\partial z_q}K(z,w).$$

The curvature of the metric K is given by the $(1,1)$ - form

$$\sum\frac{\partial^2}{\partial w_q\partial\bar{w}_p}\log K(w,w)\, dw_q\wedge d\bar{w}_p.$$

Set

$$\mathscr{K}(z,w) := \left(\frac{\partial^2}{\partial z_q\partial\bar{w}_p}\log K(z,w)\right)_{qp}.$$

Since $2K(z,w)^2\mathscr{K}(z,w)$ is of the form $G(z,w)^{\mathrm{tr}}$, it follows that $K(z,w)^2\mathscr{K}(z,w)$ defines a positive definite kernel on Ω taking values in $\mathscr{M}_m(\mathbb{C})$.

Let $\varphi : \Omega \to \Omega$ be holomorphic and $g : \Omega \to \mathbb{C}$ be the function $g(w) := K(w,w)$, $w \in \Omega$. Set $h = g\circ\varphi$. Apply the change of variable formula twice. The first time around, we have

$$(\partial_i h)(w) = \sum_\ell(\partial_\ell g)(\varphi(w))(\partial_i\varphi_\ell)(w),$$

differentiating a second time, we have

$$(\bar{\partial}_j\partial_i h)(w) = \sum_\ell\Big(\sum_k\bar{\partial}_k(\partial_\ell g)(\varphi(w))\overline{\partial_j\varphi_k(w)}\Big)(\partial_i\varphi_\ell)(w).$$

In terms of matrices, we have

$$\left(\frac{\partial^2}{\partial z_i \partial \bar{w}_j} \log K(\varphi(z), \varphi(w))\right)_{i,j}$$
$$= \left(\frac{\partial \varphi_\ell}{\partial z_i}\right)_{i,\ell} \left(\left(\frac{\partial^2}{\partial z_\ell \partial \bar{w}_k} \log K\right)(\varphi(z), \varphi(w))\right)_{\ell,k} \left(\frac{\partial \bar{\varphi}_k}{\partial \bar{z}_j}\right)_{k,j}.$$

Equivalently,

$$\mathscr{K}_{g\circ\varphi}(w) = D\varphi(w)^{\mathrm{tr}} \mathscr{K}_g(\varphi(w)) \overline{D\varphi(w)}.$$

If

$$\det{}_{\mathbb{C}} D\varphi(w)\,(g\circ\varphi)(w)\,\overline{\det{}_{\mathbb{C}} D\varphi(w)} = g(w),$$

then $\mathscr{K}_{g\circ\varphi}(w)$ equals $\mathscr{K}_g(w)$. Hence we conclude that $\mathscr{K}$ is invariant under the automorphisms φ of Ω in the sense that

$$D\varphi(w)^{\mathrm{tr}} \mathscr{K}(\varphi(w)) \overline{D\varphi(w)} = \mathscr{K}(w),\ w \in \Omega.$$

Let $Q : \Omega \to \mathscr{M}_m(\mathbb{C})$ be a real analytic function such that $Q(w)$ is positive definite for $w \in \Omega$. Let $\mathscr{H}$ be the Hilbert space of holomorphic functions on Ω which are square integrable with respect to $Q(w)dV(w)$, that is,

$$\|f\|^2 := \int_\Omega \langle Q(w)f(w), f(w)\rangle^2 dV(w),$$

where dV is the normalized volume measure on $\mathbb{C}^m$. Let $U_\varphi : \mathscr{H} \to \mathscr{H}$ be the operator

$$(U_\varphi f)(z) = m(\varphi^{-1}, z)(f \circ \varphi^{-1})(z)$$

for some cocycle m. The operator U_φ is unitary if and only if

$$\begin{aligned}
\|U_\varphi f\|^2 &= \int \langle Q(w)(U_\varphi f)(w), (U_\varphi f)(w)\rangle^2 dV(w) \\
&= \int \langle \overline{m(\varphi^{-1}, w)}^{\mathrm{tr}} Q(w) m(\varphi^{-1}, w) f(\varphi^{-1}(w)), f(\varphi^{-1}(w))\rangle^2 dV(w) \\
&= \int \langle Q(w)f(w), f(w)\rangle^2 dV(w),
\end{aligned}$$

whenever Q transforms according to the rule

$$\overline{m(\varphi^{-1}, w)}^{\mathrm{tr}} Q(w) m(\varphi^{-1}, w) = Q(\varphi^{-1}(w))|\det{}_{\mathbb{C}}(D\varphi^{-1})(w)|^2. \tag{4.13}$$

Set

$$m(\varphi^{-1}, w) = D\varphi^{-1}(w)^{\mathrm{tr}}$$

and

$$Q^{(\lambda)}(w) := b(w)^{1-\lambda} \mathscr{K}_g(w)^{-1}, \quad \lambda > 0,$$

where $b(w)$ is the restriction of the Bergman kernel to the diagonal subset of $\Omega \times \Omega$.

Then $Q^{(\lambda)}$ transforms according to the rule (4.13). If for some $\lambda > 0$, the Hilbert space $L^2_{\text{hol}}(\Omega, Q^{(\lambda)}\, dV)$ determined by the measure is nontrivial, then the corresponding reproducing kernel is of the form $b(w)^{\lambda}\mathscr{K}(w)$.

For the Euclidean ball,

$$L^2_{\text{hol}}(\mathbb{B}^m, Q^{(\lambda)}\, dV) \neq \{0\}$$

if and only if $\lambda > 1$. This means that the polarization of the real analytic function

$$\begin{aligned}(1-\langle w,w\rangle)(1-\langle w,w\rangle)^{-\lambda(n+1)}\mathscr{K}(w) &= (1-\langle w,w\rangle)^{-(\lambda-1)(n+1)}\mathscr{K}(w)\\ &= (1-\langle w,w\rangle)^{-(\lambda+1)(n+1)}\left(\langle e_p(w), e_q(w)\rangle\right)_{qp}\end{aligned}$$

must also be positive definite. Since $\left(\langle e_p(w), e_q(z)\rangle\right)_{qp}$ is positive definite, it follows that $\lambda > -1$ will ensure positivity of the kernel

$$(1-\langle w,w\rangle)^{-\lambda(n+1)}\mathscr{K}(w) = b(w)^{\lambda+2}\text{Adj}(I_n - zw^*).$$

4.2.3.2 Classification

We have already described homogeneous operators in the Cowen-Douglas class $B_1(\mathbb{D})$ using the curvature invariant. For operators in $B_k(\mathbb{D})$, $k > 1$, the curvature alone does not determine the class of the operator. Examples of irreducible homogeneous operators in $B_k(\mathbb{D})$ were given in [42] using an intertwining operator Γ. For the complete classification, we recall the description of homogeneous vector bundles via holomorphic induction, see [41]. Making this explicit in our context, we were able to construct the intertwining operator Γ in general [44]. Some of the details are reproduced below from the announcement [43].

Let $\mathfrak{t} \subseteq \mathfrak{g}^{\mathbb{C}} = \mathfrak{sl}(2,\mathbb{C})$ be the algebra $\mathbb{C}h + \mathbb{C}y$, where

$$h = \frac{1}{2}\begin{pmatrix} 1 & 0 \\ 0 & -1 \end{pmatrix}, \quad y = \begin{pmatrix} 0 & 0 \\ 1 & 0 \end{pmatrix}.$$

Linear representations (ρ, V) of the algebra $\mathfrak{t} \subseteq \mathfrak{g}^{\mathbb{C}} = \mathfrak{sl}(2,\mathbb{C})$, that is, pairs $\rho(h), \rho(y)$ of linear transformations satisfying $[\rho(h), \rho(y)] = -\rho(y)$ provide a para-metrization of the homogeneous holomorphic vector bundles.

In obtaining the classification of the homogeneous operators, it is necessary to work with the universal covering group $\tilde{G}$ of the bi-holomorphic automorphism group G of the unit disc. The $\tilde{G}$ - invariant Hermitian structures on the homogeneous holomorphic vector bundle E (making it into a homogeneous holomorphic Hermitian vector bundle), if they exist, are given by $\rho(\tilde{\mathbb{K}})$ - invariant inner products on the representation space. Here $\tilde{\mathbb{K}}$ is the stabilizer of 0 in $\tilde{G}$.

An inner product can be $\rho(\tilde{\mathbb{K}})$ - invariant if and only if $\rho(h)$ is diagonal with real diagonal elements in an appropriate basis. We are interested only in Hermitizable bundles, that is, those that admit a Hermitian structure. So, we will assume without restricting generality, that the representation space of ρ is $\mathbb{C}^n$ and that $\rho(h)$ is a real diagonal matrix.

Since $[\rho(h),\rho(y)] = -\rho(y)$, we have $\rho(y)V_\lambda \subseteq V_{\lambda-1}$, where $V_\lambda = \{\xi \in \mathbb{C}^n : \rho(h)\xi = \lambda\xi\}$. Hence $(\rho,\mathbb{C}^n)$ is a direct sum, orthogonal for every $\rho(\tilde{K})$ - invariant inner product of "elementary" representations, that is, such that

$$\rho(h) = \begin{pmatrix} -\eta I_0 & & \\ & \ddots & \\ & & -(\eta+m)I_m \end{pmatrix} \text{ with } I_j = I \text{ on } V_{-(\eta+j)} = \mathbb{C}^{d_j}$$

and

$$Y := \rho(y) = \begin{pmatrix} 0 & & & & \\ Y_1 & 0 & & & \\ & Y_2 & 0 & & \\ & & \ddots & \ddots & \\ & & & Y_m & 0 \end{pmatrix}, \quad Y_j : V_{-(\eta+j-1)} \to V_{-(\eta+j)}.$$

We denote the corresponding elementary Hermitizable bundle by $E^{(\eta,Y)}$.

The Multiplier and Hermitian structures. As in [44] we will use a natural trivialization of $E^{(\eta,Y)}$. In this, the sections of homogeneous holomorphic vector bundle $E^{(\eta,Y)}$ are holomorphic functions $\mathbb{D}$ taking values in $\mathbb{C}^n$. The $\tilde{G}$ action is given by $f \mapsto J^{(\eta,Y)}_{g^{-1}}(f\circ g^{-1})$ with multiplier

$$\left(J^{(\eta,Y)}_g(z)\right)_{p,\ell} = \begin{cases} \frac{1}{(p-\ell)!}(-c_g)^{p-\ell}(g')(z)^{\eta+\frac{p+\ell}{2}}Y_p\cdots Y_{\ell+1} & if\ p \geq \ell \\ 0 & if\ p < \ell \end{cases},$$

where c_g is the analytic function on $\tilde{G}$ which, for g near e, acting on $\mathbb{D}$ by $z \mapsto (az+b)(cz+d)^{-1}$ agrees with c.

Proposition 2.12 *We have $E^{(\eta,Y)} \equiv E^{(\eta',Y')}$ if and only if $\eta = \eta'$ and $Y' = AYA^{-1}$ with a block diagonal matrix A.*

A Hermitian structure on $E^{(\eta,Y)}$ appears as the assignment of an inner product $\langle\cdot,\cdot\rangle_z$ on $\mathbb{C}^n$ for $z \in \mathbb{D}$. We can write

$$\langle\zeta,\xi\rangle_z = \langle H(z)\zeta,\xi\rangle, \text{ with } H(z) \succ 0.$$

Homogeneity as a Hermitian vector bundle is equivalent to

$$J_g(z)H(g\cdot z)^{-1}J_g(z)^* = H(z)^{-1},\ g \in G,\ z \in \mathbb{D}.$$

The Hermitian structure is then determined by $H = H(0)$ which is a positive block diagonal matrix. We write $(E^{(\eta,Y)},H)$ for the vector bundle $E^{(\eta,Y)}$ equipped with the Hermitian structure H. We note that $(E^{(\eta,Y)},H) \cong (E^{(\eta,AYA^{-1})},A^{*-1}HA)$ for any block diagonal invertible A. Therefore every homogeneous holomorphic Hermitian vector bundle is isomorphic with one of the form $(E^{(\eta,Y)},I)$.

If $E^{(\eta,Y)}$ has a reproducing kernel K which is the case for bundles corresponding to an operator in the Cowen-Douglas class, then K satisfies

$$K(z,w) = J_g(z)K(gz,gw)J_g(w)^*$$

and induces a Hermitian structure H given by $H(0) = K(0,0)^{-1}$. *Construction of the bundles with reproducing kernel.* For $\lambda > 0$, let $\mathbb{A}^{(\lambda)}$ be the Hilbert space of holomorphic functions on the unit disc with reproducing kernel $(1-z\bar{w})^{-2\lambda}$. It is homogeneous under the multiplier $g^{\lambda}(z)$ for the action of $\tilde{G}$. This gives a unitary representation of $\tilde{G}$. Let

$$\mathbf{A}^{(\eta)} = \oplus_{j=0}^{m} \mathbb{A}^{(\eta+j)} \otimes \mathbb{C}^{d_j}.$$

For f in $\mathbf{A}^{(\eta)}$, we denote by f_j, the part of f in $\mathbb{A}^{(\eta+j)} \otimes \mathbb{C}^{d_j}$. We define $\Gamma^{(\eta,Y)}f$ as the $\mathbb{C}^n$ - valued holomorphic function whose part in $\mathbb{C}^{d_\ell}$ is given by

$$\left(\Gamma^{(\eta,Y)}f\right)_\ell = \sum_{j=0}^{\ell} \frac{1}{(\ell-j)!}\frac{1}{(2\eta+2j)_{\ell-j}} Y_\ell \cdots Y_{j+1} f_j^{(\ell-j)}$$

for $\ell \geq j$. For invertible block diagonal N on $\mathbb{C}^n$, we also define $\Gamma_N^{(\eta,Y)} := \Gamma^{(\eta,Y)} \circ N$. It can be verified that $\Gamma_N^{(\eta,Y)}$ is a $\tilde{G}$ - equivariant isomorphism of $\mathbf{A}^{(\eta)}$ as a homogeneous holomorphic vector bundle onto $E^{(\eta,Y)}$. The image $K_N^{(\eta,Y)}$ of the reproducing kernel of $\mathbf{A}^{(\eta)}$ is then a reproducing kernel for $E^{(\eta,Y)}$. A computation gives that $K_N^{(\eta,Y)}(0,0)$ is a block diagonal matrix such that its ℓ'th block is

$$K_N^{(\eta,Y)}(0,0)_{\ell,\ell} = \sum_{j=0}^{\ell} \frac{1}{(\ell-j)!}\frac{1}{(2\eta+2j)_{\ell-j}} Y_\ell \cdots Y_{j+1} N_j N_j^* Y_{j+1}^* \cdots Y_\ell^*.$$

We set

$$H_N^{(\eta,Y)} = K_N^{(\eta,Y)}(0,0)^{-1}.$$

We have now constructed a family $(E^{(\eta,Y)}, H_N^{(\eta,Y)})$ of elementary homogeneous holomorphic vector bundles with a reproducing kernel ($\eta > 0$, Y as before, N invertible block diagonal).

Theorem 2.13 *([44, Theorem 3.2] Every elementary homogeneous holomorphic vector bundle E with a reproducing kernel arises from the construction given above.*

Proof (**Sketch of proof**) As a homogeneous bundle E is isomorphic to some $E^{(\eta,Y)}$. Its reproducing kernel gives a Hilbert space structure in which the $\tilde{G}$ action on the sections of $E^{(\eta,Y)}$ is a unitary representation U. Now $\Gamma^{(\eta,Y)}$ intertwines the unitary representation of $\tilde{G}$ on $\mathbb{A}^{(\eta)}$ with U. The existence of a block diagonal N such that $\Gamma_N^{(\eta,Y)} = \Gamma^{(\eta,Y)} \circ N$ is a Hilbert space isometry follows from Schur's Lemma.

As remarked before, every homogeneous holomorphic Hermitian vector bundle is isomorphic to an $(E^{(\eta,Y)}, I)$, here Y is unique up to conjugation by a block unitary. In this form, it is easy to tell whether the bundle is irreducible: this is the case if and only if Y is not the orthogonal direct sum of two matrices of the same block type as Y. We call such a Y irreducible.

Let $\mathscr{P}$ be the set of all (η, Y) such that $E^{(\eta,Y)}$ has a reproducing kernel. Using the formula for $K_N^{(\eta,Y)}(0,0)$ we can write down explicit systems of inequalities that determine whether (η, Y) is in $\mathscr{P}$. In particular we have

Proposition 2.14 *For every Y, there exists a $\eta_Y > 0$ such that (η, Y) is in $\mathscr{P}$ if and only if $\eta > \eta_Y$.*

Finally, we obtain the desired classification.

Theorem 2.15 *([44, Theorem 4.1]) All the homogeneous holomorphic Hermitian vector bundles of rank n with a reproducing kernel correspond to homogeneous operators in the Cowen – Douglas class $\mathrm{B}_n(\mathbb{D})$. The irreducible ones are the adjoint of the multiplication operator M on the Hilbert space of sections of $(E^{(\eta,Y)}, I)$ for some (η, Y) in $\mathscr{P}$ and irreducible Y. The block matrix Y is determined up to conjugacy by block diagonal unitaries.*

Proof **(Sketch of proof)** There is a simple orthonormal system for the Hilbert space $\mathbb{A}^{(\lambda)}$. Hence we can find such a system for $\mathbf{A}^{(\eta)}$ as well. Transplant it using $\Gamma^{(\eta,Y)}$ to $E^{(\eta,Y)}$. The multiplication operator in this basis has a block diagonal form with

$$M_n := M_{|\,\mathrm{res}\,\mathscr{H}(n)} : \mathscr{H}(n) \to \mathscr{H}(n+1).$$

This description is sufficiently explicit to see:

$$M_n \sim I + 0(\frac{1}{n}).$$

Hence M is the sum of an ordinary block shift operator and a Hilbert Schmidt operator. This completes the proof.

The general case. The first examples of homogeneous operators in several variables were given in [52]. This was followed by a detailed study of these operators in the paper [5] for tube domains and in full generality in the paper [3]. First examples of homogeneous operators in $B_2(\mathbb{D})$ were given in [60]. A class of homogeneous operators in $B_n(\mathbb{D})$, which we called, generalized Wilkins operators were described in [6] using the *jet construction*. The paper [54] gives a class of homogeneous operators in $B_n(\mathscr{D})$, where $\mathscr{D}$ is one of the classical bounded symmetric domains, using the decomposition of a tensor product of two discrete series representations.

The essential ingredients from the case of the automorphism group of the unit disc is now available for an arbitrary bounded symmetric domain. In particular, the intertwining operator Γ has been found explicitly. This gives a complete classification of homogeneous operators in the Cowen-Douglas class of the ball. In general, while a description of the homogeneous holomorphic vector bundles is given in the paper

[46], it hasn't been possible to describe the operators as explicitly as in the case of the ball. These results were announced in [45] and complete proofs now appear in [46]. A different approach to finding a class of homogeneous holomorphic vector bundles is in [53].

As one might expect several questions remain open, for instance, which commuting tuple of homogeneous operators are subnormal, which of them induce a contractive, or completely contractive homomorphism. Some of these questions have been studied in [5, 3]. In a different direction, the class of quasi-homogeneous operators introduced in [39] containing all the homogeneous operators shares many of the properties of the smaller class of homogeneous operator. The Halmos question on similarity of polynomially bounded operators to a contraction has an affirmative answer for the quasi-homogeneous operators.

4.2.4 Quotient and sub-modules

In an attempt to generalize the very successful model theory of Sz.-Nagy and Foias for contractions to other settings, R. G. Douglas reformulated the question in the language of Hilbert modules over a function algebra. We describe below some aspects of this reformulation and its consequences focussing on the quotient modules for the class of quasi-free Hilbert modules $\mathscr{M}$ introduced in [26]. These Hilbert modules are obtained by taking the completion of the polynomial ring $\mathbb{C}[z]$ with respect to some inner product. We consider a class of sub-modules $\mathscr{S}_k \subseteq \mathscr{M}$ which consist of all functions in $\mathscr{M}$ that vanish to some fixed order k on a hypersurface $\mathscr{Z}$ contained in Ω. Let us recall some of the definitions (cf. [27, Section 1.2]).

(1) A *hypersurface* is a complex sub-manifold of complex dimension $m-1$, that is, a subset $\mathscr{Z} \subseteq \Omega$ is a hypersurface if for any fixed $z \in \mathscr{Z}$, there exists a neighbourhood $U \subseteq \Omega$ of z and a local defining function φ for $U \cap \mathscr{Z}$.

(2) A *local defining function* φ is a holomorphic map $\varphi : U \to \mathbb{C}$ such that $U \cap \mathscr{Z} = \{z \in U : \varphi(z) = 0\}$ and $\frac{f}{\varphi}$ is holomorphic on U whenever $f_{|U\cap\mathscr{Z}} = 0$. In particular, this implies that the gradient of φ doesn't vanish on $\mathscr{Z}$ and that any two defining functions for $\mathscr{Z}$ must differ by a unit.

(3) A function f is said to be *vanishing to order* k on the hypersurface $\mathscr{Z}$ if $f = \varphi^n g$ for some $n \geq k$, a holomorphic function g on U and a defining function φ of $\mathscr{Z}$. The order of vanishing on $\mathscr{Z}$ of a holomorphic function $f : \Omega \to \mathbb{C}$ does not depend on the choice of the local defining function. This definition can also be framed in terms of the partial derivatives normal to $\mathscr{Z}$.

We have seen that an extension of a result due to Aronszajn's provides a model for the quotient module when the sub-module consist of the maximal set of all functions vanishing on a hyper-surface. However, if the sub-module is taken to be all functions vanishing to order $k > 1$, then the situation is different and one must introduce a matrix valued kernel via the jet construction.

4.2.4.1 The jet construction

Let $\mathscr{S}_k \subseteq \mathscr{M}$ be a sub-module of a quasi-free Hilbert module over the algebra $\mathscr{A}(\Omega)$ consisting of functions vanishing to order k on a hyper-surface $\mathscr{Z} \subseteq \Omega$. Let ∂ denote the differentiation along the unit normal to the hyper-surface $\mathscr{Z}$. Let $J : \mathscr{M} \to \mathscr{M} \otimes \mathbb{C}^k$ defined by

$$h \mapsto (h, \partial h, \partial^2 h, \ldots, \partial^{k-1} h),\ h \in \mathscr{M}$$

be the jet of order k. Transplanting the inner product from $\mathscr{M}$ on $\mathscr{M} \otimes \mathbb{C}^k$ via the map J, we see that

$$\{(e_n, \partial e_n, \ldots, \partial^{k-1} e_n)_{n\geq 0} : (e_n)_{n\geq 0} \text{ is an orthonormal basis in } \mathscr{M}\}$$

is an orthonormal basis in $J(\mathscr{M}) \subseteq \mathscr{M} \otimes \mathbb{C}^k$. This makes the jet map J isometric. Now, it is not hard to see that the sub-module $J\mathscr{S}_k \subseteq J\mathscr{M}$ consisting of the maximal set of functions in $J\mathscr{M}$ vanishing on $\mathscr{Z}$ is exactly the image under the map J of the sub-module $\mathscr{S}_k \subseteq \mathscr{M}$. It is shown in [28] that the quotient module $\mathscr{Q} = \mathscr{M}/\mathscr{S}_k$ is isomorphic to $(J\mathscr{M})/(J\mathscr{S}_k)$. Thus we are reduced to the multiplicity free case and it follows that the quotient module $\mathscr{Q}$ is the restriction of $J\mathscr{M}$ to the hypersurface $\mathscr{Z}$. To complete the description, we must provide a model.

The module $J\mathscr{M}$ possesses a reproducing kernel JK, which is the infinite sum

$$JK(z,w) = \sum_{n=0}^{\infty} (Je_n)(z)(Je_n)(w)^*,\ z,w \in \Omega.$$

From this it follows that $JK : \Omega \times \Omega \to \mathscr{M}_k(\mathbb{C})$ is of the form

$$(JK)_{\ell,j}(z,w) = \big(\partial^\ell \bar{\partial}^j K\big)(z,w),\ 0 \leq \ell, j \leq k-1. \tag{4.14}$$

The module multiplication on $J\mathscr{M}$ is then naturally obtained by requiring that the map J be a module map. Thus setting Jf to be the array

$$(Jf)_{\ell,j} = \begin{cases} \binom{\ell}{j}(\partial^{\ell-j} f), & 0 \leq \ell \leq j \leq k-1 \\ 0 & \text{otherwise,} \end{cases}, \tag{4.15}$$

$f \in \mathscr{A}(\Omega)$. The module multiplication is induced by the multiplication operator M_{J_f}, where J_f is a holomorphic function defined on Ω taking values in $\mathscr{M}_k(\mathbb{C})$. The adjoint of this operator is easy to compute

$$J_f^* JK(\cdot,w) \cdot x = JK(\cdot,w)(Jf)(w)^* \cdot x,\ x \in \mathbb{C}^k. \tag{4.16}$$

We consider the Hilbert space $(J\mathscr{M})_{\text{res}}$ obtained by restricting the functions in $J\mathscr{M}$ to the hyper-surface $\mathscr{Z}$, that is,

$$(J\mathscr{M})_{\text{res}} = \big\{h_0 \text{ holomorphic on } \mathscr{Z} : h_0 = h_{|\mathscr{Z}} \text{ for some } h \in J\mathscr{M}\big\}.$$

The norm of h_0 in $(J\mathscr{M})_{\text{res}}$ is the quotient norm, that is,

$$\|h_0\| = \inf\big\{\|h\| : h_{|\mathscr{Z}} = h_0 \text{ for } h \in J\mathscr{M}\big\},$$

and the module action is obtained by restricting the map $(f,h) \to (J_f)_{|\mathscr{Z}} h_{|\mathscr{Z}}$.

We have discussed the restriction map R before and observed that it is a unitary module map in the case of Hilbert modules consisting of scalar valued holomorphic functions. However, this is true of the vector valued case as well. It shows that$JK(\cdot,w)_{\text{res}} := K(\cdot,w)_{|\mathscr{Z}}$, $w \in \mathscr{Z}$ is the kernel function for the Hilbert module $(J\mathscr{M})_{\text{res}}$.

Theorem 2.16 *([28, Theorem 3.3]) Let $\mathscr{M}$ be a Hilbert module over the algebra $\mathscr{A}(\Omega)$. If $\mathscr{S}_k$ is the sub-module of functions vanishing to order k, then the quotient module is isomorphic to $(J\mathscr{M})_{\text{res}}$ via the isometric module map JR.*

Finding invariants for the quotient module, except in the case of $k = 1$, is more challenging. The module multiplication in the quotient module involves both a semi-simple and a nilpotent part. The semi-simple part lies typically in some $B_n(\mathscr{Z})$. Now, any equivalence between two quotient modules must also intertwine the nilpotent action. In the papers [25, 27], using this additional structure, a complete invariants were found for a class of quotient modules. We describe some fascinating possibilities for finding invariant for quotient modules using the notion of module tensor products and the recent work of Harvey and Lawson.

Let $\mathscr{M}$ and $\mathscr{N}$ be any two Hilbert modules over the algebra $\mathscr{A}$. Notice that there are two possible module actions on $\mathscr{M} \otimes \mathscr{N}$, i.e., the left action: $L \otimes I : (f, h \otimes k) \mapsto f \cdot h \otimes k$ and the right action: $I \otimes R : (f, h \otimes k) \mapsto h \otimes f \cdot k$. The module tensor product $\mathscr{M} \otimes_{\mathscr{A}} \mathscr{N}$ is defined to be the module obtained by identifying these two actions. Specifically, let $\mathscr{S}$ be the closed subspace of $\mathscr{M} \otimes \mathscr{N}$ generated by vectors of the form

$$\{f \cdot h \otimes k - h \otimes f \cdot k : h \in \mathscr{M}, k \in \mathscr{N} \text{ and } f \in \mathscr{A}\}.$$

Then $\mathscr{S}$ is a submodule of $\mathscr{M} \otimes \mathscr{N}$ with respect to both the left and the right actions. The module tensor product $\mathscr{M} \otimes_{\mathscr{A}} \mathscr{N}$ is defined to be $(\mathscr{M} \otimes \mathscr{N})/\mathscr{S}$ together with the compression of either the left or the right actions, which coincide on this space. For fixed $w \in \Omega$, $\mathbb{C}$ is a module over $\mathscr{A}$ with the module action

$$(f, v) \mapsto f(w)v, \ f \in \mathscr{A}, v \in \mathbb{C}.$$

Let $\mathbb{C}_w$ denote the one dimensional module $\mathbb{C}$ with this action. We will largely confine ourselves to the module tensor product $\mathscr{M} \otimes_{\mathscr{A}} \mathbb{C}_w$, which we denote by $\mathscr{M}(w)$.

Localizing the short exact sequence

$$0 \longleftarrow \mathscr{Q} \longleftarrow \mathscr{M} \overset{X}{\longleftarrow} \mathscr{S} \longleftarrow 0$$

using the one dimensional module $\mathbb{C}_w$, one obtains a new short exact sequence of spectral sheaves (cf. [29])

$$0 \longleftarrow \mathscr{Q} \otimes_{\mathscr{A}(\Omega)} \mathbb{C}_w \longleftarrow \mathscr{M} \otimes_{\mathscr{A}(\Omega)} \mathbb{C}_w \overset{X \otimes_{\mathscr{A}(\Omega)} 1_{\mathbb{C}_w}}{\longleftarrow} \mathscr{S} \otimes_{\mathscr{A}(\Omega)} \mathbb{C}_w \longleftarrow 0.$$

Let E_0 and E be the holomorphic line bundles corresponding to the modules $\mathscr{S}$ and $\mathscr{M}$ and $\mathscr{K}_{E_0}$, $\mathscr{K}_E$ be their curvatures, respectively. It is shown in [24] that the class of the alternating sum

$$\sum_{i,j=1}^{m} \frac{\partial^2}{\partial w_i \partial \bar{w}_j} \log(X(w)^* X(w)) dw_i \wedge d\bar{w}_j - \mathscr{K}_{E_0}(w) + \mathscr{K}_E(w),$$

is the fundamental class of the hypersurface $\mathscr{Z}$. This identification makes essential use of the Poincaré-Lelong formula.

It is not clear how one can obtain such an alternating sum if the submodule $\mathscr{S}$ is not assumed to be the submodule (maximal set) of functions vanshing on $\mathscr{Z}$. A possible approach to this question using some ideas of Donaldson appearing in [22] is discussed in [28]. An adaptation of the results of Harvey and Lawson [35] to the present situation may be fruitful in the case where $\mathscr{S}$ is assumed to consist of all functions in $\mathscr{M}$ which vanish to order k. Let $\mathscr{E}$ and $\mathscr{E}_0$ be the vector bundles obtained by localization (cf. [29]) from the modules $\mathscr{M}$ and $\mathscr{M}_0$ and let φ be an ad-invariant polynomial (in particular, a Chern form) in the respective curvatures $\mathscr{K}$ and $\mathscr{K}_0$. Then the work of Harvey and Lawson [35] on singular connections gives a mechanism for studying these bundles since the natural connection on the bundle $\mathscr{M}_0$ is a singular one. They obtain a relation of the form

$$\varphi(\mathscr{K}) - \varphi(\mathscr{K}_0) = \mathrm{R}es_\varphi[\mathscr{Z}] + dT_\varphi,$$

where $\mathrm{R}es_\varphi[\mathscr{Z}]$ is a 'residue' form related to the zero set and T_φ is a transgression current. This incorporates a generalized Poincaré-Lelong formula which played a crucial role in the study of the quotient module in the rank one case; see [28].

4.2.4.2 The Clebsch-Gorden formula

Let $\mathscr{M}_1$ and $\mathscr{M}_2$ be Hilbert spaces consisting of of holomorphic functions defined on Ω possessing reproducing kernels K_1 and K_2, respectively. Assume that the natural action of $\mathbb{C}[z]$ on the Hilbert space $\mathscr{M}_1$ is continuous; that is, the map $(p,h) \to ph$ defines a bounded operator on $\mathscr{M}$ for $p \in \mathbb{C}[z]$. (We make no such assumption about the Hilbert space $\mathscr{M}_2$.) Now, $\mathbb{C}[z]$ acts naturally on the Hilbert space tensor product $\mathscr{M}_1 \otimes \mathscr{M}_2$ via the map

$$(p,(h \otimes k)) \to ph \otimes k, p \in \mathbb{C}[z],\ h \in \mathscr{M}_1,\ k \in \mathscr{M}_2.$$

The map $h \otimes k \to hk$ identifies the Hilbert space $\mathscr{M}_1 \otimes \mathscr{M}_2$ as a reproducing kernel Hilbert space of holomorphic functions on $\Omega \times \Omega$. The module action is then the point-wise multiplication $(p,hk) \to (ph)k$, where

$$((ph)k)(z_1,z_2) = p(z_1)h(z_1)k(z_2), \quad z_1,z_2 \in \Omega.$$

Let $\mathscr{M}$ be the Hilbert module $\mathscr{M}_1 \otimes \mathscr{M}_2$ over $\mathbb{C}[z]$. Let $\triangle \subseteq \Omega \times \Omega$ be the diagonal subset $\{(z,z) : z \in \Omega\}$ of $\Omega \times \Omega$. Let $\mathscr{S}$ be the maximal submodule of functions in $\mathscr{M}_1 \otimes \mathscr{M}_2$ which vanish on $\triangle$. Thus

$$0 \to \mathscr{S} \xrightarrow{X} \mathscr{M}_1 \otimes \mathscr{M}_2 \xrightarrow{Y} \mathscr{Q} \to 0$$

is a short exact sequence, where $\mathscr{Q} = (\mathscr{M}_1 \otimes \mathscr{M}_2)/\mathscr{S}$, X is the inclusion map and Y is the natural quotient map.

As we have seen earlier in Section 1.4, the theorem of Aronszajn provides a complete description of the quotient module $\mathscr{Q}$ as the restrictions of functions in $\mathscr{M}$. Now, let us investigate what happens if the submodule $\mathscr{S}$ is taken to be space of functions vanishing to order 2 on $\triangle$. Set $\mathscr{H}_1$ and $\mathscr{H}_2$ to be the submodules defined by

$$\mathscr{H}_1 = \{f \in (\mathscr{H}, K) \otimes (\mathscr{H}, K) : f|_\Delta = 0\}$$

and

$$\mathscr{H}_2 = \{f \in (\mathscr{H}, K) \otimes (\mathscr{H}, K) : f|_\Delta = \partial_1 f|_\Delta = \partial_2 f|_\Delta = \ldots = \partial_m f|_\Delta = 0\}.$$

Let $\mathscr{H}_{11} = \mathscr{H}_2^\perp \ominus \mathscr{H}_1^\perp$ We have already described the quotient module $\mathscr{H}_{00} := \mathscr{M} \ominus \mathscr{H}_1$, where $\mathscr{M} = (\mathscr{H}, K) \otimes (\mathscr{H}, K)$. This is the module $\mathscr{M}_{\text{res}}$. Set

$$\widetilde{K}(z,w) = \left(K^2(z,w)\partial_i\bar{\partial}_j \log K(z,w)\right)_{1 \le i,j \le m}.$$

We claim that the function $\tilde{K}$ taking values in $\mathscr{M}_m(\mathbb{C})$ is a non-negative definite kernel. To see this, set

$$\varphi_i(w) := K_w \otimes \bar{\partial}_i K_w - \bar{\partial}_i K_w \otimes K_w, \quad 1 \le i \le m,$$

and note that each $\varphi_i : \Omega \to \mathscr{M}$ is holomorphic. A simple calculation then shows that

$$((\langle \varphi_j(w), \varphi_i(z) \rangle_{\mathscr{M}})) = \widetilde{K}(z,w).$$

How to describe the Hilbert space, or more importantly, the Hilbert module $(\mathscr{H}, \widetilde{K})$? Maybe it is a quotient, or a sub-quotient module, of the Hilbert module $(\mathscr{H}, K) \otimes (\mathscr{H}, K)$?

Let $\mathscr{H}_0$ be the subspace of $(\mathscr{H}, K) \otimes (\mathscr{H}, K)$ given by the smallest closed subspace containing the linear span of the vectors $\{\varphi_i(w) : w \in \Omega, 1 \le i \le m\}$. From this definition, it is not clear which functions belong to the subspace. An interesting limit computation given below shows that it coincides with $\mathscr{H}_{11}$.

The point of what we have said so far is that we can explicitly describe the Hilbert modules $\mathscr{H}_2^\perp$ and $\mathscr{H}_1^\perp$, up to an isomorphism of modules. Using the jet construction followed by the restriction map, one may also describe the direct sum $\mathscr{H}_2^\perp \oplus \mathscr{H}_1^\perp$, again up to an isomorphism.

But what is the module $\mathscr{H}_{11}$? To answer this question (see [33, Section 2.4.1]), one must find the kernel function for $\mathscr{H}_{11}$. Setting K_1 to be the kernel function of the module $\mathscr{H}_1$, indeed, we have

$$\lim_{u \to z, v \to w} \frac{K_1(z,u;v,w)}{(z-u)(\bar{w}-\bar{v})} = \tfrac{1}{2} K(z,w)^2 \partial\bar{\partial} \log K(z,w).$$

This shows that $\mathscr{H}_{11}$ is isomorphic to $(\mathscr{H}, \tilde{K})$. The main challenge is to obtain an orthogonal decomposition of the Hilbert module $\mathscr{M}$ in the form of a composition series, namely, to complete the decomposition:

$$\mathscr{H} \sim \mathscr{H}_{00} \oplus \mathscr{H}_{11} \oplus \mathscr{H}_{22} \oplus \cdots,$$

where $\mathscr{H}_{00}$ is the quotient module $\mathscr{M}_{\rm res}$, as in the more familiar Clebsch-Gorden formula.

4.2.4.3 The sheaf model

For a Hilbert module $\mathscr{M}$ over a function algebra $\mathscr{A}(\Omega)$, not necessarily in the class $\mathrm{B}_1(\Omega)$, motivated by the correspondence of vector bundles with locally free sheaf, we construct a sheaf of modules $\mathscr{S}^{\mathscr{M}}(\Omega)$ over $\mathscr{O}(\Omega)$ corresponding to $\mathscr{M}$. We assume that $\mathscr{M}$ possesses all the properties for it to be in the class $\mathrm{B}_1(\Omega)$ except that the dimension of the joint kernel $\mathbb{K}(w)$ need not be constant. We note that sheaf models have occurred, as a very useful tool, in the study of analytic Hilbert modules (cf. [31]), although the model we describe below is somewhat different.

A Hilbert module $\mathscr{M} \subset \mathscr{O}(\Omega)$ is said to be in the class $\mathfrak{B}_1(\Omega)$ if it possesses a reproducing kernel K (we don't rule out the possibility: $K(w,w) = 0$ for w in some closed subset X of Ω) and the dimension of $\mathscr{M}/\mathfrak{m}_w\mathscr{M}$ is finite for all $w \in \Omega$.

Most of the examples in $\mathfrak{B}_1(\Omega)$ arises in the form of a submodule of some Hilbert module $\mathscr{H}(\subseteq \mathscr{O}(\Omega))$ in the Cowen-Douglas class $\mathrm{B}_1(\Omega)$. Are there others?

Let $\mathscr{S}^{\mathscr{M}}(\Omega)$ be the subsheaf of the sheaf of holomorphic functions $\mathscr{O}(\Omega)$ whose stalk at $w \in \Omega$ is

$$\big\{(f_1)_w\mathscr{O}_w + \cdots + (f_n)_w\mathscr{O}_w : f_1,\ldots,f_n \in \mathscr{M}\big\},$$

or equivalently,

$$\mathscr{S}^{\mathscr{M}}(U) = \Big\{\sum_{i=1}^{n} (f_{i|U})g_i : f_i \in \mathscr{M}, g_i \in \mathscr{O}(U)\Big\}$$

for U open in Ω.

Proposition 2.17 *The sheaf $\mathscr{S}^{\mathscr{M}}(\Omega)$ is coherent.*

Proof The sheaf $\mathscr{S}^{\mathscr{M}}(\Omega)$ is generated by the set of functions $\{f : f \in \mathscr{M}\}$. Let $\mathscr{S}_J^{\mathscr{M}}(\Omega)$ be the subsheaf generated by the set of functions

$$J = \{f_1,\ldots,f_\ell\} \subseteq \mathscr{M} \subseteq \mathscr{O}(\Omega).$$

Thus $\mathscr{S}_J^{\mathscr{M}}(\Omega)$ is coherent. An application of Noether's Lemma [34] then guarantees that

$$\mathscr{S}^{\mathscr{M}}(\Omega) = \cup_{J\,\mathrm{finite}}\mathscr{S}_J^{\mathscr{M}}(\Omega)$$

is coherent.

We note that the coherence of the sheaf implies, in particular, that the stalk $(\mathscr{S}^{\mathscr{M}})_w$ at $w \in \Omega$ is generated by a finite number of elements $g_1,\ldots,g_n$ from $\mathscr{O}(\Omega)$.

If K is the reproducing kernel for $\mathscr{M}$ and $w_0 \in \Omega$ is a fixed but arbitrary point, then for w in a small neighborhood U of w_0, we obtain the following decomposition theorem.

Theorem 2.18 *Suppose g_i^0, $1 \le i \le n$, is a minimal set of generators for the stalk $(\mathcal{S}^{\mathcal{M}})_0 := (\mathcal{S}^{\mathcal{M}})_{w_0}$. Then we have*

$$K(\cdot, w) := K_w = g_1^0(w)K_w^{(1)} + \cdots + g_n^0(w)K_w^{(n)},$$

where $K^{(p)} : U \to \mathcal{M}$ defined by $w \mapsto K_w^{(p)}$, $1 \le p \le n$, is anti-holomorphic. Moreover, the elements $K_{w_0}^{(p)}$, $1 \le p \le n$, are linearly independent in $\mathcal{M}$, they are eigenvectors for the adjoint of the action of $\mathcal{A}(\Omega)$ on the Hilbert module $\mathcal{M}$ at w_0, and they are uniquely determined by these generators.

We also point out that the Grammian

$$G(w) = ((\langle K_w^{(p)}, K_w^{(q)} \rangle))_{p,q=1}^n$$

is invertible in a small neighborhood of w_0 and is independent of the generators $g_1, \ldots, g_n$. Thus

$$t : w \mapsto (K_{\bar{w}}^{(1)}, \ldots, K_{\bar{w}}^{(n)})$$

defines a holomorphic map into the Grassmannian $\mathrm{G}(\mathcal{H}, n)$ on the open set U^*. The pull-back E_0 of the canonical bundle on $\mathrm{G}(\mathcal{H}, n)$ under this map then defines a holomorphic Hermitian bundle on U^*. Clearly, the decomposition of K given in our Theorem is not canonical in any way. So, we can't expect the corresponding vector bundle E_0 to reflect the properties of the Hilbert module $\mathcal{M}$. However, it is possible to obtain a canonical decomposition following the construction in [21]. It then turns out that the equivalence class of the corresponding vector bundle E_0 obtained from this canonical decomposition is an invariant for the isomorphism class of the Hilbert module $\mathcal{M}$. These invariants are by no means easy to compute. At the end of this subsection, we indicate, how to construct invariants which are more easily computable. For now, the following corollary to the decomposition theorem is immediate.

Corollary 2.19 *The dimension of the joint kernel $\mathbb{K}(w)$ is greater or equal to the number of minimal generators of the stalk $(\mathcal{S}^{\mathcal{M}})_w$ at $w \in \Omega$.*

Now is the appropriate time to raise a basic question. Let $\mathfrak{m}_w \subseteq \mathcal{A}(\Omega)$ be the maximal ideal of functions vanishing at w. Since we have assumed $\mathfrak{m}_w \mathcal{M}$ is closed, it follows that the dimension of the joint kernel $\mathbb{K}(w)$ equals the dimension of the quotient module $\mathcal{M}/(\mathfrak{m}_w \mathcal{M})$. However it is not clear if one may impose a natural hypothesis on $\mathcal{M}$ to ensure

$$\dim \mathcal{M}/(\mathfrak{m}_w \mathcal{M}) = \dim \mathbb{K}(w) = \dim (\mathcal{S}^{\mathcal{M}})_w / (\mathfrak{m}(\mathcal{O}_w)(\mathcal{S}^{\mathcal{M}})_w),$$

where $\mathfrak{m}(\mathcal{O}_w)$ is the maximal ideal in $\mathcal{O}_w$, as well.

More generally, suppose $p_1, \ldots, p_n$ generate $\mathcal{M}$. Then $\dim \mathbb{K}(w) \le n$ for all $w \in \Omega$. If the common zero set V of these is $\{0\}$ then $(p_1)_0, \ldots, (p_n)_0$ need not be a minimal set of generators for $(\mathcal{S}^{\mathcal{M}})_0$. However, we show that they do if we assume

$p_1,\ldots,p_n$ are homogeneous of degree k, say. Furthermore, basis for $\mathbb{K}(0)$ is the set of vectors:

$$\{p_1(\bar{\partial})\}K(\cdot,w)_{|w=0},\ldots,p_n(\bar{\partial})\}K(\cdot,w)_{|w=0}\},$$

where $\bar{\partial}=(\bar{\partial}_1,\ldots,\bar{\partial}_m)$.

Going back to the example of $H_0^2(\mathbb{D}^2)$, we see that it has two generators, namely z_1 and z_2. Clearly, the joint kernel

$$\mathbb{K}(w):=\ker D_{(M_1^*-\bar{w}_1,M_2^*-\bar{w}_2)}$$

at $w=(w_1,w_2)$ is spanned by

$$\{z_1\otimes_{\mathscr{A}(\mathbb{D}^2)}1_w,z_2\otimes_{\mathscr{A}(\mathbb{D}^2)}1_w\}=\{w_1K_{H_0^2(\mathbb{D}^2)}(z,w),w_2K_{H_0^2(\mathbb{D}^2)}(z,w)\}$$

which consists of two vectors that are linearly dependent except when $w=(0,0)$. We also easily verify that

$$\left(\mathscr{S}^{H_0^2(\mathbb{D}^2)}\right)_w\cong\begin{cases}\mathscr{O}_w & w\neq(0,0)\\ \mathfrak{m}(\mathscr{O}_0) & w=(0,0).\end{cases}$$

Since the reproducing kernel

$$K_{H_0^2(\mathbb{D}^2)}(z,w)\quad=\quad K_{H^2(\mathbb{D}^2)}(z,w)-1=\frac{z_1\bar{w}_1+z_2\bar{w}_2-z_1z_2\bar{w}_1\bar{w}_2}{(1-z_1\bar{w}_1)(1-z_2\bar{w}_2)},$$

we find there are several choices for $K_w^{(1)}$ and $K_w^{(2)}$, $w\in U$. However, all of these choices disappear if we set $\bar{w}_1\theta_1=\bar{w}_2$ for $w_1\neq 0$, and take the limit:

$$\lim_{(w_1,w_2)\to 0}\frac{K_{H_0^2(\mathbb{D}^2)}(z,w)}{\bar{w}_1}=K_0^{(1)}(z)+\theta_1K_0^{(2)}(z)=z_1+\theta_1z_2$$

because $K_0^{(1)}$ and $K_0^{(2)}$ are uniquely deteremined by Theorem 1. Similarly, for $\bar{w}_2\theta_2=\bar{w}_1$ for $w_2\neq 0$, we have

$$\lim_{(w_1,w_2)\to 0}\frac{K_{H_0^2(\mathbb{D}^2)}(z,w)}{\bar{w}_2}=K_0^{(2)}(z)+\theta_2K_0^{(1)}(z)=z_2+\theta_2z_1.$$

Thus we have a Hermitian line bundle on the complex projective space $\mathbb{P}^1$ given by the frame $\theta_1\mapsto z_1+\theta_1z_2$ and $\theta_2\mapsto z_2+\theta_2z_1$. The curvature of this line bundle is then an invariant for the Hilbert module $H_0^2(\mathbb{D}^2)$ as shown in [28]. This curvature is easily calculated and is given by the formula $\mathscr{K}(\theta)=(1+|\theta|^2)^{-2}$. The decomposition theorem yields similar results in many other examples.

Let $\mathscr{I}$ be an ideal in the polynomial ring $\mathbb{C}[\mathbf{z}]$. The characteristic space of an ideal $\mathscr{I}$ in $\mathbb{C}[\mathbf{z}]$ at the point w is the vector space

$$\mathbb{V}_w(\mathscr{I}):=\{q\in\mathbb{C}[\mathbf{z}]:q(D)p|_w=0,\,p\in\mathscr{I}\}.$$

The envelope $\mathscr{I}_w^e$ of the ideal $\mathscr{I}$ is

$$\{p \in \mathbb{C}[\mathbf{z}] : q(D)p|_w = 0, q \in \mathbb{V}_w(\mathscr{I})\}.$$

If the zero set of the ideal $\mathscr{I}$ is $\{w\}$ then $\mathscr{I}_w^e = \mathbb{V}_w(\mathscr{I})$.

This describes an ideal by prescribing conditions on derivatives. We stretch this a little more. Let $\tilde{\mathbb{V}}_w(\mathscr{I})$ be the auxiliary space $\mathbb{V}_w(\mathfrak{m}_w\mathscr{I})$. We have

$$\dim(\cap \operatorname{Ker}(M_j - w_j)^*) = \dim \tilde{\mathbb{V}}_w(\mathscr{I})/\mathbb{V}_w(\mathscr{I}).$$

Actually, we have something much more substantial.

Lemma 2.20 *Fix $w_0 \in \Omega$ and polynomials $q_1, \dots, q_t$. Let $\mathscr{I}$ be a polynomial ideal and K be the reproducing kernel corresponding to the Hilbert module $[\mathscr{I}]$, which is assumed to be in $\mathfrak{B}_1(\Omega)$. Then the vectors*

$$q_1(\bar{D})K(\cdot, w)|_{w=w_0}, \dots, q_t(\bar{D})K(\cdot, w)|_{w=w_0}$$

form a basis of the joint kernel $\cap_{j=1}^m \ker(M_j - w_{0j})^*$ if and only if the classes $[q_1^*], \dots, [q_t^*]$ form a basis of $\tilde{\mathbb{V}}_{w_0}(\mathscr{I})/\mathbb{V}_{w_0}(\mathscr{I})$.

However, it is not clear if we can choose the polynomials $\{q_1, \dots, q_t\}$ to be a generating set for the ideal $\mathscr{I}$. Nonetheless, the following theorem produces a new set $\{q_1, \dots, q_v\}$ of generators for $\mathscr{I}$, which is more or less *"canonical"*. Indeed, it is uniquely determined modulo linear transformations.

Theorem 2.21 *([10, Proposition 2.10]) Let $\mathscr{I} \subset \mathbb{C}[\mathbf{z}]$ be a homogeneous ideal and $\{p_1, \dots, p_v\}$ be a minimal set of generators for $\mathscr{I}$ consisting of homogeneous polynomials. Let K be the reproducing kernel corresponding to the Hilbert module $[\mathscr{I}]$, which is assumed to be in $\mathfrak{B}_1(\Omega)$. Then there exists a set of generators $q_1, ..., q_v$ for the ideal $\mathscr{I}$ such that the set*

$$\{q_i(\bar{D})K(\cdot, w)|_{w=0} : 1 \le i \le v\}$$

is a basis for the joint kernel $\cap_{j=1}^m \ker M_j^*$.

4.2.5 The flag structure

We have already discussed the class of operators $\mathscr{F}B_2(\Omega)$ which is contained in the Cowen-Douglas class $B_2(\Omega)$. A natural generalization is given below. The main point of introducing this smaller class is to show that a complete set of tractable unitary invariants exists for this class.

Definition 2.22 *We let $\mathscr{F}B_n(\Omega)$ be the set of all bounded linear operators T defined on some complex separable Hilbert space $\mathscr{H} = \mathscr{H}_0 \oplus \cdots \oplus \mathscr{H}_{n-1}$, which are of the form*

$$T = \begin{pmatrix} T_0 & S_{0,1} & S_{0,2} & \cdots & S_{0,n-1} \\ 0 & T_1 & S_{1,2} & \cdots & S_{1,n-1} \\ \vdots & \ddots & \ddots & \ddots & \vdots \\ 0 & \cdots & 0 & T_{n-2} & S_{n-2,n-1} \\ 0 & \cdots & \cdots & 0 & T_{n-1} \end{pmatrix},$$

where the operator $T_i : \mathscr{H}_i \to \mathscr{H}_i$, defined on the complex separable Hilbert space $\mathscr{H}_i$, $0 \le i \le n-1$, is assumed to be in $B_1(\Omega)$ and $S_{i,i+1} : \mathscr{H}_{i+1} \to \mathscr{H}_i$, is assumed to be a non-zero intertwining operator, namely, $T_i S_{i,i+1} = S_{i,i+1} T_{i+1}$, $0 \le i \le n-2$.

The set of operators in $\mathscr{F}B_n(\Omega)$, as before, is also contained in the Cowen-Douglas class $B_n(\Omega)$. To show this starting with the base case of $n = 2$, using induction, we don't even need the intertwining condition. A rigidity theorem for this class of operators was proved in [38, Theorem 3.5], which is reproduced below.

Theorem 2.23 (Rigidity) *Any two operators T and $\tilde{T}$ in $\mathscr{F}B_n(\Omega)$ are unitarily equivalent if and only if there exists unitary operators U_i, $0 \le i \le n-1$, such that $U_i T_i = \tilde{T}_i U_i$ and $U_i S_{i,j} = \tilde{S}_{i,j} U_j$, $i < j$.*

From the rigidity, it is easy to extract unitary invariants for the operators in the class $\mathscr{F}B_n(\Omega)$, see [38, Theorem 3.6]. These invariants come from the intrinsic complex geometry inherent in the definition of the operators in the class $\mathscr{F}B_n(\Omega)$. Also, all the operators in this class are irreducible.

4.2.5.1 An application to module tensor products

In the early development of the quotient and sub-modules, it was expected that the localization using modules of rank > 1 might provide new insight. This was discussed in some detail in the paper [28] but the outcome was not conclusive. The results from [38, Section 4] completes the study initiated in [28]. Let f be a polynomial in one variable. Set

$$\mathscr{J}_\mu(f)(z) = \begin{pmatrix} \mu_{1,1} f(z) & 0 & \cdots & 0 \\ \mu_{2,1} \frac{\partial}{\partial z} f(z) & \mu_{2,2} f(z) & \cdots & 0 \\ \vdots & \vdots & \ddots & \vdots \\ \mu_{k,1} \frac{\partial^{k-1}}{\partial z^{k-1}} f(z) & \mu_{k-1,1} \frac{\partial^{k-2}}{\partial z^{k-2}} f(z) & \cdots & \mu_{k,k} f(z) \end{pmatrix}$$

where $\mu = (\mu_{i,j})$ is a lower triangular matrix of complex numbers with $\mu_{i,i} = 1$, $1 \le i \le k$. Lemma 4.1 of [38] singles out those μ which ensure $\mathscr{J}_\mu(fg) = \mathscr{J}_\mu(f)\mathscr{J}_\mu(g)$. Fix one such μ. For $\mathbf{x}$ in $\mathbb{C}^k$, and f in the polynomial ring $\mathbb{C}[z]$, define the module multiplication on $\mathbb{C}^k$ as follows:

$$f \cdot \mathbf{x} = \mathscr{J}_\mu(f)(w)\mathbf{x}.$$

The finite dimensional Hilbert space $\mathbb{C}^k$ equipped with this module multiplication is denoted by $\mathbb{C}^k_w[\mu]$.

Let $\mathscr{M}$ be a Hilbert module possessing a reproducing kernel. We construct a module of k - jets by setting

$$J\mathscr{M} = \Big\{ \sum_{l=0}^{k-1} \frac{\partial^i}{\partial z^i} h \otimes \varepsilon_{i+1} : h \in \mathscr{M} \Big\},$$

where $\varepsilon_{i+1}, 0 \leq i \leq k-1$, are the standard basis vectors in $\mathbb{C}^k$. Recall that there is a natural module action on $J\mathscr{M}$, namely,

$$\Big(f, \sum_{l=0}^{k-1} \tfrac{\partial^i}{\partial z^i} h\Big) \mapsto \mathscr{J}(f)\Big(\sum_{l=0}^{k-1} \tfrac{\partial^i}{\partial z^i} h \otimes \varepsilon_{i+1}\Big), f \in \mathbb{C}[z], h \in \mathscr{M},$$

where

$$\mathscr{J}(f)_{i,j} = \begin{cases} \binom{i-1}{j-1} \partial^{i-j} f & \text{if } i \geq j, \\ 0 & \text{otherwise.} \end{cases} \tag{4.17}$$

The module tensor product $J\mathscr{M} \otimes_{\mathbb{C}[z]} \mathbb{C}^k_w[\mu]$ is easily identified with the quotient module $\mathscr{N}^\perp$, where $\mathscr{N} \subseteq \mathscr{M}$ is the sub-module spanned by the vectors

$$\Big\{\sum_{l=1}^{k} (J_f \cdot \mathbf{h}_l \otimes \varepsilon_l - \mathbf{h}_l \otimes (\mathscr{J}_\mu(f))(w) \cdot \varepsilon_l) : \mathbf{h}_l \in J\mathscr{M}, \varepsilon_l \in \mathbb{C}^k, f \in [z]\Big\}.$$

Theorem 2.24 *The Hilbert modules corresponding to the localizations $J\mathscr{M} \otimes_{\mathbb{C}[z]} \mathbb{C}^k_w[\mu_i]$, $i = 1, 2$, are in $\mathscr{F}B_k(\Omega)$ and they are isomorphic if and only if $\mu_1 = \mu_2$.*

Acknowledgment

The author thanks Soumitra Ghara, Dinesh Kumar Keshari, Md. Ramiz Reza and Subrata Shyam Roy for going through a preliminary draft of this article carefully and pointing out several typographical errors. The author gratefully acknowledges the financial support from the Department of Science and Technology in the form of the J. C. Bose National Fellowship and from the University Grants Commission, Centre for Advanced Study.

References

[1] M. B. Abrahamse and R. G. Douglas, A class of subnormal operators related to multiply connected domains, *Adv. Math.* **19** (1976), 106 – 148.

[2] O. P. Agrawal and N. Salinas, Sharp kernels and canonical subspaces (revised), *Amer. J. Math.* **110** (1988), 23 – 47.

[3] J. Arazy and G. Zhang, Homogeneous multiplication operators on bounded symmetric domains, *J. Func. Anal.* **202** (2003), 44 – 66.

[4] N. Aronszajn, Theory of reproducing kernels, *Trans. Amer. Math. Soc.* **68** (1950), 337 – 404.

[5] B. Bagchi and G. Misra, Homogeneous tuples of multiplication operators on twisted Bergman space, *J. Func. Anal.* **136** (1996), 171 – 213.

[6] ______, Homogeneous operators and projective representations of the Möbius group: a survey, *Proc. Ind. Acad. Sc. (Math. Sci.)* **111** (2001), 415 – 437.

[7] I. Biswas and G. Misra, $\widetilde{\mathrm{SL}}(2,\mathbb{R})$-homogeneous vector bundles, *Internat. J. Math.* **19** (2008), 1 – 19.

[8] S. Biswas, G. Ghosh, G. Misra, and S. Shyam Roy, On reducing sub-modules of hilbert modules with $\mathfrak{S}_n$-invaraint kernels, *J. Funct. Anal.* **276** (2019), 751-784.

[9] S. Biswas, D. K. Keshari, and G. Misra, Infinitely divisible metrics and curvature inequalities for operators in the Cowen-Douglas class, *J. Lond. Math. Soc. (2)* **88** (2013), no. 3, 941 – 956.

[10] S. Biswas and G. Misra, Resolution of singularities for a class of Hilbert modules, *Indiana Univ. Math. J.* **61** (2012), 1019 – 1050.

[11] S. Biswas, G. Misra, and M. Putinar, Unitary invariants for Hilbert modules of finite rank, *J. Reine Angew. Math.* **662** (2012), 165 – 204.

[12] Li Chen, A dual geometric theory for bundle shifts, *J. Func. Anal.* **263** (2012), 846 – 868.

[13] X. Chen and R. G. Douglas, Localization of Hilbert modules, *Mich. Math. J.* **39** (1992), 443 – 454.

[14] X. Chen and K. Guo, *Analytic Hilbert Modules*, Chapman and Hall/CRC, 2003.

[15] D. N. Clark and G. Misra, Curvature and similarity, *Mich. Math. J.* **30** (1983), 361 – 367.

[16] ______, On weighted shifts, curvature and similarity, *J. London Math. Soc.* **31** (1985), 357 – 368.

[17] M. J. Cowen and R. G. Douglas, Complex geometry and operator theory, *Acta Math.* **141** (1978), 187 – 261.

[18] ______, On operators possessing an open set of eigenvalues, Memorial Conf. for Féjer-Riesz, *Colloq. Math. Soc. J. Bolyai*, 1980, 323 – 341.

[19] ______, On moduli for invariant subspaces, in *Invariant Subspaces and Other Topics* (Timisoara/Herculane, 1981), pp. 65–73, *Operator Theory: Adv. Appl.* **6**, Birkhäuser Basel, Basel, 1982.

[20] ______, Equivalence of connections, *Adv. Math.* **56** (1985), 39 – 91.

[21] R. E. Curto and N. Salinas, Generalized Bergman kernels and the Cowen-Douglas theory, *Amer J. Math.* **106** (1984), 447 – 488.

[22] S. K. Donaldson, Anti self-dual Yang-Mills Connection over complex algebraic surfaces and stable vector bundles, *Proc. London Math. Soc.* **50** (1985), 1 – 26.

[23] R. G. Douglas, H.-K. Kwon, and S. Treil, Similarity of n-hypercontractions and backward Bergman shifts, *J. London Math. Soc.* **88** (2013), 237 – 648.

[24] R. G. Douglas and G. Misra, Geometric invariants for resolutions of Hilbert modules, *Operator Theory: Advances and Applications*, Birkhauser, 1993, 83 – 112.

[25] ———, Equivalence of quotient Hilbert modules, *Proc. Indian Acad. Sc. (Math. Sc.)* **113** (2003), 281 – 291.

[26] ———, Quasi-free resolutions of Hilbert modules, *Integral Equations and Operator Theory* **47** (2003), 435 – 456.

[27] ———, Equivalence of quotient modules - II, *Trans. Amer. Math. Soc.* **360** (2008), 2229 – 2264.

[28] R. G. Douglas, G. Misra, and C. Varughese, On quotient modules - the case of arbitrary multiplicity, *J. Func. Anal.* **174** (2000), 364 – 398.

[29] R. G. Douglas and V. I. Paulsen, *Hilbert Modules over Function Algebras*, Pitman Research Notes in Mathematics, no. **217**, Longman Scientific and Technical, 1989.

[30] R. G. Douglas, V. I. Paulsen, C.-H. Sah, and K. Yan, Algebraic reduction and rigidity for Hilbert modules, *Amer. J. Math.* **117** (1995), 75 – 92.

[31] J. Eschmeier and M. Putinar, *Spectral Decompositions and Analytic Sheaves*, Oxford University Press, London, 1996.

[32] J. Eschmeier and J. Schmitt, Cowen-Douglas operators and dominating sets, *J. Operator Theory* **72** (2014), 277 – 290.

[33] S. Ghara, Decomposition of the tensor product of Hilbert modules via the jet construction and weakly homogeneous operators, PhD thesis, Indian Institute of Science, 2018.

[34] H. Grauert and R. Remmert, *Coherent Analytic Sheaves*, Springer-Verlag, Berlin, 1984.

[35] F. R. Harvey and H. B. Lawson, A theory of characteristic currents associated with a singular connection, *Asterisque* **213** (1993), 1 – 268.

[36] C. Jiang, Similarity Classification of Cowen-Douglas Operators, *Canad. J. Math.* **56** (2004), 742 – 775.

[37] C. Jiang and K. Ji, Similarity classification of holomorphic curves, *Adv. Math.* **215** (2007), 446 – 468.

[38] K. Ji, C. Jiang, D. K. Keshari, and G. Misra, Rigidity of the flag structure for a class of Cowen-Douglas operators, *J. Func. Anal.* **272** (2017), 2899 – 2932.

[39] C. Jiang, K. Ji, and G. Misra, Classification of quasi-homogeneous holomorphic curves and operators in the Cowen-Douglas class, *J. Func. Anal.* **273** (2017), 2870 – 2915.

[40] C. Jiang and Z. Wang, *Strongly Irreducible Operators on Hilbert Space*, Pitman Research Notes in Mathematics Series, vol. 389, Longman, Harlow, 1998.

[41] A. Kirillov, *Elements of the Theory of Representations*, Springer-Verlag, 1976.

[42] A. Korányi and G. Misra, Homogeneous operators on Hilbert spaces of holomorphic functions, *J. Func. Anal.* **254** (2008), 2419 – 2436.

[43] ______, A classification of homogeneous operators in the Cowen - Douglas class, *Integral Equations and Operator Theory* **63** (2009), 595 – 599.

[44] ______, A classification of homogeneous operators in the Cowen-Douglas class, *Adv. Math.* **226** (2011), 5338 – 5360.

[45] ______, Homogeneous Hermitian holomorphic vector bundles and the Cowen-Douglas class over bounded symmetric domains, *C. R. Math. Acad. Sci. Paris* **354** (2016), 291 – 295.

[46] ______, Homogeneous hermitian holomorphic vector bundles and the Cowen-Douglas class over bounded symmetric domains, arXiv:1806.01955, 2018.

[47] G. Misra, Curvature and the backward shift operator, *Proc. Amer. Math. Soc.* **91** (1984), 105 – 107.

[48] ______, Curvature inequalities and extremal properties of the bundle shifts, J. Operator Theory **11** (1984), 305 – 317.

[49] G. Misra and A. Pal, Contractivity, complete contractivity and curvature inequalities, arXiv:1410.7493, to appear in *J. d'Analyse Mathematique* (2018).

[50] G. Misra and Md. Ramiz Reza, Curvature inequalities and extremal operators, 2018.

[51] G. Misra and S. Shyam Roy, The curvature invariant for a class of homogeneous operators, *Proc. London Math. Soc.* **99** (2009), 557 – 584.

[52] G. Misra and N. S. N. Sastry, Homogeneous tuples of operators and holomorphic discrete series representation of some classical groups, *J. Operator Theory* **24** (1990), 23 – 32.

[53] G. Misra and H. Upmeier, Homogeneous vector bundles and intertwining operators for symmetric domains, *Adv. Math.* **303** (2016), 1077 – 1121.

[54] L. Peng and G. Zhang, Tensor products of holomorphic representations and bilinear differential operators, *J. Func. Anal.* **210** (2004), 171 – 192.

[55] Md. Ramiz Reza, Curvature inequalities for operators in the Cowen-Douglas class of a planar domain, *Indiana Univ. Math. J.* **67** (2018), 1255 – 1279.

[56] W. Rudin, Analytic functions of class H_p, *Trans. Amer. Math. Soc.* **78** (1955), 46 – 66.

[57] D. Sarason, The H^p spaces of an annulus, *Mem. Amer. Math. Soc.* No. **56** (1965), 78 pages.

[58] N. Suita, On a metric induced by analytic capacity. II, *Kōdai Math. Sem. Rep.* **27** (1976), 159 – 162.

[59] K. Wang, and G. Zhang, Curvature inequalities for operators of the Cowen-Douglas class, *Israel J. Math.* **222** (2017), 279 – 296.

[60] D. R. Wilkins, Homogeneous vector bundles and Cowen-Douglas operators, *Intern. J. Math.* **4** (1993), 503 – 520.

[61] D. V. Yakubovich, Dual piecewise analytic bundle shift models of linear operators, *J. Func. Anal.* **136** (1996), 294 – 330.

[62] K. Zhu, Operators in Cowen-Douglas classes, *Illinois J. Math.* **44** (2000), 767 – 783.

Chapter 5

Toeplitz Operators and Toeplitz C^*-Algebras in Several Complex Variables

Harald Upmeier
Fachbereich Mathematik, Universität Marburg, Marburg, 35032, Germany
upmeier@Mathematik.Uni-Marburg.de

CONTENTS

5.1 Introduction

The theory of Toeplitz operators on the Hardy or Bergman space over the unit disk is one of the central topics in operator theory. In this survey article we describe some of the main features concerning the multi-variable case. The main topics are the analysis of Hilbert spaces of holomorphic functions in several complex variables, the C^*-algebraic structure of Toeplitz operators and aspects of geometric quantization such as the Berezin transform. These topics have been studied for three main classes of domains, namely strongly pseudoconvex domains, bounded symmetric domains and Reinhardt domains. The common intersection of these classes is the unit ball. We also discuss recent developments concerning Hilbert quotient modules, related to holomorphic functions vanishing on analytic subvarieties of the underlying domain.

5.2 Toeplitz operators on Hilbert spaces of multi-variable holomorphic functions

In the following let $Z \approx \mathbf{C}^n$ be a hermitian vector space, with inner product denoted by $(z|w) = \sum_i z_i \overline{w}_i$. Consider the Wirtinger "gradients"

$$\partial := (\frac{\partial}{\partial z_i}, \ldots, \frac{\partial}{\partial z_n}), \qquad \overline{\partial} := (\frac{\partial}{\partial \overline{z}_i}, \ldots, \frac{\partial}{\partial \overline{z}_n}).$$

Then

$$(\partial|\partial) = \sum_i \frac{\partial}{\partial z_i} \frac{\partial}{\partial \overline{z}_i} \tag{5.1}$$

is the Laplace operator on Z. Let Ω be a domain in Z. Denote by $\mathscr{O}(\Omega)$ the commutative algebra of holomorphic functions on Ω. By restriction, $\mathscr{O}(\Omega)$ contains the algebra $\mathscr{P}(Z)$ of all holomorphic polynomials on Z. Let μ be a smooth probability measure on Ω having full support. The (generalized) **Bergman space**

$$H^2_\mu(\Omega) = \mathscr{O}(\Omega) \cap L^2_\mu(\Omega)$$

is the space of all square-integrable holomorphic functions, endowed with the inner product

$$(\varphi|\psi)^\mu_\Omega := \int_\Omega d\mu(z)\ \overline{\varphi(z)}\psi(z)$$

chosen to be conjugate-linear in the first variable. As in the 1-dimensional setting, $H^2_\mu(\Omega)$ is a closed subspace of $L^2_\mu(\Omega)$ and thus a Hilbert space. For each $z \in \Omega$ the linear evaluation map

$$H^2_\mu(\Omega) \to \mathbf{C}, \quad \varphi \mapsto \varphi(z) \tag{5.2}$$

is bounded. Let $\mathscr{K}_z \in H^2_\mu(\Omega)$ be the adjoint. The sesqui-holomorphic function $\mathscr{K}^\mu_\Omega(z,w) := \mathscr{K}_w(z)$ is called the **reproducing kernel** of $H^2_\mu(\Omega)$ because of the reproducing property

$$\varphi(z) = (\mathscr{K}_z|\varphi)^\mu_\Omega = \int_\Omega d\mu(w)\ \mathscr{K}^\mu_\Omega(z,w)\ \varphi(w)$$

for all $\varphi \in H^2_\mu(\Omega)$. Denoting the evaluation map (5.2) by $\mathscr{K}_z^* : H^2_\mu(\Omega) \to \mathbf{C}$, the orthogonal projection $P^\mu_\Omega : L^2_\mu(\Omega) \to H^2_\mu(\Omega)$ is expressed as an integral

$$P^\mu_\Omega = \int_\Omega d\mu(w)\ \mathscr{K}_w \mathscr{K}_w^*$$

of rank 1 operators. In fact, we have

$$\int_\Omega d\mu(w)\ \mathcal{K}_w\mathcal{K}_w^*\varphi(z) = \int_\Omega d\mu(w)\ \mathcal{K}_w(z)(\mathcal{K}_w|\varphi)_\Omega^\mu$$

$$= \int_\Omega d\mu(w)\ \mathcal{K}_w(z)\ \varphi(w) = \int_\Omega d\mu(w)\ \mathcal{K}_\Omega^\mu(z,w)\ \varphi(w) = (P_\Omega^\mu\varphi)(z).$$

For any bounded measurable function f on Ω, the (bounded) **Toeplitz operator** $\tau_\Omega^\mu f$ acting on $H_\mu^2(\Omega)$ is defined by

$$(\tau_\Omega^\mu f)\varphi := P_\Omega^\mu(f\cdot\varphi)$$

for all $\varphi \in H_\mu^2(\Omega)$. Its integral representation is

$$\tau_\Omega^\mu f = \int_\Omega d\mu(w)\ f(w)\ \mathcal{K}_w\mathcal{K}_w^*.$$

In case Ω is a **bounded domain** we have the (normalized) Lebesgue measure $d\mu(z) = dz/Vol(\Omega)$ and obtain the **standard Bergman space** denoted by $H^2(\Omega)$. The associated Toeplitz operators $\tau_\Omega f$ are the Bergman-Toeplitz operators. On the other hand, the **flat case** $\Omega = Z$, with Gaussian probability measure

$$d\mu_Z(z) = e^{-(z|z)}\ \frac{dz}{\pi^n},$$

yields the **Fock (or Segal-Bargmann) space** denoted by $H^2(Z)$, consisting of all square-integrable entire functions $\varphi : Z \to \mathbf{C}$. The inner product is

$$(\varphi|\psi)_Z := \frac{1}{\pi^n}\int_Z dz\ e^{-(z|z)}\ \overline{\varphi(z)}\,\psi(z). \tag{5.3}$$

The normalized monomials $z^\alpha/\sqrt{\alpha!}$, for multi-indices $\alpha \in \mathbf{N}^n$, form an orthonormal basis. The polynomials $\mathcal{P}(Z)$ are a dense subspace, and the reproducing kernel is

$$\mathcal{K}_Z(z,w) = e^{(z|w)}.$$

The unitary group $U(Z) \approx U(n)$ acts unitarily on $H^2(Z)$ by substitution, and there is an orthogonal decomposition

$$\mathcal{P}(Z) = \sum_{m\in\mathbf{N}} \mathcal{P}_m(Z), \tag{5.4}$$

where $\mathcal{P}_m(Z)$ denotes the irreducible subspace consisting of all m-homogeneous polynomials. This finite-dimensional Hilbert space has the reproducing kernel

$$\mathcal{K}_Z^m(z,w) := \frac{(z|w)^m}{m!}.$$

Assume in the following that $H^2_\mu(\Omega)$ contains all polynomials. Then $\mathcal{K}^\mu_\Omega(w,w) > 0$ for all $w \in \Omega$ and we may form the **normalized kernel** $\kappa_w \in H^2_\mu(\Omega)$ defined by

$$\kappa_w(z) := \frac{\mathcal{K}^\mu_\Omega(z,w)}{\mathcal{K}^\mu_\Omega(w,w)^{1/2}}.$$

A bounded operator T on $H^2_\mu(\Omega)$ has a **Berezin symbol** $\sigma^\mu_\Omega T : \Omega \to \mathbf{C}$ defined by

$$(\sigma^\mu_\Omega T)(z) := (\kappa_z|T\kappa_z)^\mu_\Omega = \frac{1}{\mathcal{K}^\mu_\Omega(z,z)}(\mathcal{K}_z|T\mathcal{K}_z)^\mu_\Omega.$$

In terms of the normalized kernels, a Toeplitz operator can be written as

$$\tau^\mu_\Omega f = \int_\Omega d\mu_0(z)\ f(z)\ \kappa_z\kappa^*_z,$$

where $d\mu_0(z) = \mathcal{K}^\mu_\Omega(z,z)\ d\mu(z)$ is an (unbounded) 'invariant' measure. For symbol functions f on Ω and operators T on $H^2_\mu(\Omega)$ we have the formal duality

$$(f|\sigma^\mu_\Omega T)_{\mu_0} = (\tau^\mu_\Omega f|T)_{HS},$$

for the L^2-norm with respect to μ_0 and the Hilbert-Schmidt inner product. Combining the Toeplitz map and the symbol map, we arrive at the fundamental **Berezin transform**

$$f \mapsto \beta^\mu_\Omega f := \sigma^\mu_\Omega(\tau^\mu_\Omega f)$$

which is a smoothing operator acting on symbol functions. The Berezin transform arises naturally in the process of **geometric quantization**, cf. [18]. Let ρ be a plurisubharmonic function on Ω, called a **Kähler potential**, such that the $(1,1)$-form

$$\omega := \frac{i}{2}\partial\overline{\partial}\rho$$

defines a Kähler metric on Ω. Let $\omega^n/n!$ be the associated Liouville measure and put

$$d\mu^\nu_\Omega(z) := e^{-\nu\rho(z)}\ \frac{\omega^n(z)}{n!} \tag{5.5}$$

(up to normalization). The corresponding **weighted Bergman space** will be denoted by $H^2_\nu(\Omega)$. Here $\nu > 0$ is a deformation parameter interpreted as the inverse Planck constant. Letting $\nu \to \infty$ is called the 'classical limit'.

In the 'flat' situation $\Omega = Z$, with $\rho(z) = (z|z)$, the Liouville and Lebesgue measures agree (up to normalization) and the **weighted Fock space** $H^2_\nu(Z)$ is associated with the probability measure

$$d\mu^\nu_Z(z) = \frac{\nu^n}{\pi^n}e^{-\nu(z|z)}dz$$

where $\nu > 0$. The reproducing kernel is

$$\mathcal{K}^\nu_Z(z,w) = e^{\nu(z|w)}$$

and the Berezin transform is essentially the **heat semigroup**

$$\beta_Z^\nu f = e^{(\partial|\partial)/\nu} f$$

induced by the Laplace operator (5.1) [1, 29]. This simple relationship holds only in the flat setting, but the bounded symmetric domains discussed below allow for an interesting generalization.

5.3 Strongly pseudoconvex domains

In this section we consider bounded domains Ω with smooth boundary $S = \partial\Omega$. Let ρ be a smooth 'defining' function on the closure $\overline{\Omega}$ satisfying $\rho < 0$ on Ω, $\rho = 0$ on S and $d\rho \neq 0$ near S. The domain Ω is called **pseudoconvex** if at every boundary point $u \in S$ the **Levi form**

$$\sum_{i,j} \frac{\partial^2 \rho}{\partial z_i \partial \overline{z}_j}(u)\, dz_i d\overline{z}_j \tag{5.6}$$

is positive semi-definite, when restricted to the holomorphic tangent space $T_u^{1,0}(S) := (\overline{\partial}\rho)^\perp$. If the Levi form (5.6) is positive definite, then Ω is called **strongly pseudoconvex**. The basic example of a strongly pseudoconvex domain is the n**-ball**

$$\mathbf{B} := \{z \in \mathbf{C}^n : (z|z) < 1\}$$

where $\rho(z) = (z|z) - 1$. Its boundary is the odd-dimensional sphere $\mathbf{S} := \mathbf{S}^{2n-1}$ and the holomorphic tangent space at $u \in \mathbf{S}$ is the orthogonal complement $u^\perp$, since $(\overline{\partial}\rho)(u) = u$.

For a strongly pseudoconvex smooth domain, let $j : S \to \overline{\Omega}$ be the inclusion map. The 'contact form' $\alpha := \frac{1}{2i} j^*(\partial\rho - \overline{\partial}\rho)$ on S gives rise to a $(2n-1)$-form $\alpha \wedge (d\alpha)^{n-1}$ which induces a measure on S [8, 9]. Let $L^2(S)$ be the associated Lebesgue space and define the **Hardy space** $H^2(S)$ as the closure of the space of smooth functions on S which admit a holomorphic continuation onto Ω. The orthogonal projection $P_S : L^2(S) \to H^2(S)$ is the **Cauchy-Szegö projection**. Given a 'symbol' function $f \in L^\infty(S)$ define the (bounded) **Hardy-Toeplitz** operator $\tau_S f$ on $H^2(S)$ by

$$\tau_S f := P_S\, M_f\, P_S,$$

where M_f is the multiplication operator.

Recall that an associative complex Banach $*$-algebra $\mathscr{A}$ is called a C^***-algebra** if $\|a^*a\| = \|a\|^2$ for all $a \in \mathscr{A}$. The Banach $*$-algebra $\mathscr{B}(H)$ of all bounded operators on a complex Hilbert space H is a C^*-algebra, and so is every operator-norm closed $*$-subalgebra. Conversely, by the Gelfand-Naimark theorem, every C^*-algebra can be realized in this way. The **compact operators** $\mathscr{K}(H)$ form a C^*-subalgebra of $\mathscr{B}(H)$.

Theorem 3.1 *For the Hardy space $H^2(S)$ of a strongly pseudoconvex bounded domain Ω, with smooth boundary S, the associated* **Hardy-Toeplitz C^*-algebra**

$$\mathscr{T}(S) := C^*(\tau_S f : f \in \mathscr{C}(S)), \tag{5.7}$$

generated by all Hardy-Toeplitz operators with continuous symbol, acts irreducibly on $H^2(S)$ and contains $\mathscr{K}(H^2(S))$ as its commutator ideal. Moreover, there is a **symbol homomorphism**

$$\mathscr{T}(S) \xrightarrow{\sigma} \mathscr{C}(S)$$

satisfying $\sigma(\tau_S f)(u) = f(u)$ for all $u \in S$ and $f \in \mathscr{C}(S)$. We have $ker(\sigma) = \mathscr{K}(H^2(S))$, resulting in a short exact sequence

$$0 \to \mathscr{K}(H^2(S)) \xrightarrow{\iota} \mathscr{T}(S) \xrightarrow{\sigma} \mathscr{C}(S) \to 0 \tag{5.8}$$

called the **Toeplitz extension**. *Here ι is the inclusion map.*

A similar result holds for the Bergman space $H^2(\Omega)$ and the **Bergman-Toeplitz C^*-algebra**

$$\mathscr{T}(\Omega) := C^*(\tau_\Omega f : f \in \mathscr{C}(\overline{\Omega})),$$

generated by all Bergman-Toeplitz operators with continuous symbol f on the closure $\overline{\Omega}$. The symbol homomorphism is still supported on the boundary, yielding a short exact sequence

$$0 \to \mathscr{K}(H^2(\Omega)) \xrightarrow{\iota} \mathscr{T}(\Omega) \xrightarrow{\sigma} \mathscr{C}(S) \to 0,$$

where $\sigma : \mathscr{T}(\Omega) \to \mathscr{C}(S)$ maps $\tau_\Omega f$ to the restriction $f|_S$.

Theorem 3.1 is proved in [12] for the n-ball. The general case is treated in [33, 44, 59], cf. [44, Theorem 2.2] (for the Hardy space) and [44, Theorem 3.1] (for the Bergman space). In the 1-dimensional case we recover the Gohberg-Krein theorem for Toeplitz operators on the unit disk. An important ingredient in the theory is the concept of 'peaking functions' developed in [27] and the solution of the $\overline{\partial}$-Neumann problem on strongly pseudoconvex domains [36]. For a more detailed account of operator theory in this context, see [60].

Theorem 3.1 has interesting consequences (and refinements) for **index theory** and **trace formulas**. Consider a strongly pseudoconvex domain Ω with smooth boundary $S = \partial\Omega$. By (5.8) a Toeplitz operator with smooth symbol is Fredholm if and only if the symbol is invertible. This holds also for smooth matrix-valued symbols $F = (f_{ij}) : S \to \mathbf{C}^{N \times N}$ and the associated operator matrix $\tau_S F = (\tau_S f_{ij})$ acting on the vector-valued Hardy space $H^2(S) \otimes \mathbf{C}^N$. In higher dimension matrix-valued symbols are needed in order to obtain Fredholm operators with non-zero index. As a special case of [8, Theorem 2] (cf. also [15]) we have

$$Index\ \tau_S F = -\frac{(n-1)!}{(2n-1)!(2\pi i)^n} \int_S tr(F^{-1} dF)^{2n-1} \tag{5.9}$$

for $F : S \to GL_N(\mathbf{C})$, in direct generalization of the classical index formula

$$Index\ \tau_{\mathbf{T}} f = -\frac{1}{2\pi i} \int_{\mathbf{T}} f^{-1} df$$

for the unit circle $\mathbf{T}$, yielding the negative winding number of a smooth symbol $f : \mathbf{T} \to GL_1(\mathbf{C})$. For the n-ball $\mathbf{B}$ and $N \geq n$ the index formula (5.9) yields $(-1)^n$ times the topological degree of F. This result was proved earlier in [57].

In terms of K-theory the short exact sequence (5.8) yields a six-term exact sequence

$$\begin{array}{ccccc}
\mathbf{Z} & \xrightarrow{\iota_0} & K_0(\mathscr{T}(S)) & \xrightarrow{\sigma_0} & K^0(S) \\
\uparrow{\scriptstyle \delta_1} & & & & \downarrow{\scriptstyle \delta_0} \\
K^1(S) & \xleftarrow[\sigma_1]{} & K_1(\mathscr{T}(S)) & \xleftarrow[\iota_1]{} & 0
\end{array}$$

The connecting homomorphism δ_1 is the so-called index map. It is shown in [10] that

$$\sigma_0 : K_0(\mathscr{T}(S)) \approx K^0(S)$$

is an isomorphism, and $K_1(\mathscr{T}(S)) \oplus \mathbf{Z} \approx K^1(S)$. In case S is homotopic to the sphere $\mathbf{S}$, we have $K_0(\mathscr{T}(S)) \approx K^0(S) \approx \mathbf{Z}$ and δ_1 is an isomorphism $K^1(S) \approx \mathbf{Z}$, so that $K_1(\mathscr{T}(S)) = \{0\}$. For more general strongly pseudoconvex domains, the index map δ_1 need not be injective and the K-theory is more complicated. For example, the strongly pseudoconvex boundary S defined by $\sum_{i=1}^n (|z_i| - 1)^2 = r^2$, for $0 < r < \frac{1}{2}$, has $K_0(\mathscr{T}(S)) = K^0(S) = \mathbf{Z}^{2^n}$ and $K_1(\mathscr{T}(S)) \neq \{0\}$ [10, Proposition 6]. A similar example is mentioned in [8, p. 269].

Theorem 3.1 shows in particular that for a strongly pseudoconvex domain Ω commutators of Toeplitz operators with smooth symbol are compact. A natural question arises whether this property can be strengthened to membership in a norm ideal such as a Schatten p-class. Here the multi-variable case $n > 1$ exhibits new features conpared to the 1-dimensional case. Given smooth functions $f_1, \ldots, f_{2n}$ on the closed n-ball $\overline{\mathbf{B}}$, a classical result of Helton-Howe [32, Theorem 7.2] asserts that the **complete anti-symmetrization** $[\tau_{\mathbf{B}} f_1, \ldots, \tau_{\mathbf{B}} f_{2n}]$ of Bergman-Toeplitz operators is of trace-class and its trace is given by

$$tr[\tau_{\mathbf{B}} f_1, \ldots, \tau_{\mathbf{B}} f_{2n}] = \frac{1}{Vol(\mathbf{B})} \int_{\mathbf{B}} df_1 \wedge \ldots \wedge df_{2n}.$$

On the other hand, a non-trivial n-fold product of Toeplitz commutators is not of trace-class if $n > 1$, as pointed out in [24]. In order to obtain a trace formula, one has to pass to the so-called **Dixmier trace**. Let $\mathscr{K}$ denote the compact operators and let $\mu_1(T) \geq \mu_2(T) \geq \cdots$ denote the singular values of $T \in \mathscr{K}$. For $p > 1$, we say that $T \in \mathscr{K}^{p,\infty}$ if the sequence $\left(j^{1/p} \mu_j(T)\right)$ is bounded. The **Dixmier class** $\mathscr{K}^{1,\infty}$ consists of all $T \in \mathscr{K}$ such that the sequence

$$\left(\frac{1}{\log(j)} \sum_{i=1}^{j} \mu_i(T) \right)$$

is bounded. For such operators the Dixmier trace $tr_\omega(T)$ is defined (initially for $T \geq 0$) by

$$tr_\omega(T) = \omega - \lim_{j\to\infty} \frac{1}{\log(j)} \sum_{i=1}^{j} \mu_i(T)$$

whenever this limit exists. Here ω denotes a positive linear functional on $\ell^\infty(\mathbf{N})/c_0(\mathbf{N})$. For the so-called 'measurable' operators, $tr_\omega(T)$ is independent of ω. For more details, see [14]. The following result has been obtained in [21] for the n-ball and generalized in [24].

Theorem 3.2 *Let Ω be a strongly pseudoconvex domain with smooth boundary S. Then, for smooth symbols f_i, g_i, the commutator product $[\tau_S f_1, \tau_S g_1]\ldots[\tau_S f_n, \tau_S g_n]$ on the Hardy space belongs to the Dixmier class $\mathcal{K}(H^2(S))^{1,\infty}$ and has Dixmier trace*

$$tr_\omega[\tau_S f_1, \tau_S g_1]\ldots[\tau_S f_n, \tau_S g_n] = \frac{1}{n!}\int_S d\sigma \prod_{i=1}^{n} \{f_i, g_i\}_\flat.$$

Here $d\sigma$ is the normalized surface measure on S and $\{f,g\}_\flat$ denotes the so-called **boundary Poisson bracket** *of smooth functions f, g.*

For the n-ball the boundary Poisson bracket has the explicit form

$$\{f,g\}_\flat = (\partial g|\partial \overline{f}) - (\partial f|\partial \overline{g}) + (\partial f|\overline{z})\,(\overline{z}|\partial \overline{g}) - (\partial g|\overline{z})\,(\overline{z}|\partial \overline{f}).$$

A crucial ingredient in the proof of the above theorem is the general theory of Toeplitz operators with pseudodifferential symbol developed in [9] and a classical formula for the Dixmier trace of pseudodifferential operators of order $-n$ on an n-dimensional compact manifold [14].

5.4 Symmetric domains and Jordan triples

Many interesting domains do not have a smooth boundary. An important class are the so-called **bounded symmetric domains**. By a deep theorem of H. Cartan [43, Chapter 9] the group

$$G = Aut(\Omega) = \{g : \Omega \to \Omega \text{ biholomorphic}\}$$

of a bounded domain Ω is a real Lie group. Ω is called **symmetric** if around each point $o \in \Omega$ there exists an involutive **symmetry** $s_o \in G$ fixing o. One can show that in this case G acts **transitively** on Ω. Thus $\Omega = G/K$, where the isotropy subgroup

$$K = \{g \in G :\ g(o) = o\}$$

at a given base point $o \in \Omega$ is compact. Harish-Chandra has shown that a symmetric domain Ω can be realized as a convex circular domain, i.e. as the unit ball

$$\Omega = \{z \in Z : \ \|z\| < 1\}$$

of a complex vector space $Z \approx \mathbf{C}^n$ for the so-called **spectral norm**. Choosing the base point $0 \in \Omega$, it follows that K consists of linear transformations. For symmetric domains of rank $r > 1$ the spectral norm is not a Hilbert norm.

Example 4.1 *For $1 \leq r \leq s$ the* **matrix unit ball**

$$\Omega = \{z \in \mathbf{C}^{r\times s} : \ 1_r - zz^* > 0\} = \{z \in \mathbf{C}^{r\times s} : \ 1_s - z^*z > 0\}$$

is a symmetric domain, with respect to the **pseudounitary group**

$$\begin{aligned} G &= U(r,s) \\ &= \left\{ \begin{pmatrix} a & b \\ c & d \end{pmatrix} \in GL_{r+s}(\mathbf{C}) : \ \begin{pmatrix} a & b \\ c & d \end{pmatrix} \begin{pmatrix} 1 & 0 \\ 0 & -1 \end{pmatrix} \begin{pmatrix} a^* & c^* \\ b^* & d^* \end{pmatrix} = \begin{pmatrix} 1 & 0 \\ 0 & -1 \end{pmatrix} \right\}, \end{aligned}$$

acting on Ω via **Moebius transformations**

$$\begin{pmatrix} a & b \\ c & d \end{pmatrix}(z) = (az+b)(cz+d)^{-1}.$$

In this case $\|z\| = \sup \sigma(zz^*)^{1/2}$ coincides with the operator norm. The maximal compact subgroup is

$$K = \left\{ \begin{pmatrix} a & 0 \\ 0 & d \end{pmatrix} : \ a \in U(r), \ d \in U(s) \right\}$$

*with the linear action $z \mapsto azd^*$. For $r = 1, s = n$ we obtain the n-***ball**

$$\mathbf{B} = SU(1,n)/U(n).$$

Example 4.2 *For $n \geq 3$ the* **Lie ball**

$$\Omega_n := \{z \in \mathbf{C}^n : \ z\cdot\bar{z} < 1, \ 1 - 2z\cdot\bar{z} + |z\cdot z|^2 > 0\} \tag{5.10}$$

is an irreducible symmetric domain of rank $r = 2$. It is also defined by

$$z\cdot\bar{z} + \sqrt{(z\cdot\bar{z})^2 - |z\cdot z|^2} < 1,$$

but the description (5.10) via two polynomials (since $r = 2$) is more systematic [40, p. 4.16]. For $n = 2$ the Lie ball Ω_2 is reducible and isomorphic to the **bidisk**. *This case has been studied intensively by Rongwei Yang and others.*

A vector space Z endowed with a ternary composition

$$Z \times Z \times Z \to Z, \quad (x,y,z) \mapsto \{x;y;z\} = \{z;y;x\} \tag{5.11}$$

is called a **Jordan triple** if the 'multiplication' endomorphisms $L(x,y)z := \{x;y;z\}$ of Z satisfy the commutator identity (a kind of Jacobi identity)

$$[L(x,y),L(u,v)] = L(\{x;y;u\},v) - L(u,\{v;x;y\}).$$

A complex Jordan triple Z is called **hermitian** if (5.11) is anti-linear in the inner variable y and the hermitian form

$$(x,y) \mapsto tr\, L(x,y) \tag{5.12}$$

is positive definite. A basic theorem of M. Koecher (cf. [35, 40]) states that every symmetric domain can be realized as the unit ball of a unique hermitian Jordan triple (and conversely). Moreover, K consists of all linear transformations preserving the Jordan triple product. For the structure and analysis on Jordan algebras and Jordan triples, see [25, 40].

Example 4.3 *The matrix space $Z = \mathbf{C}^{r\times s}$ carries the hermitian Jordan triple product*

$$\{x;y;z\} = xy^*z + zy^*x \tag{5.13}$$

for all $x,y,z \in Z$. Note that (5.13) *makes sense for rectangular matrices. In the rank 1 case,* (5.13) *reduces to*

$$\{x;y;z\} = (x|y)z + (z|y)x$$

for vectors $x,y,z \in \mathbf{C}^n$.

Example 4.4 *The Lie ball* (5.10) *belongs to the so-called* **spin factor** *with hermitian Jordan triple product*

$$\{xy^*z\} = (x\cdot\overline{y})\, z + (z\cdot\overline{y})\, x - (x\cdot z)\overline{y} \tag{5.14}$$

for vectors $x,y,z \in \mathbf{C}^n$. This triple product can also be realized via matrices, but of higher order 2^n related to the spin representation of $SO(n)$.

Closely related to Jordan triples are **Jordan algebras**. A vector space X endowed with a non-associative product $(x,y) \mapsto x\circ y = y\circ x$ is called a Jordan algebra if the multiplication endomorphisms $L(x)y := x\circ y$ on X satisfy the commutator identity

$$[L(x),L(x^2)] = 0.$$

A real Jordan algebra X is called **euclidean** if the symmetric bilinear form

$$(x|y) := tr\, L(x\circ y)$$

is positive definite. A euclidean Jordan algebra X with unit element e has a canonical **determinant polynomial** $N : X \to \mathbf{R}$, normalized by $N(e) = 1$, such that the inverse in X is given by Cramer's rule

$$x^{-1} = \frac{grad_x N}{N(x)}.$$

Moreover, the set of squares $\{x^2 : x \in X\}$ is a closed convex cone and its interior

$$\Lambda := \{x^2 : x \in X,\ N(x) > 0\}$$

becomes a **symmetric (self-dual homogeneous) cone** [26]. As a basic example, the real vector space $\mathscr{H}_r(\mathbf{K})$ of self-adjoint $r \times r$-matrices over $\mathbf{K} = \mathbf{R}, \mathbf{C}$ or $\mathbf{H}$ (quaternions) becomes a euclidean Jordan algebra under the anti-commutator product

$$x \circ y = \frac{1}{2}(xy + yx).$$

The associated cone $\Lambda = \mathscr{H}_r^+(\mathbf{K})$ consists of all positive definite matrices. For $\mathbf{K} = \mathbf{R}, \mathbf{C}$ we have $N(x) = \det x$, whereas for quaternions $N(x)$ is related to the Pfaffian determinant. For $r = 3$ one can also take the (non-associative) division algebra $\mathbf{O}$ of octonions (Cayley numbers), yielding the **exceptional Jordan algebra** $\mathscr{H}_3(\mathbf{O})$ of dimension 27.

The **complexification** $Z = X \otimes \mathbf{C}$ of a euclidean Jordan algebra, endowed with the extended Jordan product $z \circ w$ and the involution $z \mapsto z^*$ with self-adjoint part X, becomes a hermitian Jordan triple under the triple product

$$\{u; v; w\} := (u \circ v^*) \circ w + (w \circ v^*) \circ u - (u \circ w) \circ v^*$$

which formally resembles (5.14). These Jordan triples, and the associated bounded symmetric domains Ω, are said to be of **tube type**, since the Jordan theoretic **Cayley transform** maps Ω onto the symmetric tube domain $\Lambda \times iX \subset Z$ over the cone Λ. Via this transformation Toeplitz operators on Ω are related to the so-called Wiener-Hopf operators on the symmetric cone Λ [42, 52].

The **classification** of irreducible hermitian Jordan triples [40] comprises the following types: The first type are the 'matrix triples' $Z = \mathbf{C}^{r \times s}$ described in Example 4.3 and the Jordan subtriples

$$Z_+ := \{z \in \mathbf{C}^{r \times r} : z^t = z\}$$

and

$$Z_- := \{z \in \mathbf{C}^{s \times s} : z^t = -z\}$$

consisting of symmetric matrices and anti-symmetric matrices, respectively. The second type are the spin factors described in Example 4.4. Finally, there are two **exceptional Jordan triples** of dimension 16 and 27, respectively, which are related to the (non-associative) Cayley numbers and cannot be realized via matrix spaces.

In view of the 1-1 correspondence between hermitian Jordan triples and bounded symmetric domains, the Jordan theoretic classification yields a classification of (irreducible) bounded symmetric domains into matrix balls (first type), Lie balls (second type) and two exceptional domains (third type). The Jordan theoretic approach to symmetric domains allows one to treat the classical domains and the exceptional domains in a uniform manner.

As a consequence of the classification, an irreducible hermitian Jordan triple Z is uniquely determined by its rank r and two numerical invariants a, b (**characteristic multiplicities**) such that

$$n = \dim Z = r(1 + \frac{a}{2}(r-1) + b).$$

The matrix triple (Example 4.3) has $a = 2, b = s - r$. In particular, the n-ball $\mathbf{B}$ has $r = 1, a = 2, b = n - 1$. The spin factor (Example 4.4) has $r = 2, a = n - 2, b = 0$. The hermitian Jordan triples of tube type correspond to $b = 0$. Another important numerical invariant is the **genus**

$$p := 2 + a(r-1) + b. \tag{5.15}$$

Thus $p = r + s$ for the matrix ball, $p = n + 1$ for the n-ball and $p = n$ for the spin factor.

For a hermitian Jordan triple Z the **Bergman endomorphism** $B(z, w) \in End(Z)$ is defined by

$$B(z,w)v = v - \{z;w;v\} + \frac{1}{4}\{z;\{w;v;w\};z\}$$

for $z, w, v \in Z$. Let

$$(u|v) = \frac{1}{p}\, tr\, L(u,v) \tag{5.16}$$

be the normalized K-invariant inner product (cf. (5.12)), where p is the genus (5.15). Then the G-invariant **Bergman metric** is given by

$$(u|v)_z = (B(z,z)^{-1}u|v)$$

for all $u, v \in Z$, identified with the holomorphic tangent space at $z \in \Omega$. For the matrix case $Z = \mathbf{C}^{r\times s}$, we have

$$(u|v) = \frac{1}{r+s}\, tr\, L(u,v) = tr\, uv^*$$

and

$$B(z,w)v = v - zw^*v - vw^*z + z(wv^*w)^*z = (1 - zw^*)v(1 - w^*z).$$

Therefore the Bergman metric at $z \in \Omega$ is

$$(u,v)_z = tr\,(1 - zz^*)^{-1}\, u(1 - z^*z)^{-1}\, v^*.$$

As a fundamental fact there exists a sesqui-polynomial called the **Jordan triple determinant** $D : Z \times Z \to \mathbf{C}$ (holomorphic in z and anti-holomorphic in w), such that

$$\det B(z,w) = D(z,w)^p.$$

In the matrix case $Z = \mathbf{C}^{r\times s}$ we have

$$\det B(z,w) = \det(1_r - zw^*)^{r+s}.$$

It follows that

$$D(z,w) = \det(1_r - zw^*) = \det(1_s - w^*z).$$

For the n-ball ($r = 1, s = n$) we have $D(z,w) = 1 - (z|w)$ for all $z,w \in Z = \mathbf{C}^n$. For Jordan triples of tube type, the Jordan triple determinant $D(z,w)$ is related to the Jordan algebra determinant (which depends only on one vector variable).

As a convex domain, a symmetric domain Ω is pseudoconvex (in the sense of the Cartan-Thullen theorem [43, Chapter 7]). If $r > 1$, then Ω is not strongly pseudoconvex and the boundary $\partial\Omega$ is not smooth. The fine structure of the boundary is best described using the Jordan theoretic approach. An element u in a hermitian Jordan triple Z is called a **tripotent** if $\{u;u;u\} = 2u$. For matrices $Z = \mathbf{C}^{r\times s}$ the tripotents are the partial isometries characterized by $u = uu^*u$. The set S_ℓ of all tripotents of fixed rank $\ell \le r$ is a compact K-homogeneous manifold. S_1 consists of all minimal tripotents and $S_r = S$ is the so-called **Shilov boundary** consisting of all extreme boundary points. In holomorphic terms, the Shilov boundary S is the minimal closed set of uniqueness for the function algebra $\mathscr{A}(\Omega)$ of all holomorphic functions on Ω which are continuous on the closure $\overline{\Omega}$. For the n-ball we have $S = S_1 = \mathbf{S}$.

For hermitian Jordan triples of tube type, the Jordan algebra determinant N is an 'inner' function, satisfying $|N(u)| = 1$ for all $u \in S$, and there is a decomposition

$$S = \mathbf{T}\cdot S', \tag{5.17}$$

where $\mathbf{T}$ is the 1-torus and the 'reduced Shilov boundary' $S' := \{u \in S:\ N(u) = 1\}$ is simply-connected. For the (square) matrix space $Z = \mathbf{C}^{r\times r}$ we have $S = U(r),\ S' = SU(r)$. For the spin factor, we obtain the **cosphere bundle**

$$S_1 \approx S^*(\mathbf{S}^{n-1}) = \{(x,\xi) : \|x\| = 1,\ \|\xi\| = 1,\ (x|\xi) = 0\},\ z = \frac{x+i\xi}{2} \tag{5.18}$$

and the **Lie sphere**

$$S = S_2 = \mathbf{T}\cdot\mathbf{S}^{n-1}. \tag{5.19}$$

Here $S' = \mathbf{S}^{n-1}$ is simply-connected since $n \ge 3$.

Every tripotent u induces a **Peirce decomposition**

$$Z = Z_u^2 \oplus Z_u^1 \oplus Z_u^0,$$

where

$$Z_u^\alpha = \{z \in Z:\ \{u;u;z\} = 2\alpha z\}$$

is an eigenspace of $L(u,u)$. For $u \in S_\ell$ the Peirce 0-space $Z_u := Z_u^0$ is an irreducible hermitian Jordan subtriple of rank $r - \ell$. Hence its unit ball

$$\Omega_u := \Omega \cap Z_u \tag{5.20}$$

is a bounded symmetric domain of rank $r-\ell$. For $u \neq 0$ the translated set $u+\Omega_u \subset \partial\Omega$ is called a **boundary component** of Ω. One can show that these boundary components are pairwise disjoint. Taking the disjoint union

$$\partial_\ell\Omega = \bigcup_{u\in S_\ell} u+\Omega_u \tag{5.21}$$

of all boundary components of fixed rank $r-\ell$, one obtains a **stratification**

$$\partial\Omega = \bigcup_{\ell=1}^{r} \partial_\ell\Omega. \tag{5.22}$$

The Shilov boundary $S = S_r = \partial_r\Omega$ agrees with the top stratum, since $u \in S$ has a trivial Peirce 0-space $Z_u^0 = \{0\}$ and the boundary component $u+\Omega_u = u+\{0\} = \{u\}$ reduces to an extreme point.

5.5 Holomorphic function spaces on symmetric domains

An irreducible hermitian Jordan triple Z of rank r and dimension n carries the normalized K-invariant inner product (5.16). The associated Fock space $H^2(Z)$ is the closure of the polynomial algebra $\mathscr{P}(Z)$. Since K is a subgroup of $U(Z)$ (which is proper if $r > 1$) the elementary decomposition (5.4) is refined by the orthogonal **Hua-Schmid-Kostant decomposition** [46]

$$\mathscr{P}(Z) = \sum_{m\in\mathbf{N}_+^r} \mathscr{P}_m(Z). \tag{5.23}$$

Here $\mathbf{N}_+^r$ denotes the set of all **partitions** $m = (m_1, m_2, \ldots, m_r)$ of integers

$$m_1 \geq m_2 \geq \ldots \geq m_r \geq 0.$$

Partitions correspond to the well-known **Young diagrams**. In [49] the highest weight vector N_m of the finite-dimensional irreducible K-module $\mathscr{P}_m(Z)$ is described in terms of Jordan theoretic determinants. Denote by $\mathscr{K}_Z^m(z,w)$ the reproducing kernel of the finite-dimensional Hilbert space $\mathscr{P}_m(Z) \subset H^2(Z)$. Then

$$\frac{(z|w)^m}{m!} = \sum_{|m|=m} \mathscr{K}_Z^m(z,w),$$

for all $m \in \mathbf{N}$, since $\mathscr{P}_m(Z)$ consists of polynomials with total homogeneity $|m| := m_1+\ldots+m_r$. In terms of harmonic analyis, the functions $\mathscr{K}_Z^m$ are closely related to the **Jack polynomials**, which are of major importance in combinatorics [47]. Using the multi-variable **Pochhammer symbol**

$$(\nu)_m = \prod_{j=1}^{r} \frac{\Gamma(\nu + m_j - \frac{a}{2}(j-1))}{\Gamma(\nu - \frac{a}{2}(j-1))}, \tag{5.24}$$

for a vector parameter $m = (m_1, \ldots, m_r)$ and $\nu \in \mathbf{C}$, we define the multi-variable **hypergeometric series**

$$\mathscr{F}_Z^{p,q}\begin{pmatrix} \alpha_1, \ldots, \alpha_p \\ \beta_1, \ldots, \beta_q \end{pmatrix}(z,w) := \sum_m \frac{(\alpha_1)_m \cdots (\alpha_p)_m}{(\beta_1)_m \cdots (\beta_q)_m} \mathscr{K}_Z^m(z,w)$$

summing over all partitions m. Here $\alpha_1, \ldots, \alpha_p, \beta_1, \ldots, \beta_q$ are complex parameters. Then (5.23) shows that the Segal-Bargmann kernel is

$$\mathscr{K}_Z(z,w) = e^{(z|w)} = \sum_m \mathscr{K}_Z^m(z,w) = \mathscr{F}_Z^{0,0}(z,w). \tag{5.25}$$

The irreducible symmetric domain Ω associated with Z satisfies $\Omega = G/K$, where K is compact. Hence there exists a G-invariant measure μ_0 on Ω, which, up to a constant multiple, is given by

$$d\mu_0(z) = D(z,z)^{-p}\,\frac{dz}{\pi^n} = \det B(z,z)^{-1}\,\frac{dz}{\pi^n},$$

where dz denotes Lebesgue measure. For any parameter $\nu > p-1$ one shows that

$$d\mu_\Omega^\nu(z) := \left(\nu - \frac{n}{r}\right)_{n/r} D(z,z)^{\nu - p}\,\frac{dz}{\pi^n} \tag{5.26}$$

is a probability measure. Here $n/r := (n/r, \ldots, n/r)$. The Hilbert space $H_\nu^2(\Omega)$ of all holomorphic functions $\varphi : \Omega \to \mathbf{C}$ which are square-integrable with respect to (5.26) is called the **weighted Bergman space**. The inner product is

$$(\varphi|\psi)_\Omega^\nu := \left(\nu - \frac{n}{r}\right)_{n/r} \int_\Omega \frac{dz}{\pi^n}\, D(z,z)^{\nu-p}\,\overline{\varphi(z)}\,\psi(z). \tag{5.27}$$

Since the domain Ω is bounded and circular, $H_\nu^2(\Omega)$ contains the polynomials $\mathscr{P}(Z)$ as a dense subspace. The reproducing kernel of $H_\nu^2(\Omega)$ is given by

$$\mathscr{K}_\Omega^\nu(z,w) = D(z,w)^{-\nu} = \det B(z,w)^{-\nu/p}.$$

For $\nu = p$ one obtains the standard Bergman space relative to the (normalized) Lebesgue measure, with (standard) Bergman kernel

$$\mathscr{K}_\Omega(z,w) = D(z,w)^{-p} = \det\, B(z,w)^{-1}.$$

For the n-ball $\mathbf{B}$ of rank $r = 1$ we have $p = n+1$ and

$$d\mu_{\mathbf{B}}^\nu(z) = \frac{\Gamma(\nu)}{\Gamma(\nu - n)}(1 - (z|z))^{\nu - n - 1}\,\frac{dz}{\pi^n}$$

for $\nu > n$. The reproducing kernel becomes

$$\mathscr{K}_{\mathbf{B}}^{\nu}(z,w) = (1-(z|w))^{-\nu}.$$

The missing numerical factors for the kernel functions are due to the fact that we use probability measures.

The weighted Bergman spaces are of fundamental importance in harmonic analysis: They constitute the so-called **scalar holomorphic discrete series** of the semi-simple Lie group G under the irreducible unitary (projective) representation

$$(U_{\Omega}^{\nu}(g^{-1})\varphi)(z) := \det g'(z)^{\nu/p}\ \varphi(g(z))$$

for $g \in G$ and $\varphi \in H_{\nu}^{2}(\Omega)$. The 'analytic continuation' of this series of representations gives rise to the **Wallach set** W of Ω, consisting of all parameters ν such that the $N \times N$-matrix $(\mathscr{K}_{\Omega}^{\nu}(z_i,z_j))_{i,j}$ is positive for any choice of points $z_1,\ldots,z_N$ in Ω, and thus defines a reproducing kernel Hilbert space $\mathscr{H}_{\nu}$ with an irreducible unitary (projective) G-representation.

The well-known **Faraut-Korányi binomial formula** [25]

$$D(z,w)^{-\nu} = \sum_{m} (\nu)_m\ \mathscr{K}_{Z}^{m}(z,w) = \mathscr{F}_{Z}^{1,0}\begin{pmatrix}\nu\\ -\end{pmatrix}(z,w)$$

is valid for all $\nu \in \mathbf{C}$, with compact convergence on $\Omega \times \Omega$. As a consequence, the inner product (5.27) and the Fock space inner product (5.3) are related by

$$(\varphi|\psi)_Z = (\nu)_m\ (\varphi|\psi)_{\Omega}^{\nu} \tag{5.28}$$

whenever $\varphi, \psi \in \mathscr{P}_m(Z)$. A parameter ν belongs to the Wallach set if and only if the Pochhammer symbols $(\nu)_m$ are non-negative. The **continuous part**, where we have strict positivity, is given by the half-line

$$W_c = \{\nu \in \mathbf{R}:\ \nu > \frac{a}{2}(r-1)\}.$$

The resulting Hilbert space $\mathscr{H}_{\nu}$ is the completion of $\mathscr{P}(Z)$ under the inner product determined by (5.28) on each homogeneous component. The remaining **discrete part** consists of r equidistant points

$$W_d = \{\frac{a}{2}\ell:\ \ell = 0,1,\ldots,r-1\}. \tag{5.29}$$

Here $(\nu)_m = 0$ for all partitions satisfying $m_{\ell+1} > 0$. Therefore the resulting Hilbert space $\mathscr{H}_{\ell a/2}$ is the closure of

$$\mathscr{P}_{\ell}(Z) := \sum_{m \in \mathbf{N}_{+}^{\ell}} \mathscr{P}_{m,0\ldots,0}(Z) \tag{5.30}$$

corresponding to partitions of length ℓ. The parameter $\nu = 0$, where $\mathscr{P}_0(Z) = \mathbf{C}$ are the constant functions, yields the trivial representation of G. For deeper results on the discrete Wallach points, cf. [2, 60].

The continuous part W_c contains a special set of r equidistant points

$$\nu_\ell := \frac{n}{r} + \frac{a}{2}\ell \tag{5.31}$$

for integers $0 \leq \ell < r$. In particular,

$$\nu_0 = \frac{n}{r} = 1 + \frac{a}{2}(r-1) + b \tag{5.32}$$

gives rise to the **Hardy space** $H^2(S)$ over the Shilov boundary S of Ω, defined as the closed subspace of $L^2(S)$ (for the normalized K-invariant measure du on S) spanned by boundary values of holomorphic functions on Ω. The Hardy space does not have a reproducing kernel in the strict sense, but the orthogonal **Cauchy-Szegö projection** $P_S : L^2(S) \to H^2(S)$ can be realized as an integral operator

$$(P_S\varphi)(z) = \int_S dv\, D(z,v)^{-n/r}\, \varphi(v)$$

for all $\varphi \in L^2(S)$ and $z \in \Omega$. In a distribution sense, this formula holds also at points $z \in S$. By [2] the parameters (5.31) can also be realized by measures supported on the boundary strata $\partial_{r-\ell}\Omega$ (cf. (5.21)). The other continuous Wallach parameters, in particular the 'desert' points

$$\frac{a}{r}(r-1) < \nu < \frac{n}{r}$$

are quite difficult to realize in a geometric way.

For the n-ball $\Omega = \mathbf{B}$, the only discrete Wallach point is $\nu = 0$, and for the continuous Wallach set $\nu > 0$ one obtains the Hilbert spaces

$$\mathscr{H}_\nu = \{\varphi \in \mathscr{O}(\mathbf{B}) : \sum_{\alpha \in \mathbf{N}^n} \frac{1}{\alpha!(\nu)_{|\alpha|}} \left| \frac{\partial^\alpha \varphi}{\partial z^\alpha}(0) \right|^2 < \infty\}$$

with reproducing kernel $(1-(z|w))^{-\nu}$. The parameter $\nu = n+1$ yields the standard Bergman space, for $\nu = n$ we obtain the Hardy space $H^2(S)$ on the boundary $\mathbf{S} = \mathbf{S}^{2n-1}$ and $\nu = 1$, with $(1)_{|\alpha|} = |\alpha|!$, yields the **Drury-Arveson space** with reproducing kernel $(1-(z|w))^{-1}$. For symmetric domains of higher rank, the Drury-Arveson parameter is $\nu = 1 + \frac{a}{2}(r-1)$, which is smaller than the Hardy space parameter (5.32) (and hence belongs to the 'desert') if the domain is not of tube type.

For hermitian Jordan triples of tube type, the decomposition (5.17) induces a Hilbert space tensor product decomposition

$$L^2(S) = L^2(\mathbf{T}) \otimes L^2(S')$$

and the Cauchy-Szegö projection has the form $P_S = P_{\mathbf{T}} \otimes 1$, where $P_{\mathbf{T}}$ is the Cauchy projection for the circle $\mathbf{T}$. It follows that the Hardy space over S has a tensor product decomposition

$$H^2(S) = H^2(\mathbf{T}) \otimes L^2(S') \tag{5.33}$$

involving the classical Hardy space $H^2(\mathbf{T})$. For the spin factor, (5.33) specializes to

$$H^2(\mathbf{T} \cdot \mathbf{S}^{n-1}) = H^2(\mathbf{T}) \otimes L^2(\mathbf{S}^{n-1}). \tag{5.34}$$

The Jordan algebra determinant N is invariant under the commutator subgroup K'. For any $m \in \mathbf{N}$, the space $\mathcal{P}_{m,\ldots,m}(Z)$ is 1-dimensional and consists of all multiples of N^m. Using the notation (5.30), one shows that

$$\mathcal{P}_{r-1}(Z) = \{\psi \in \mathcal{P}(Z) : \partial_N \psi = 0\}$$

consists of all **harmonic polynomials**. Since

$$\mathcal{P}_m(Z) = \{N(z)^{m_r}\ \psi(z) : \ \psi \in \mathcal{P}_{m-m_r}(Z)\},$$

one obtains a tensor product decomposition [49]

$$\mathcal{P}(Z) = \mathbf{C}[N] \otimes \mathcal{P}_{r-1}(Z), \ \varphi(z) = \sum_{\ell \geq 0} N(z)^\ell\ \psi_\ell(z)$$

into 'invariant' and 'harmonic' polynomials with respect to K'. Note that these properties hold only in the tube type case. For the spin factor, we have $N(z) = \sum_i z_i^2$ and $\mathcal{P}_1(Z)$ consists of holomorphic polynomials which are harmonic in the classical sense.

5.6 Toeplitz C^*-algebras on symmetric domains

By Gelfand's theorem, every commutative unital C^*-algebra $\mathcal{A}$ is isometrically isomorphic to the continuous functions $\mathcal{C}(\mathcal{A}^\sharp)$, where the compact Hausdorff space $\mathcal{A}^\sharp$ is the so-called **spectrum** of $\mathcal{A}$, consisting of all characters $\chi : \mathcal{A} \to \mathbf{C}$ or, equivalently, of all maximal ideals of $\mathcal{A}$. For a non-commutative C^*-algebra $\mathcal{A}$, the spectrum is defined as the space of all irreducible $*$-representations $\pi : \mathcal{A} \to \mathcal{B}(H_\pi)$ on a complex Hilbert space H_π. For the Toeplitz C^*-algebra over symmetric domains, it is therefore an important problem to classify all irreducible representations, and obtain a non-commutative version of the Gelfand isomorphism. Note that the well-known Gelfand-Naimark theorem achieves a similar goal in a weaker form, by considering all, not necessarily irreducible, $*$-representations induced by (positive unital) states φ of $\mathcal{A}$, and proving that the canonical evaluation map from $\mathcal{A}$ into the direct product of the associated field $\mathcal{B}(H_\varphi)$ is a C^*-embedding (in general not surjective). The irreducible representations correspond to the pure states (extreme points) in the sense of convex analysis.

The main results of [50, 51] show that, for a symmetric domain Ω of arbitrary rank r, the Toeplitz C^*-algebra has all its irreducible representations realized via 'little' Toeplitz C^*-algebras on the boundary components of the underlying domain, which are themselves symmetric domains of lower rank. Moreover, the non-Hausdorff topology on the spectrum has an explicit description in terms of the stratification of the full boundary $\partial\Omega$, which, as mentioned above, is not smooth if $r > 1$. On the algebraic level, the stratification corresponds to a composition series of ideals which has length r and generalizes the short exact sequence (5.8) for $r = 1$. In this sense the symmetric domains serve as a model class for more general multi-variable domains with a non-smooth stratified boundary.

Consider first the Hardy-Toeplitz C^*-algebra $\mathscr{T}(S)$ over the Shilov boundary S of an irreducible symmetric domain Ω of rank r, defined as in (5.7). For each tripotent $u \in S_\ell$ let

$$S_u = S_{r-\ell} \cap Z_u$$

be the Shilov boundary of the symmetric domain Ω_u defined in (5.20). Then $u + S_u \subset S$. Let $H^2(S_u)$ denote the Hardy space and consider the Hardy-Toeplitz C^*-algebra $\mathscr{T}(S_u)$. The main result of [50, 51] can now be formulated as follows.

Theorem 6.1 *For each tripotent $u \in Z$ there exists a C^*-homomorphism σ_u : $\mathscr{T}(S) \to \mathscr{T}(S_u)$ such that*

$$\sigma_u(\tau_S f) := \tau_{S_u} f_u$$

for all $f \in \mathscr{C}(S)$, where $f_u \in \mathscr{C}(S_u)$ is defined by $f_u(w) := f(u+w)$. The corresponding irreducible representations are pairwise inequivalent, and constitute the full spectrum of $\mathscr{T}(S)$.

It is possible to express σ_u more directly via the following formula. Let $h_u(z) = \exp(z|u)$ and consider the sequence of 'peaking functions'

$$h_u^n(z) := \frac{1}{c_n}(h_u(z))^n$$

normalized by $\|h_u^n\|_S = 1$. Then for $T \in \mathscr{T}(S)$ the symbol $\sigma_u T$ acting on $H^2(S_u)$ is uniquely determined by the formula

$$\|h_u^n \cdot ((\sigma_u T)q) - T(q \cdot h_u^n)\|_S \to 0$$

as $n \to \infty$, where $q \in \mathscr{P}(Z_u)$ is any polynomial in the Peirce 0-space Z_u.

Corollary 6.2 *The spectrum of the Toeplitz C^*-algebra $\mathscr{T}(S)$ is the set*

$$\Sigma = \bigcup_{0 \le \ell \le r} S_\ell \tag{5.35}$$

of all tripotents of Z (including 0), viewed as a 'non-commutative space' endowed with a non-Hausdorff topology related to the stratification (5.22). *More precisely, the closure of S_ℓ is given by $\bigcup_{k \ge \ell} S_k$.*

Thus the Shilov boundary $S = S_r$ is a closed subset, corresponding to the characters of $\mathscr{T}(S)$, and $S_0 = \{0\}$ is a dense point in the spectrum, corresponding to the faithful representation on $H^2(S)$. The Toeplitz C^*-algebra is a basic example of 'non-commutative geometry' in the sense that its spectrum (5.35) is a stratified, non-Hausdorff space and, formally,

$$\mathscr{T}(S) = \sum_{u \in \Sigma}^{\oplus} \mathscr{K}(H^2(S_u)) \tag{5.36}$$

corresponds to the global sections of the field $(\mathscr{K}(H^2(S_u)))_{u \in \Sigma}$ of compact operators.

Corollary 6.3 *Define ideals $\mathscr{I}_\ell$ of $\mathscr{T}(S)$ by*

$$\mathscr{I}_\ell = \bigcap_{u \in S_\ell} ker(\sigma_u).$$

Then there is a composition series

$$\mathscr{K}(H^2(S)) = \mathscr{I}_1 \subset \mathscr{I}_2 \subset \ldots \subset \mathscr{I}_r = [\mathscr{T}(S), \mathscr{T}(S)] \subset \mathscr{I}_{r+1} = \mathscr{T}(S)$$

such that each subquotient $\mathscr{I}_{\ell+1}/\mathscr{I}_\ell$ is (stably) isomorphic to $\mathscr{C}(S_\ell)$.

The special case of the Lie ball (5.10) had been treated before in [6]. For the case $S = U(n)$, cf. [58].

An analogous result [54] holds for the weighted Bergman spaces $H^2_\nu(\Omega)$ and the Bergman-Toeplitz C^*-algebra $\mathscr{T}_\nu(\Omega)$ generated by weighted Bergman-Toeplitz operators with continuous symbol on the closure $\overline{\Omega}$. Here a shift

$$\nu \mapsto \nu' := \nu - \frac{a}{2}\ell \tag{5.37}$$

in the deformation parameter is necessary since the 'little' symmetric domain Ω_u has lower rank $r - \ell$. For $\nu > p - 1$, let $H^2_{\nu'}(\Omega_u)$ denote the weighted Bergman space on the boundary component (5.20), and consider the associated Bergman-Toeplitz C^*-algebra $\mathscr{T}_{\nu'}(\Omega_u)$.

Theorem 6.4 *For each tripotent $u \in Z$ there exists a C^*-homomorphism σ_u : $\mathscr{T}_\nu(\Omega) \to \mathscr{T}_{\nu'}(\Omega_u)$ such that*

$$\sigma_u(\tau^\nu_\Omega f) = \tau^{\nu'}_{\Omega_u} f_u$$

for all $f \in \mathscr{C}(\overline{\Omega})$, where $f_u \in \mathscr{C}(\overline{\Omega}_u)$ is defined by $f_u(w) := f(u + w)$. The corresponding irreducible representations are pairwise inequivalent, and constitute the full spectrum of $\mathscr{T}_\nu(\Omega)$.

The parameter shift (5.37) is compatible with the discrete Wallach points (5.29) and the special continuous Wallach points (5.31). In particular, the Hardy space for Ω is mapped to the Hardy space for Ω_u. Thus Theorem 6.1 can be viewed as a special case of the parameter dependent Theorem 6.4. On the other hand, the standard Bergman space parameter $\nu = 2+a(r-1)+b$ is not preserved, since the shifted parameter $\nu' = 2+a(r-1-\ell/2)+b$ exceeds the standard Bergman parameter $2+a(r-\ell-1)+b$ for Ω_u. This means that the image representation becomes 'more classical'.

The **index theory** for Toeplitz operators on symmetric domains is much less developed than in the strongly pseudoconvex case. If $\Omega \subset Z$ is an irreducible bounded symmetric domain of tube type, with Jordan determinant $N : Z \to \mathbf{C}$, the Shilov boundary $S = K/L$ is a compact symmetric space and the reduced Shilov boundary $S' := \{u \in S : N(u) = 1\}$ is simply connected. By [3, Theorem 3.3] every continuous non-vanishing function $f : S \to GL_1(\mathbf{C})$ has a factorization

$$f(u) = N(u)^k \, e^{i\varphi(u)}$$

where $\varphi : S \to \mathbf{R}$ is continuous and $k \in \mathbf{Z}$ is a sort of winding number. Moreover, the Toeplitz operator $\tau_S f$ is of Fredholm type, with index $-k$, in the more abstract setting developed in [41]. For the Lie sphere (5.19) of rank 2 the reduced Shilov boundary $S' = \mathbf{S}^{n-1}$ is a sphere and the decomposition (5.34) induces an isomorphism

$$\mathscr{I}_2 = [\mathscr{T}(S), \mathscr{T}(S)] \approx \mathscr{K}(H^2(\mathbf{T})) \otimes CZ(\mathbf{S}^{n-1})$$

between the commutator ideal and the C^*-algebra tensor product of compact operators with the **Calderon-Zygmund operators** (pseudodifferential operators of order 0). Since the index theory of pseudodifferential operators is well-developed, this allows for a somewhat more explicit index theorem [6, 39]. For a deeper study in the higher rank case, cf. [53, 54].

In the symmetric case the **Berezin transform**, for the weighted Fock spaces $H^2_\nu(Z)$ and weighted Bergman spaces $H^2_\nu(\Omega)$, can be defined as in Section 2. In the flat setting we now have

$$\beta^\nu_Z f = e^{(\partial|\partial)/\nu} f = \mathscr{F}^{0,0}_Z(\partial/\nu, \partial)$$

for symbol functions f on Z, where $(\partial|\partial)$ is the Laplace operator on Z. In the bounded setting, the Berezin transform is not a semigroup anymore, but can be expressed [55] as a function

$$\beta^\nu_\Omega = F_\nu(\Delta_1, \ldots, \Delta_r)$$

(in the sense of spectral theory) of the G-**invariant differential operators** $\Delta_1, \ldots, \Delta_r$ on Ω, with Δ_1 the Laplace-Beltrami operator (which is different from the 'flat' Laplacian). As shown in [1] the Berezin transform has also an **asymptotic expansion** into a series of differential operators

$$(\beta^\nu_\Omega f)(0) = \sum_{m \in \mathbf{N}^r_+} \frac{1}{(\nu)_m} (\mathscr{K}^m_Z(\partial, \partial) f)(0) = \mathscr{F}^{0,1}_Z \binom{-}{\nu} (\partial, \partial) f(0)$$

as $\nu \to \infty$, involving the Pochhammer symbols (5.24) and the Fischer-Fock kernels (5.25). For the n-ball this simplifies to

$$(\beta_{\mathbf{B}}^{\nu} f)(0) = \sum_{m \in \mathbf{N}} \frac{1}{(\nu)_m} \Big(\frac{(\partial|\partial)^m}{m!} f \Big)(0).$$

The evaluation at other points $z \in \Omega$ uses the transitive G-action.

The Berezin symbol is of great importance in measuring the 'growth' of an operator. For example, the **Axler-Zheng theorem** for the unit disk has been generalized to bounded symmetric domains [19] in the following form: Let T be a finite sum of finite products $\tau_{\Omega}^{\nu} f_1 \ldots \tau_{\Omega}^{\nu} f_k$ of Bergman-Toeplitz operators with L^{∞}-symbol. Then, assuming that the parameter ν is large enough, T is compact if and only if its Berezin symbol $\sigma_{\Omega}^{\nu} T$ vanishes on the boundary. Concerning the regularity of the Bergman projection P_{Ω} for bounded symmetric domains, a deep result [4] states that for $f \in L^2(\Omega)$ the densely defined operator $[M_f, P_{\Omega}]$ is bounded (resp. compact) if and only if f satisfies a kind of BMO-condition (resp. VMO-condition) with respect to the Bergman metric.

5.7 Hilbert quotient modules and Kepler varieties

Operators on Hilbert spaces of holomorphic functions are also of interest when a domain in $\mathbf{C}^n$ is replaced by a complex subvariety. This is related to the so-called **Hilbert module program** initiated by R. Douglas. For a polynomial ideal $\mathscr{I}$ consider the algebraic variety

$$\mathscr{I}^{\circ} := \{ z \in \mathbf{C}^n : \ f(z) = 0 \, \forall \, f \in \mathscr{I} \}.$$

Conversely, an algebraic variety $V \subset \mathbf{C}^n$ defines a radical ideal

$$V^{\circ} := \{ f \in \mathbf{C}[z_1, \ldots, z_n] : \ f|_V = 0 \}.$$

By Hilbert's Nullstellensatz, the 'double' $\mathscr{I}^{\circ\circ}$ is the radical of $\mathscr{I}$. Now consider a reproducing kernel Hilbert space $\mathscr{H}$ of holomorphic functions on a bounded domain $\Omega \subset \mathbf{C}^n$, on which the coordinate functions z_i, and hence all polynomials, act as bounded multiplication operators. A standard example is the (unweighted) Bergman space $H^2(\Omega)$. The closed subspace

$$\mathscr{H}_V^{\perp} := \{ \varphi \in \mathscr{H} : \ \varphi|_{V \cap \Omega} = 0 \}$$

is a submodule, and the **Hilbert quotient module**

$$\mathscr{H}_V := \mathscr{H} / \mathscr{H}_V^{\perp} = \mathscr{H} \ominus \mathscr{H}_V^{\perp}$$

corresponds to holomorphic functions 'living' on the variety $V \cap \Omega$. Its module structure is given by the 'sub-Toeplitz operators'

$$S_i := P T_i P, \tag{5.38}$$

where $T_i := M_{z_i}$ and $P : \mathscr{H} \to \mathscr{H}_V$ is the orthogonal projection. The **Arveson-Douglas conjecture** (in its weaker geometric form dealing only with radical ideals) concerns the commutators of sub-Toeplitz operators (5.38) for homogeneous subvarieties of a strongly pseudoconvex domain Ω. More precisely, the conjecture states that

$$[S_i^*, S_j] \in \mathscr{K}^p(\mathscr{H}_V), \quad \forall\, p > \dim V. \tag{5.39}$$

Here $\mathscr{K}^p$ denotes the Schatten p-class and $\dim V$ is the complex dimension of V. The Arveson-Douglas conjecture for the n-ball is an open problem in general but has been proved in various special cases, e.g. for principal varieties defined by a single polynomial or for low dimension $n \leq 3$ [30]. For varieties which are smooth away from the origin, major progress has been made (independently) in [20, 17].

Theorem 7.1 *Let V be a homogeneous algebraic variety in the n-ball $\mathbf{B}$ such that $V \setminus \{0\}$ is smooth. Consider the Hilbert module $\mathscr{H} = H_\nu^2(\mathbf{B})$ for $\nu > 0$. Then the sub-Toeplitz operators S_i, $1 \leq i \leq n$, on the quotient module $\mathscr{H}_V$ satisfy* (5.39).

Algebraic varieties and Hilbert quotient modules are also of interest for symmetric domains. Let Z be an irreducible hermitian Jordan triple of rank r. For fixed $0 \leq \ell \leq r$ the direct sum

$$\mathscr{P}_\ell^\perp(Z) := \sum_{m_{\ell+1} > 0} \mathscr{P}_m(Z) \tag{5.40}$$

is an ideal in the polynomial algebra, which is generated by the Jordan algebraic minors of degree $> \ell$. The associated algebraic variety

$$V_\ell = \{z \in Z : \; rank(z) \leq \ell\} \subset Z$$

is called the **Kepler variety** associated with Z. Its smooth part

$$\mathring{V}_\ell = \{z \in V_\ell : \; rank(z) = \ell\} = V_\ell \setminus V_{\ell-1}$$

is a $K^{\mathbf{C}}$-homogeneous manifold called a **Kepler manifold**. These Kepler manifolds are precisely the 'Matsuki duals' of the G-orbits (5.21) in the boundary of Ω. For the matrix triple $Z = \mathbf{C}^{r \times s}$ we obtain the 'determinantal varieties' defined by vanishing of all $(\ell+1) \times (\ell+1)$-minors. For the spin factor $Z = \mathbf{C}^n$ the Jordan determinant is given by

$$N(z) = z \cdot z = z_1^2 + \ldots + z_n^2.$$

Therefore

$$V_1 := \{z \in \mathbf{C}^n : \; z \cdot z = 0\} \tag{5.41}$$

is the complex light cone. In terms of symplectic geometry

$$\mathring{V}_1 \approx T^*(\mathbf{S}^{n-1}) \setminus 0 = \{(x, \xi) : \|x\| = 1, \; \xi \neq 0, \; (x|\xi) = 0\}, \; z = \frac{x + i\xi}{2}$$

gives the cotangent bundle of $\mathbf{S}^{n-1}$. The corresponding contact manifold is the cosphere bundle (5.18).

In [23], it is shown that V_ℓ is a **normal variety having only rational singularities**. This is well-known for the classical case (5.41) and was known before for non-exceptional types. Our proof is uniform, based on Kempf's collapsing vector bundle theorem. As an important consequence, the second Riemann extension theorem holds for V_ℓ (except for the trivial case of tube domains and $\ell = r$): Every holomorphic function on the Kepler manifold $\mathring{V}_\ell$ has a unique extension to the closure V_ℓ.

For each tripotent $u \in S_\ell$, the Peirce 2-space Z_u^2 becomes a Jordan algebra of rank ℓ, with unit element u and multiplication $(x,y) \mapsto \{x;u;y\}$. Let Λ_u denote the symmetric cone in the self-adjoint part of Z_u^2. Similar to the boundary strata (5.21) of a symmetric domain Ω, the Kepler manifold admits a fibration (or polar decomposition)

$$\mathring{V}_\ell = \bigcup_{u \in S_\ell} \Lambda_u.$$

The compact group K acts on this fibration. As a consequence, every K-invariant measure on $\mathring{V}_\ell$ has a polar decomposition whose **radial part** is a measure on the symmetric cone Λ_c of rank ℓ, where $c \in S_\ell$ is a chosen base point. Using the Jordan determinant N_c of the Jordan algebra Z_c^2 of rank ℓ, one defines a differential operator

$$\mathscr{D}_c := N_c^{\frac{a}{2}(\ell-r)} \partial_{N_c}^b \, N_c^{\frac{a}{2}(r-\ell-1)+b+1} \left(\partial_{N_c} N_c^{\frac{a}{2}} \partial_{N_c}^{a-1}\right)^{r-\ell} N_c^{\frac{a}{2}(r-\ell+1)-1} \tag{5.42}$$

of order $\ell((r-\ell)a+b)$ on the symmetric cone Λ_c. For $\ell = 1$, $\Lambda_c = \mathbf{R}^+ \cdot c$ is the open half-line and $N_c(tc) = t$ for $t \in \mathbf{R}$. Then $\mathscr{D}_c$ becomes a polynomial differential operator of order $(r-1)a + b = p-2$ on the half-line $t > 0$.

We now define Hilbert quotient modules of holomorphic functions for Kepler manifolds. In view of (5.40) these are completions of the quotient space

$$\mathscr{P}(Z)/\mathscr{P}_\ell^\perp(Z) \approx \mathscr{P}_\ell(Z)$$

defined in (5.30). Its elements are uniquely determined by the restriction to V_ℓ.

Given a pluri-subharmonic potential function ρ on a subvariety V of dimension k, define a measure

$$d\mu_V^\nu(z) := e^{-\nu\rho(z)} \frac{1}{k!}\left(\frac{i}{2}\partial\overline{\partial}\rho\right)^k$$

on V similar to (5.5). Let $H_\nu^2(V)$ denote the associated Hilbert quotient module. The following results [23] express reproducing kernel functions in terms of hypergeometric series.

Theorem 7.2 *Consider the pluri-subharmonic function $\rho(z) = (z|z)$ on the Kepler manifold V_ℓ, of dimension n_ℓ. Then the Hilbert space $H_\nu^2(V_\ell)$ has the reproducing kernel*

$$\mathscr{K}_{V_\ell}^\nu(t,c) = \mathscr{D}_c \, \mathscr{F}_c^{1,1}\binom{n_\ell/\ell}{n/\ell}(\nu t) \tag{5.43}$$

for t in the radial cone Λ_c. Here $\mathscr{D}_c$ is the differential operator defined in (5.42) *and $\mathscr{F}_c^{1,1}$ is the 'confluent' hypergeometric function relative to Z_c^2.*

A similar result holds in the bounded setting.

Theorem 7.3 *Consider the pluri-subharmonic function*

$$\rho(z) = \log D(z,z)^{-p} = \log\det B(z,z)^{-1}$$

on the intersection

$$\mathring{V}_\ell \cap \Omega := \{z \in \mathring{V}_\ell : \|z\| < 1\}$$

with the bounded symmetric domain Ω (Kepler domain). Then the Hilbert space $H^2_\nu(V_\ell \cap \Omega)$ has the reproducing kernel

$$\mathscr{K}^\nu_{V_\ell\cap\Omega}(t,c) = \mathscr{D}_c\, \mathscr{F}^{2,1}_c \begin{pmatrix} n_\ell/\ell;\nu \\ n_\ell/\ell \end{pmatrix}(t) \tag{5.44}$$

on the radial 'interval' $\Lambda_c \cap (c - \Lambda_c)$. Here $\mathscr{F}^{2,1}_c$ is the Gauss hypergeometric function relative to Z^2_c.

In the simplest case $\ell = 1$ (5.43) and (5.44) yield explicit expansion formulas on the half-line $t > 0$ and the interval $0 < t < 1$, respectively.

In order to obtain **Dixmier trace formulas** for symmetric domains Ω of higher rank, one has to consider Hilbert quotient modules since the usual Toeplitz operators on Ω are not essentially normal. For $\ell = 1$ the Kepler variety V_1 has dimension

$$n_1 = p - 1 = 1 + a(r-1) + b.$$

Consider the associated Hilbert quotient module

$$H^2_1(S) := H^2(S)/V_1^\circ = \sum_{m\in\mathbf{N}} \mathscr{P}_{m,0,\dots,0}(Z)$$

of the Hardy space $H^2(S)$ over the Shilov boundary $S = S_r$. Denote the associated sub-Toeplitz operators by $\tau^1_S f$. The following results have been obtained in [56].

Theorem 7.4 *Let f,g be polynomials on Z. Then $[\tau^1_S f, \tau^1_S \overline{g}] \in \mathscr{K}^{n_1,\infty}(H^2_1(S))$. Hence*

$$[\tau^1_S f_1, \tau^1_S \overline{g}_1]\cdots[\tau^1_S f_{n_1}, \tau^1_S \overline{g}_{n_1}] \in \mathscr{K}^{1,\infty}(H^2_1(S))$$

for polynomials f_i, g_i on Z.

In order to compute the Dixmier trace, we pass to the strongly pseudoconvex Kepler ball $V_1 \cap \Omega$. It has a singularity at the origin, but its boundary S_1, the set of all minimal tripotents, is a smooth K-homogeneous manifold. Denote by $L^2(S_1)$ the L^2-space with respect to the normalized K-invariant measure. The Hardy space $H^2(S_1)$ is the closure of the algebra $\mathscr{P}(Z)$ of all polynomials on Z, restricted to S_1. For $\ell > 1$, the minors N_ℓ vanish on S_1. It follows that

$$H^2(S_1) = \sum_{m\geq 0} \mathscr{P}_{m,0,\dots,0}(Z)|_{S_1}.$$

By [56, Lemma 4.5] the inner products $(\varphi|\psi)_S$ and $(\varphi|\psi)_{S_1}$ are related by

$$(\varphi|\psi)_S = d_m \frac{m!}{(d/r)_m} (\varphi|\psi)_{S_1}$$

for all $\varphi, \psi \in \mathscr{P}_{m,0,\dots,0}(Z)$, where $d_m := \dim \mathscr{P}_{m,0,\dots,0}(Z)$. This leads to an isometric isomorphism $U : H^2_1(S) \to H^2(S_1)$ which, by [56, Proposition 4.6] satisfies

$$U[\tau^1_S f, \tau^1_S \overline{g}]U^* - [\tau_{S_1}(f|_{S_1}), \tau_{S_1}(\overline{g}|_{S_1})] \in \mathscr{K}^{n_1/2,\infty}(H^2(S_1))$$

for linear functionals f, g on Z. Applying Theorem 3.2 above to S_1 one obtains:

Theorem 7.5 *Let $f_i, g_i \in \mathscr{P}(Z)$. Then the Toeplitz commutator product has the Dixmier trace*

$$tr_\omega[\tau^1_S f_1, \tau^1_S \overline{g}_1] \cdots [\tau^1_S f_{n_1}, \tau^1_S \overline{g}_{n_1}] = C \int_{S_1} d\sigma \prod_{j=1}^{n_1} \{\hat{f}_j, \hat{\overline{g}_j}\}_\flat,$$

where $d\sigma$ is the normalized K-invariant measure, $\hat{f}$ is the Poisson extension of f and $\{f, g\}_\flat$ denotes the boundary Poisson bracket of $f, g \in \mathscr{C}^\infty(S_1)$. The constant C can be computed in terms of invariants of the underlying domain.

The previous two results handle Toeplitz commutator products for real-analytic polynomials. It is desirable to extend them to all smooth functions.

5.8 Toeplitz operators on Reinhardt domains

The Hardy and Bergman spaces associated with symmetric domains are sometimes called 'non-commutative' function spaces, because of the natural action under the non-commutative Lie groups K and G. The class of **Reinhardt domains** is more closely related to the commutative n-torus $\mathbf{T}^n$, but in the multi-variable case these domains exhibit interesting behavior not present in the 1-dimensional case. A domain $\Omega \subset \mathbf{C}^n$ is called a Reinhardt (or polycircular) domain if it is invariant under the $\mathbf{T}^n$-action

$$z = (z_1, \dots, z_n) \mapsto (\vartheta_1 z_1, \dots, \vartheta_n z_n)$$

for $\vartheta_1, \dots, \vartheta_n \in \mathbf{T}$. In short, $\mathbf{T}^n \cdot \Omega = \Omega$. Put $|z| := (|z_1|, \dots, |z_n|) \in \mathbf{R}^n_+$. Then

$$|\Omega| := \{|z| : z \in \Omega\}$$

is a domain in $\mathbf{R}^n_+$ called the 'absolute domain'. By definition, we have $\Omega = \mathbf{T}^n \cdot |\Omega|$. In case Ω is bounded, we may assume by rescaling that Ω is contained in the closed polydisk. Then $|\Omega|$ is a domain in the n-cube $[0,1]^n$. Define

$$\log |z| := (\log |z_1|, \dots, \log |z_n|).$$

Then

$$\log|\Omega| := \{\log|z| : z \in \Omega\} \tag{5.45}$$

becomes a domain in $[-\infty, 0[^n$. Here we put $\log 0 := -\infty$. A basic theorem [43, Chapter 7] asserts that a Reinhardt domain Ω is a domain of holomorphy (i.e. pseudoconvex) if and only if Ω is polybalanced and (5.45) is convex. This we assume from now on.

Basic examples of Reinhardt domains are the polydisk and the n-ball $\mathbf{B}$. A more sophisticated example is the so-called Hartogs' wedge $\Omega \subset \mathbf{C}^2$ determined by

$$\log|\Omega| = \{(x,y) \in \mathbf{R}^2 : x < 0, \ \vartheta_2 x < y < \vartheta_1 x\} \tag{5.46}$$

where $0 < \vartheta_1 < \vartheta_2$ are irrational. Symmetric domains of higher rank are not Reinhardt domains.

Let $H^2(\Omega)$ denote the (standard) Bergman space over a bounded Reinhardt domain Ω, assumed to be polybalanced and log-convex. The torus action leads to a kind of Fourier analysis of $H^2(\Omega)$. Each $\varphi \in H^2(\Omega)$ admits a Taylor expansion

$$\varphi(z) = \sum_{\alpha \in \mathbf{N}^n} \varphi_\alpha \, z^\alpha$$

where the coefficient family $(\varphi_\alpha)_{\alpha \in \mathbf{N}^n}$ belongs to a weighted ℓ^2-space. For $1 \leq i \leq n$ and $\alpha \in \mathbf{N}^n$ define weights

$$w_i(\alpha) := \frac{\|z^{\alpha+\varepsilon_i}\|}{\|z^\alpha\|}. \tag{5.47}$$

Here $\varepsilon_i = (0,\ldots,0,1,0\ldots,0)$ with 1 at position i. Then the Toeplitz operator $T_i := \tau_\Omega z_i$ satisfies $T_i \, e_\alpha = w_i(\alpha) \, e_{\alpha+\varepsilon_i}$ for the normalized monomials e_α. More generally,

$$T^\alpha := T_1^{\alpha_1} \ldots T_n^{\alpha_n} \tag{5.48}$$

satisfies $T^\alpha e_\beta = w_\alpha(\beta) \, e_{\alpha+\beta}$, where $w_\alpha(\beta)$ can be computed from (5.47) via a cocycle property. Define operators U_α, W_α on $\ell^2(\mathbf{N}^n)$ by

$$(U_\alpha \zeta)(\beta) := \zeta(\beta - \alpha), \quad (W_\alpha \zeta)(\beta) := w_\alpha(\beta) \, \zeta(\beta) \tag{5.49}$$

for all $\zeta \in \ell^2(\mathbf{N}^n)$. Then the previous arguments yield the polar decomposition

$$T^\alpha = U_\alpha \, W_\alpha$$

of the operators (5.48) which form a representation of the semigroup $\mathbf{N}^n$ by weighted shifts [11]. Hence the Toeplitz C^*-algebra $\mathscr{T}(\Omega)$, generated by the operators (5.48), is (properly) contained in the unital C^*-algebra $C^*(U,W)$ generated by the operators (5.49). As shown in [11] the latter C^*-algebra can be realized as a **groupoid C^*-algebra**, making the well-developed theory of groupoids and their C^*-algebras [45] available for complex analysis on Reinhardt domains.

The underlying groupoid is defined as follows: For $\alpha \in \mathbf{Z}^n$ define a bounded function $\tau_\alpha w_i$ on $\mathbf{Z}^n$ by

$$(\tau_\alpha w_i)(\beta) = \begin{cases} w_i(\beta - \alpha) & \beta - \alpha \in \mathbf{N}^n \\ 0 & \beta - \alpha \notin \mathbf{N}^n \end{cases}.$$

These functions generate a non-unital C^*-subalgebra $\mathscr{A} \subset \ell^\infty(\mathbf{Z}^n)$. Its **spectrum** (character space) $\mathscr{A}^\sharp$ is a locally compact, non-compact Hausdorff space, endowed with a natural translation action of $\mathbf{Z}^n$, since $\mathscr{A}$ is translation invariant. Consider the transformation groupoid $\mathbf{Z}^n \times \mathscr{A}^\sharp$ [42, 45]. Let $\chi(\alpha)h := h(\alpha)$, for $h \in \mathscr{A}$ and $\alpha \in \mathbf{Z}^n$, denote the 'elementary' characters. Then $C^*(U,W)$ can be identified with the groupoid C^*-algebra associated with the 'reduction' $(\mathbf{Z}^n \times \mathscr{A}^\sharp)|_{\overline{\chi(\mathbf{N}^n)}}$ to the closure of $\chi(\mathbf{N}^n)$.

For a more refined analysis, consider the **faces** F of the compact convex set

$$C := \log|\overline{\Omega}| \subset [-\infty, 0]^n.$$

Let π_F denote the orthogonal projection onto the tangent space TF. If F is a maximal face, we may write (using $(x|y)$ for the inner product)

$$F = \{x \in C : \ (x|u) = \sup(C|u)\} = u^\perp + u \ \sup(C|u),$$

where $u \in \mathbf{R}^n$ is a unit vector, with orthogonal complement $u^\perp$. By [48] there is a continuous embedding $\chi : \overline{\pi_F(\mathbf{N}^n)} \to \overline{\chi(\mathbf{N}^n)}$ given by

$$\chi(s) := w^* - \lim_{k \to \infty} \chi(v_k),$$

where $v_k \in \mathbf{N}^n$ satisfies $\|v_k\| \to \infty$, $v_k/\|v_k\| \to u$ and $\pi_F v_k \to s$. This can be generalized to all faces. One can show that the compact space $\overline{\chi(\mathbf{N}^n)}$ is the union of the compact subsets $\chi(F) := \chi(\overline{\pi_F(\mathbf{N}^n)})$, as F ranges over all faces of C. Moreover, the subsets corresponding to the interior $\mathring{F}$ relative to TF are pairwise disjoint. Thus we have as a disjoint union

$$\overline{\chi(\mathbf{N}^n)} = \bigcup_{F \subset C} \chi(\mathring{F})$$

similar to (5.22). The trivial face C itself forms the 'interior' $\chi(\mathring{C}) = \chi(\mathbf{N}^n)$. The other faces belong to the 'boundary' consisting of w^*-limits of unbounded sequences $(\chi(v_k))_{k \in \mathbf{N}}$ of elementary characters.

The groupoid C^*-algebra $C^*(U,W)$ contains the compact operators on $\ell^2(\mathbf{Z}^n)$ which can be realized via a further reduction $(\mathbf{Z}^n \times \mathscr{A}^\sharp)|_{\chi(\mathbf{N}^n)}$ to the interior. All other faces have groupoid C^*-algebras corresponding to the reduction $(\mathbf{Z}^n \times \mathscr{A}^\sharp)|_{\chi(\mathring{F})}$. However, in contrast to the symmetric case these boundary parts in general are not just the compact operators (on suitable Hilbert spaces defined on boundary components) but are realized as **foliation C^*-algebras** $C^*(\mathscr{F}_F)$ for the Kronecker type foliation $\mathscr{F}_F$ of $\mathbf{T}^n$ whose leaves are given by

$$\mathscr{L}_s := \{e(s+t) : \ t \in TF\}$$

for $s \in \mathbf{R}^n$. Here we put $e(x) := (e^{2\pi i x_1}, \ldots, e^{2\pi i x_n})$. Thus, similar to (5.36), we can formally write

$$C^*(U,W) = \sum_{F \subset C}^{\oplus} C^*(\mathscr{F}_F).$$

In particular, the Toeplitz C^*-algebra associated with a Reinhardt domain is not of type I if the boundary contains faces of 'irrational slope'. This facial structure is already interesting in the 2-dimensional case, where domains such as the Hartogs wedge (5.46) or the L-shaped domains considered in [11] give rise to irrational rotation C^*-algebras instead of the compact operators. In [48] the structure and index theory of Toeplitz C^*-algebras over Reinhardt domains, with non-trivial applications to complex analysis (existence of proper holomorphic mappings), has been studied. In principle it is possible to unify symmetric domains and Reinhardt domains in the class of 'non-commutative' Reinhardt domains [38] and develop a uniform theory of Toeplitz C^*-algebras in this more general setting.

References

[1] J. Arazy and B. Orsted, Asymptotic expansions of Berezin transforms, *Indiana Univ. Math. J.* **49** (2000), 7-30.

[2] J. Arazy and H. Upmeier, Boundary measures for symmetric domains and integral formulas for the discrete Wallach points, *Int. Equ. Op. Th.* **47** (2003), 375-434.

[3] A. B. Badi, Non-vanishing functions and Toeplitz operators on tube type domains, *J. Funct. Anal.* **258** (2010), 3841-3854.

[4] D. Bekollé, C. Berger, L. Coburn, and K. Zhu, BMO in the Bergman metric on bounded symmetric domains, *J. Funct. Anal.* **93** (1990), 310-350.

[5] C. Berger and L. Coburn, Wiener-Hopf operators on U_2, *Int. Eq. Op. Th.* **2** (1979), 139-173.

[6] C. Berger, L. Coburn, and A. Korányi, Opérateurs de Wiener-Hopf sur les sphères de Lie, *C.R. Acad. Sci. Paris* **290** (1980), 989-991.

[7] L. Brown, R. Douglas, and P. Fillmore, Extensions of C^*-algebras and K-homology, *Ann. Math.* **105** (1977), 265-324.

[8] L. Boutet de Monvel, On the index of Toeplitz operators in several complex variables, *Invent. Math.* **50** (1979), 249-272.

[9] L. Boutet de Monvel and V. Guillemin, *The Spectral Theory of Toeplitz Operators*, Ann. Math. Studies **99** (1981), Princeton University Press, Princeton, New Jersey.

[10] G. Cao, Toeplitz algebras on strongly pseudoconvex domains, *Nagoya Math. J.* **185** (2007), 171-186.

[11] R. Curto and P. Muhly, C^*-algebras of multiplication operators on Bergman spaces, *J. Funct. Anal.* **64** (1985), 315-329.

[12] L. Coburn, Singular integral operators and Toeplitz operators on odd spheres, *Indiana Univ. Math. J.* **23** (1973), 433-439.

[13] A. Connes, A survey of foliations and operator algebras, *Proc. Symp. Pure Math.* **38.1** (1981), 521-628.

[14] A. Connes, *Noncommutative Geometry*, Academic Press, San Diego, 1994.

[15] A. Dynin, An algebra of pseudodifferential operators on the Heisenberg group, *Soviet Math. Dokl.* **17** (1976), 508-512.

[16] A. Dynin, Multivariable Wiener-Hopf operators I. *Int. Equ. Op. Th.* **9** (1986), 537-556.

[17] R. Douglas, X. Tang, and G. Yu, An analytic Grothendieck-Riemann-Roch theorem, preprint, arXiv:1404.4396 (2014).

[18] M. Englis, Berezin transformation and reproducing kernels on complex domains, *Trans. Amer. Math. Soc.* **348** (1996), 411-479.

[19] M. Englis, Compact Toeplitz operators via the Berezin transform on bounded symmetric domains, *Int. Equ. Op. Th.* **33** (1999), 426-455.

[20] M. Englis and J. Eschmeier, Geometric Arveson-Douglas conjecture, *Adv. Math.* **9** (2015), 606–630.

[21] M. Englis, K. Guo, and G. Zhang, Toeplitz and Hankel operators and Dixmier traces on the unit ball of $\mathbf{C}^n$, *Proc. Amer. Math. Soc.* **137** (2009), 3669-3678.

[22] M. Englis and R. Rochberg, The Dixmier trace of Hankel operators on the Bergman space, *J. Funct. Anal.* **257** (2009), 1445-1479.

[23] M. Englis and H. Upmeier, Reproducing kernel functions and asymptotic expansions on Jordan-Kepler manifolds, preprint, arXiv: 1708.03388 (2017).

[24] M. Englis and G. Zhang, Hankel operators and the Dixmier trace on strictly pseudoconvex domains, *Documenta Math.* **15** (2010), 601-622.

[25] J. Faraut and A. Korányi, Function spaces and reproducing kernels on bounded symmetric domains, *J. Funct. Anal.* **88** (1990), 64-89.

[26] J. Faraut and A. Korányi, *Analysis on Symmetric Cones*, Clarendon Press, Oxford (1994).

[27] J. Fornaess and S. Krantz, Continuously varying peaking functions, *Pac. J. Math.* **83** (1979), 341-347.

[28] S. Gindikin, Invariant generalized functions in homogeneous domains, *Funct. Anal. Appl.* **9** (1975), 50-53.

[29] V. Guillemin, Toeplitz operators in n-dimensions, *Int. Equ. Op. Th.* **7** (1984), 145-205.

[30] K. Guo and K. Wang, Essentially normal Hilbert modules and K-homology, *Math. Ann.* **340** (2008), 907-934.

[31] K. Guo, K. Wang, and G. Zhang, Trace formulas and p-essentially normal properties of quotient modules on the bidisk, *J. Operator Th.* **67** (2012), 511-535.

[32] W.J. Helton and R. Howe, Traces of commutators of integral operators, *Acta Math.* **135** (1975), 271-305.

[33] J. Janas, Toeplitz operators related to certain domains in $\mathbf{C}^n$, *Studia Math.* **54** (1975), 73-79.

[34] N. Jewell and S. Krantz, Toeplitz operators and related function algebras on certain pseudoconvex domains, *Trans. Amer. Math. Soc.* **252** (1979), 297-311.

[35] M. Koecher, *An Elementary Approach to Bounded Symmetric Domains*, Rice University, Houston (1969).

[36] J. J. Kohn, Harmonic integrals on strongly pseudoconvex manifolds I, *Ann. Math.* **78** (1963), 112-124.

[37] S. Krantz, *Function Theory of Several Complex Variables*, Wiley, New York (1982).

[38] M. Lassalle, L'espace de Hardy d'un domain de Reinhardt generalisé, *J. Funct. Anal.* **60** (1985), 309-340.

[39] M. Lesch, K-theory and Toeplitz C^*-algebras, a survey, *Séminaire de theorie spectrale et géometrie* **9** (1990-91), 119-132.

[40] O. Loos, *Bounded Symmetric Domains and Jordan Pairs*, University of California, Irvine (1977).

[41] G. J. Murphy, Topological and analytic indices in C^*-algebras, *J. Funct. Anal.* **234** (2006), 261-276.

[42] P. Muhly and J. Renault, C^*-algebras of multivariable Wiener-Hopf operators, *Trans. Amer. Math. Soc.* **274** (1982), 1-44.

[43] R. Narasimhan, *Several Complex Variables*, University of Chicago Press, Chicago (1971).

[44] I. Raeburn, On Toeplitz operators associated with strongly pseudoconvex domains, *Studia Math.* **63** (1979), 253-258.

[45] J. Renault, *A Groupoid Approach to C^*-algebras*, Springer Lect. Notes in Math. **793** (1980).

[46] W. Schmid, Die Randwerte holomorpher Funktionen auf hermitesch symmetrischen Räumen, *Invent. Math.* **9** (1969), 61-80.

[47] R. Stanley, Some combinatorial properties of Jack symmetric functions, *Adv. Math.* **77** (1989), 76-115.

[48] N. Salinas, A. Sheu, and H. Upmeier, Toeplitz operators on pseudoconvex domains and foliation C^*-algebras, *Ann. Math.* **130** (1989), 531-565.

[49] H. Upmeier, Jordan algebras and harmonic analysis on symmetric spaces, *Amer. J. Math.* **108** (1986), 1-25.

[50] H. Upmeier, Toeplitz operators on bounded symmetric domains, *Trans. Amer. Math. Soc.* **280** (1983), 221-237.

[51] H. Upmeier, Toeplitz C^*-algebras on bounded symmetric domains, *Ann. Math. 119* (1984), 549-576.

[52] H. Upmeier, Toeplitz operators on symmetric Siegel domains, *Math. Ann.* **271** (1985), 401-414.

[53] H. Upmeier, Index theory for multivariable Wiener-Hopf operators, *J. reine angew. Math.* **384** (1988), 57-79.

[54] H. Upmeier, Toeplitz Operators and Index Theory in Several Complex Variables, *Operator Theory* **81**, Birkhäuser, Basel-Boston-Berlin (1996).

[55] A. Unterberger and H. Upmeier, The Berezin transform and invariant differential operators, *Commun. Math. Phys.* **164** (1994), 563-597.

[56] H. Upmeier and K. Wang, Dixmier trace for Toeplitz operators on symmetric domains, *J. Funct. Anal.* **271** (2016), 532-565.

[57] U. Venugopalkrishna, Fredholm operators associated with strongly pseudoconvex domains in $\mathbf{C}^n$, *J. Funct. Anal.* **9** (1972), 349-373.

[58] A. Wassermann, Algèbres d'opérateurs de Toeplitz sur les groupes unitaires, *C.R. Acad. Sci. Paris* **299** (1984), 871-874.

[59] K. Yabuta, A remark on a paper of Janas: Toeplitz operators related to certain domains in $\mathbf{C}^n$, *Studia Math.* **62** (1978), 73-74.

[60] G. Zhang, Tensor products of minimal holomorphic representations, *Representation Theory* **5** (2001), 164-190.

[61] K. Zhu, *Operator Theory in Function Spaces*, Marcel Dekker, New York/Basel (1990).

Chapter 6

Möbius Invariant $\mathcal{Q}_p$ and $\mathcal{Q}_K$ Spaces

Hasi Wulan
Department of Mathematics, Shantou University, Guangdong 515063, China
wulan@stu.edu.cn

CONTENTS

6.1 Introduction

As a natural extention of the BMO-theory, R. Aulaskari, J. Xiao and R. Zhao introduced the $\mathcal{Q}_p$ spaces in [AXZ1995]. The study of these spaces has attracted considerable attention in recent years and has resulted in a new code 30H25 in the 2010 AMS subject classifications. The $\mathcal{Q}_p$ spaces contain many well-known function spaces, including the Dirichlet space, BMOA and the Bloch space. Two monographs of J. Xiao [Xi2001, Xi2006] contain most of the results about $\mathcal{Q}_p$ spaces.

The spaces $\mathcal{Q}_K$ were introduced by the author and his collaborators [EW2000, WW2001] at the beginning of this century. One of the main motivations for the introduction of $\mathcal{Q}_K$ spaces is to understand the gap between BMOA and the Bloch space. For example, a natural question is whether or not there exist Möbius invariant spaces between them. The $\mathcal{Q}_p$ theory is clearly not fine enough for this, but the theory of $\mathcal{Q}_K$ spaces answers this question in the affirmative and provides a large number of other examples.

After more than two decades of research, the theory of $\mathcal{Q}_p$ and $\mathcal{Q}_K$ spaces has reached a certain level of maturity. This article will briefly outline the development of these spaces from their origins to the present and summarize some interesting results based on the interest of the author. This should not be viewed as a complete survey. I hope that the article will be accessible not only to workers in the field but also to analysts in general. In addition to Xiao's monographs, we also draw the reader's attention to the recent book [WZ2017] on $\mathcal{Q}_K$ spaces.

6.2 Background

6.2.1 BMO

F. John and L. Nirenberg [JN1961] introduced the space of functions of bounded mean oscillation (abbreviated BMO) and they applied it to smoothness problems in partial differential equations. Their results show that BMO functions cannot become unbounded too rapidly and that BMO is situated between L^∞ and L^p for all $0 < p < \infty$. Later on, C. Fefferman and E. Stein [FS1972] obtained the principal result on the connection between BMO and conjugate harmonic functions. This concludes that BMO consists exactly of all sums of pairs of analytic functions, one of which has a bounded real part and another one has a bounded imaginary part. C. Fefferman's famous equation, $(H^1)^* = BMOA$, describes a deep relation between BMO and the Hardy space [Fe1971]. Here BMOA denotes the set of analytic functions $f(z)$ in the unit disc $\mathbb{D}$ on the complex plane $\mathbb{C}$ whose boundary function $f(e^{i\theta})$ belongs to BMO of the unit circle, that is,

$$\|f\|_{\mathrm{BMO}} := \sup_I \frac{1}{|I|}\int_I \left| f(e^{i\theta}) - \frac{1}{|I|}\int_I f(e^{it})dt \right| d\theta < \infty, \tag{6.1}$$

where the supremum in (6.1) is taken over all intervals I on the unit circle $\mathbb{T} = \partial\mathbb{D}$ and $|I|$ denotes the arc length of the interval I. BMOA may be defined, equivalently, as the family of functions f in BMO whose Poisson extensions to $\mathbb{D}$ are analytic. BMO functions play a very important role in function theory and it is quite natural to study such functions in both the real variable theory and the complex analytic theory. The number of references concerning BMO is huge; some examples are [Ba1980, Ca1980, Du1970, Ga1981, Zhu2007].

6.2.2 Definition of $\mathcal{Q}_p$ spaces

For the benefit of readers who do not work in the field, here are a few of the basic definitions. Let $H(\mathbb{D})$ denote the space of analytic functions in $\mathbb{D}$, let

$$g(z,a) = \log \frac{1}{|\varphi_a(z)|}$$

be Green's function of $\mathbb{D}$ with logarithmic singularity at a, and let $dA(z)$ be the Euclidean area element on $\mathbb{D}$, where

$$\varphi_a(z) = \frac{a-z}{1-\bar{a}z}$$

is a Möbius transformation of $\mathbb{D}$. A function $f \in H(\mathbb{D})$ is said to belong to the space $\mathcal{Q}_p$ for $0 \le p < \infty$ if

$$\|f\|_{\mathcal{Q}_p} = \left\{ \sup_{a\in\mathbb{D}} \int_{\mathbb{D}} |f'(z)|^2 (g(z,a))^p \, dA(z) \right\}^{\frac{1}{2}} < \infty. \tag{6.2}$$

Note that $\mathcal{Q}_1 = BMOA$ and $\mathcal{Q}_0$ is the classical Dirichlet space $\mathcal{D} = \mathcal{D}_0$, which is a special case of the Dirichlet type spaces $\mathcal{D}_p$ defined by

$$||f||_{\mathcal{D}_p} = \left\{ \int_{\mathbb{D}} |f'(z)|^2 (1-|z|^2)^p \, dA(z) \right\}^{\frac{1}{2}}, -1 < p < \infty.$$

R. Aulaskari and P. Lappan [AL1994] showed that for all $1 < p < \infty$ the spaces $\mathcal{Q}_p$ are the same and equal to the Bloch space $\mathcal{B}$ consisting of functions $f \in H(\mathbb{D})$ such that

$$\|f\|_{\mathcal{B}} = \sup_{z \in \mathbb{D}} (1-|z|^2)|f'(z)| < \infty. \tag{6.3}$$

The space $\mathcal{Q}_p$ is a Banach space with the norm $|f(0)| + \|f\|_{\mathcal{Q}_p}$. It is clear that $\mathcal{Q}_p$ is Möbius invariant, that is, $\|f \circ \varphi_a\|_{\mathcal{Q}_p} = \|f\|_{\mathcal{Q}_p}$ for all $f \in \mathcal{Q}_p$ and $a \in \mathbb{D}$. An observation of L. Rubel and R. Timoney [RT1979] shows that the Bloch space $\mathcal{B}$ is the largest Möbius invariant Banach space of analytic functions on $\mathbb{D}$. The smallest $\mathcal{Q}_p$ space is the Dirichlet space $\mathcal{D}$ since $\mathcal{Q}_p$ has the nesting property that

$$\mathcal{D} \subsetneqq \mathcal{Q}_{p_1} \subsetneqq \mathcal{Q}_{p_2} \subsetneqq \text{BMOA} \subsetneqq \mathcal{B}$$

for $0 < p_1 < p_2 < 1$ (cf. [AXZ1995]). It is easy to check that $\mathcal{Q}_p$ is the Möbius invariant space induced by the Dirichlet type space $\mathcal{D}_p$ for $0 < p < 1$; that is

$$\|f\|_{\mathcal{Q}_p} \approx \sup_{a \in \mathbb{D}} \|f \circ \varphi_a - f(a)\|_{\mathcal{D}_p}. \tag{6.4}$$

Similarly, BMOA is the Möbius invariant space induced by the Hardy space H^2 and the Bloch space $\mathcal{B}$ is the Möbius invariant space induced by the Bergman space A^2; see [Ax1986] and [Ba1980]. More general Möbius invariant spaces $F(p,q,s)$ are discussed by R. Zhao in [Zha1996].

6.3 Basic properties of $\mathcal{Q}_p$ spaces

6.3.1 Connection between $\mathcal{Q}_p$ and Hardy spaces

Recall that for $0 < p \leq \infty$ the Hardy space H^p consists of those functions $f \in H(\mathbb{D})$ for which

$$\|f\|_{H^p} = \sup_{0<r<1} M_p(r,f) < \infty,$$

where

$$M_p(r,f) = \left(\frac{1}{2\pi} \int_0^{2\pi} |f(re^{i\theta})|^p d\theta \right)^{1/p}, 0 < p < \infty,$$

and

$$M_\infty(r,f) = \max_{0 \leq \theta \leq 2\pi} |f(re^{i\theta})|.$$

We know that all bounded analytic functions belong to BMOA, but the situation is different for the spaces $\mathcal{Q}_p$. M. Essén and J. Xiao's examples [EX1997] show that

$$H^\infty \setminus \bigcup_{0<p<1} \mathcal{Q}_p \neq \emptyset, \qquad \bigcap_{0<p<1} \mathcal{Q}_p \setminus H^\infty \neq \emptyset.$$

6.3.2 $\mathcal{Q}_p$ and some classical function spaces

Let $\mathscr{A}$ denote the space of all analytic functions in $\mathbb{D}$ which are continuous on $\overline{\mathbb{D}}$. It is clear that $\mathscr{A} \subsetneqq H^\infty$. However, $\mathscr{A} \not\subset \cup_{0<p<1} \mathcal{Q}_p$; see [Rä2001].

Recall that the Hadamard product of

$$f(z) = \sum_{n=0}^{\infty} a_n z^n$$

and

$$g(z) = \sum_{n=0}^{\infty} b_n z^n$$

is defined by

$$f * g(z) = \sum_{n=0}^{\infty} a_n b_n z^n.$$

For $0 < p < 1$, R. Aulaskari, D. Girela and H. Wulan [AGW2000] showed that the space $\mathcal{Q}_p$ admits the following characterization:

$$\|f\|_{\mathcal{Q}_p} \approx \sup_{a \in \mathbb{D}} \|(f \circ \varphi_a) * g_p - f(a)\|_{H^2}, \tag{6.5}$$

where

$$g_p(z) = 1 + \sum_{n=1}^{\infty} n^{\frac{1-p}{2}} z^n.$$

Note that (6.5) and (6.4) are the same for $p = 1$.

For $1 \leq q < \infty$ and $0 < \alpha \leq 1$, we say that $f \in H(\mathbb{D})$ is in the mean Lipschitz space $\Lambda(q, \alpha)$ if

$$M_q(r, f') = O((1-r)^{\alpha-1}), \qquad 0 < r < 1. \tag{6.6}$$

A result of R. Aulaskari, D. Stegenga and J. Xiao [ASX1996] and an example of M. Essén and J. Xiao [EX1997] assert that

$$\Lambda\left(q, \frac{1}{q}\right) \subsetneqq \mathcal{Q}_p$$

for $2 < q < \infty$ and $1 - \frac{2}{q} < p$. Furthermore, the lower bound $1 - \frac{1}{q}$ is sharp in the sense that

$$\Lambda\left(q, \frac{1}{q}\right) \not\subset \mathcal{Q}_{1-\frac{1}{q}}$$

for $q > 2$. Some results of R. Aulaskari, D. Girela and H. Wulan [AGW2001] can also be considered as a complement to the above results. Some discussions on the preduals of $\mathcal{Q}_p$ spaces can be found in [AlCP2007-2, WX2003, Xi2006].

6.4 Carleson measures

6.4.1 Classical Carleson measures

In his work on interpolation by bounded analytic functions on $\mathbb{D}$, L. Carleson [Ca1958, Ca1962] obtained the following well-known result. For a positive Borel measure μ on $\mathbb{D}$ and $0 < p < \infty$ the inequality

$$\left(\int_{\mathbb{D}} |f(z)|^p d\mu(z)\right)^{1/p} \leq C\|f\|_{H^p} \tag{6.7}$$

holds for all $f \in H^p$ if and only if there exists a constant $C > 0$ such that

$$\mu(S(I)) \leq C|I| \tag{6.8}$$

for all *Carleson boxes*

$$S(I) = \{z \in \mathbb{D} : z/|z| \in I, 1 - |I| \leq |z| < 1\}.$$

In this case, μ is called a *Carleson measure* on $\mathbb{D}$.

Carleson measures have been developed into a crucial tool in complex analysis, harmonic analysis and partial differential equations. Vanishing Carleson measures first appeared in [Po1980], where S. Power established a connection between Carleson measures and Hankel matrices. A well-known characterization of BMOA in terms of Carleson measures is that $f \in H(\mathbb{D})$ belongs to BMOA if and only if the differential form

$$d\mu_f(z) = |f'(z)|^2(1 - |z|^2)\, dA(z)$$

is a Carleson measure on $\mathbb{D}$; see [Ga1981]. There has been much additional recent progress on Carleson measures.

6.4.2 α-Carleson measures

For $0 < \alpha < \infty$, a positive Borel measure μ defined on $\mathbb{D}$ is said to be an *α-Carleson measure* if

$$\sup\{\mu(S(I))/|I|^\alpha : I \subset \partial\mathbb{D}\} < \infty. \tag{6.9}$$

The early work of H. Wulan [Wul1993], also [ASX1996], gives a useful characterization of α-Carleson measures, namely, μ is α-Carleson on $\mathbb{D}$ if and only if

$$\sup_{a \in \mathbb{D}} \int_{\mathbb{D}} \left(\frac{1 - |a|^2}{|1 - \bar{a}z|^2}\right)^\alpha d\mu(z) < \infty. \tag{6.10}$$

The relationship between $\mathcal{Q}_p$ spaces and α-Carleson measures has been described by R. Aulaskari, D. Stegenga and J. Xiao [ASX1996] as follows. For $0 < p < 1$, a function $f \in H(\mathbb{D})$ belongs to $\mathcal{Q}_p$ if and only if

$$|f'(z)|^2(1-|z|^2)^p dA(z) \tag{6.11}$$

is a p-Carleson measure. R. Aulaskari, M. Novake and R. Zhao [ANZ1998] obtained a version in terms of higher order derivatives; that is, for any positive integer n, (6.11) can be replaced by

$$|f^{(n)}(z)|^2(1-|z|^2)^{2n+p-2} dA(z). \tag{6.12}$$

The above results, together with (6.10), tell us that the Green function $g(z,a)$ in the definition (6.2) can be replaced by the expression $1-|\varphi_a(z)|^2$ since $|f'(z)|^2$ is subharmonic in $\mathbb{D}$. In fact, a more general case is still true. Let $u \geq 0$ be subharmonic in $\mathbb{D}$ and $0 < p < \infty$. Then the following conditions are equivalent.

(i) $\sup_{a\in\mathbb{D}} \int_{\mathbb{D}} u(z)(g(z,a))^p dA(z) < \infty$.

(ii) $\sup_{a\in\mathbb{D}} \int_{\mathbb{D}} u(z)(1-|\varphi_a(z)|^2)^p dA(z) < \infty$.

R. Aulaskari, H. Wulan and R. Zhao [AWZ2000] and [Wul1998] showed by an example that this is not true if u is not subharmonic.

Now let us return to the Carleson inequality (6.7) and review some related results. For $0 < p \leq q < \infty$, P. Duren [Du1969] gave an extension of (6.7) for q/p-Carleson measure μ. D. Luecking [Lu1985] considered the derivatives of H^p functions to describe $q/p+nq$-Carleson measure; that is, μ is a $(q/p+nq)$-Carleson measure if and only if there exists a constant $C > 0$ such that

$$\left(\int_{\mathbb{D}} |f^{(n)}(z)|^q d\mu(z)\right)^{1/q} \leq C\|f\|_{H^p}.$$

for all $f \in H^p$

6.4.3 $\mathscr{D}_p$-Carleson measures

For different function spaces, we can also define the notion of Carleson measures using a condition analagous to (6.7). For example, if $0 < p < \infty$, a positive Borel measure μ on $\mathbb{D}$ is said to be a $\mathscr{D}_p$-Carleson measure if there exists a constant $C > 0$ such that for all $f \in \mathscr{D}_p$

$$\left(\int_{\mathbb{D}} |g(z)|^2 d\mu(z)\right)^{1/2} \leq C\|f\|_{\mathscr{D}_p}. \tag{6.13}$$

For $1 \leq p < \infty$, the notions of $\mathscr{D}_p$-Carleson measures and p-Carleson measures are the same. But for $0 \leq p < 1$ they are different. For any subset E of the unit circle $\partial\mathbb{D}$, let $\tau(E)$ denote the corresponding subset in the interval $[-\pi,\pi]$ obtained by the natural identification of this interval with the line. A remarkable work by D. Stegenga (see [St1980-1]) shows that for $0 < p < 1$ the estimate that $\mu(S(I)) \leq C|I|^p$ for all $S(I)$ is not enough to obtain (6.13) for all $f \in \mathscr{D}_p$. For this case $0 \leq p < 1$, (6.13) is equivalent to

$$\mu\left(\bigcup_{j=1}^{n}(S(I_j))\right) \leq C\Gamma_p\left(\tau\left(\bigcup_{j=1}^{n} I_j\right)\right)$$

whenever $I_1, \ldots, I_n$ are disjoint arcs on $\partial\mathbb{D}$. Here

$$\Gamma_p(I) \approx \begin{cases} |I|^p, & 0 < p < 1, \\ (\log|I|^{-1}))^{-1}, & p = 0. \end{cases}$$

N. Arcozzi, R. Rochberg and E. Sawyer [ARS2000] described these measures by a different condition; that is, a positive Borel measure μ is $\mathscr{D}_p$-Carleson if and only if

$$\sup_{a\in\mathbb{D}} \frac{1}{\mu(S(a))} \int_{\widetilde{S}(a)} \frac{(\mu(S(a)\cap S(z)))^2}{(1-|z|^2)^{2+p}} dA(z) < \infty, \tag{6.14}$$

where

$$\widetilde{S}(a) = \{z \in \mathbb{D} : 1-|z| \leq 2(1-|a|), \quad |\arg(\bar{z}a)| \leq \pi(1-|a|)\}$$

and

$$S(a) = \{z \in \mathbb{D} : 1-|z| \leq 1-|a|, \quad |\arg(\bar{z}a)| \leq \pi(1-|a|)\}.$$

6.4.4 $\mathscr{Q}_p$-Carleson measures

Similar to (6.13) we hope to describe positive Borel measures μ such that

$$\int_{\mathbb{D}} |f'(z)|^2 d\mu(z) \leq C\|f\|^2_{\mathscr{Q}_p}, \qquad f \in \mathscr{Q}_p, \tag{6.15}$$

or

$$\int_{\mathbb{D}} |f(z)|^2 d\mu(z) \leq C\|f\|^2_{\mathscr{Q}_p}, \qquad f \in \mathscr{Q}_p. \tag{6.16}$$

In the case $p > 1$, J. Arazy, S. Fisher and J. Peetre [AFP1985] gave a result on (6.15) for the Bloch-Carleson measure.

Question 4.1 *For $0 < p \leq 1$, if (6.15) or (6.16) holds for all functions $f \in \mathscr{Q}_p$, what geometric conditions must the measure μ satisfy?*

Question 4.1 is still open. The strongest partial result is due to J. Xiao [Xi2008] who gave a local characterization for the question. For a positive Borel measure μ on $\mathbb{D}$ and $0 < p < 1$,

$$\sup_{I\subset\partial\mathbb{D}} |I|^{-p} \int_{S(I)} |f(z)|^2 d\mu(z) \leq C\|f\|^2_{\mathscr{Q}_p}$$

holds for all $f \in \mathcal{Q}_p$ if and only if for all $I \subset \mathbb{T}$ we have

$$\mu(S(I)) \le |I|^p (\log \frac{2}{|I|})^2. \tag{6.17}$$

Returning to p-Carleson measures, we have the following conjecture.

Question 4.2 *For $0 < p \le 1$, a positive Borel measure μ on $\mathbb{D}$ is p-Carleson if and only if there exists a constant $C > 0$ such that*

$$\sup_{a\in\mathbb{D}} \int_{\mathbb{D}} |f(z) - f(a)|^2 \left(\frac{1-|a|^2}{|1-\bar{a}z|^2} \right)^p d\mu(z) \le C\|f\|^2_{\mathcal{Q}_p} \tag{6.18}$$

holds for all $f \in \mathcal{Q}_p$.

6.5 The boundary value characterizations

The boundary value definition of BMOA given in (6.1) is actually equivalent to

$$\|f\|_{\mathrm{BMO}(\mathbb{T})} := \left\{ \sup_{I\subset\mathbb{T}} \frac{1}{|I|^2} \int_I \int_I |f(z) - f(w)|^2 |dz||dw| \right\}^{1/2} < \infty. \tag{6.19}$$

Thus, $\mathcal{Q}_1 = BMOA = BMO(\mathbb{T}) \cap H^2$.

6.5.1 Boundary $\mathcal{Q}_p$ spaces

By using a deep result of D. Stegenga [St1980-1], M. Essén and J. Xiao [EX1997] proved the following theorem. For $0 < p < 1$ and $f \in H^2$, $f \in \mathcal{Q}_p$ if and only if

$$\sup_{I\subset\mathbb{T}} \frac{1}{|I|^p} \int_I \int_I \frac{|f(z) - f(w)|^2}{|z-w|^{2-p}} |dz||dw| < \infty. \tag{6.20}$$

It is not possible to consider the case $p > 1$ for above result since Bloch functions in general have no boundary values.

However, (6.19) and (6.20) lead quite naturally to introduce a boundary counterpart of $\mathcal{Q}_p$ in which the analytic property of functions is not needed. For $p \in (-\infty, \infty)$ define the class $\mathcal{Q}_p(\mathbb{T})$ as the set of all functions $f \in L^2(\mathbb{T})$ satisfying (6.20). It is clear that for a special case this space corresponds with BMO; that is, $\mathcal{Q}_2(\mathbb{T}) = BMO(\mathbb{T})$. For general cases, J. Xiao [Xi2000-1] gave the following properties of $\mathcal{Q}_p(\mathbb{T})$:

- $\mathcal{Q}_p(\mathbb{T})$ is nondecreasing in $p \in (-\infty, \infty)$.
- If $p \in (-\infty, -1]$, then $\mathcal{Q}_p(\mathbb{T})$ contains constants only.

- If $-1 < p_1, p_2 \leq 1$, then $\mathcal{Q}_{p_1}(\mathbb{T}) \neq BMO(\mathbb{T})$ and $\mathcal{Q}_{p_1}(\mathbb{T}) \neq \mathcal{Q}_{p_2}(\mathbb{T})$ for $p_1 \neq p_2$.
- If $p \in (1,\infty)$, then $\mathcal{Q}_p(\mathbb{T}) = BMO(\mathbb{T})$.

There is a close connection between $\mathcal{Q}_p(\mathbb{T})$ and the Sobolev space $\mathcal{L}^2_p(\mathbb{T})$ of all measurable functions on $\mathbb{T}$ for which

$$\|f\|_{\mathcal{L}^2_p(\mathbb{T})} = \left\{ \int_{\mathbb{T}} \int_{\mathbb{T}} \frac{|f(z)-f(w)|^2}{|z-w|^{2-p}} |dz|\,|dw| \right\}^{1/2} < \infty. \tag{6.21}$$

By [Xi2000-1] and [ZB2015] we know that $\mathcal{L}^2_p(\mathbb{T})$ contains constants only for $p \leq -1$ and $\mathcal{L}^2_p(\mathbb{T}) = L^2(\mathbb{T})$ for $p \in [1,2]$. If $p \in (0,\infty)$, then $f \in \mathcal{Q}_p(\mathbb{T})$ if and only if

$$\sup_{a\in\mathbb{D}} \|f \circ \varphi_a\|_{\mathcal{L}^2_p(\mathbb{T})} < \infty.$$

It shows that $\mathcal{Q}_p(\mathbb{T})$ can be identified with the Möbius invariant function space corresponding to the Sobolev space $\mathcal{L}^2_p(\mathbb{T})$. Moreover, if $f \in H(\mathbb{D})$, we have that $f \in \mathcal{Q}_1(\mathbb{T})$ if and only if

$$\sup_{a\in\mathbb{D}} \int_{\mathbb{D}} |f'(z)|^2 (1-|\varphi_a(z)|^2) \log \frac{1}{1-|\varphi_a(z)|^2}\, dA(z) < \infty. \tag{6.22}$$

Comparing the definition of $\mathcal{Q}_1$ and (6.22), we deduce that

$$H(\mathbb{D}) \cap \mathcal{Q}_1(\mathbb{T}) \subsetneqq \mathcal{Q}_1 = \text{BMOA}.$$

A complete description of multipliers between Möbius invariant spaces $\mathcal{Q}^s_p(\mathbb{T})$ and the spectra of multiplier operators on $\mathcal{Q}^s_p(\mathbb{T})$ is given in [BP2016].

6.5.2 Geometric characterizations

Using the logarithmic capacity, W. Hayman and Ch. Pommenrenke [HP1978] gave a sufficient geometric condition, discovered independently by D. Stegenga [St1980-2], on the domain G which contains the image of a function analytic on the unit disc which guarantees that the boundary function belongs to BMO. In the converse direction, they proved that if G does not satisfy this thick complement condition, then universal covering maps from $\mathbb{D}$ onto G do not belong to BMOA. Thus there is no weaker omitted set condition which will always guarantee that $f \in BMOA$. W. Hayman and Ch. Pommenrenke's result [HP1978] is now stated as follows: The domain $G \subset \mathbb{C}$ has the property that every function $f(z)$ analytic in $\mathbb{D}$ with values in G belongs to BMOA if and only if there exist constants R and $\delta > 0$ such that

$$\text{cap}(E \cap \{|w-w_0| \leq R\}) > \delta \quad (w_0 \in G), \tag{6.23}$$

where $E = \mathbb{C} \setminus G$ and cap denotes the logarithmic capacity.

We know that $f \in \mathcal{B}$ if and only if the Riemann surface of f contains no large schlicht disks. This leads to the problem of finding a geometric condition on the Riemann surface of f which is both necessary and sufficient for f to belong to BMOA. Such a condition can been found in [SS1981].

A closed Jordan curve Γ in $\mathbb{C}$ is said to be quasi-conformal if

$$\max_{w\in\Gamma(w_1,w_2)} \frac{|w_2-w|+|w-w_1|}{|w_2-w_1|} \tag{6.24}$$

is bounded on Γ and quasi-smooth if $L(w_1,w_2)/|w_1-w_2|$, $w_1,w_2\in\Gamma$, is bounded on Γ, where $\Gamma(w_1,w_2)$ is the shorter arc of Γ between the points w_1 and w_2 on Γ and $L(w_1,w_2)$ is the length of this arc. Let g be analytic and univalent in $\mathbb{D}$ and $\Gamma=\partial g(\mathbb{D})$ be a Jordan curve. Consider the representations of the form

$$f(z)=b\log g'(z) \quad (b\in\mathbb{C}, b\neq 0).$$

Then (cf. [Po1977])

(i) $f\in BMOA$ if and only if Γ is quasi-smooth;

(ii) $f\in\mathcal{B}$ if and only if Γ is quasi-conformal.

Question 5.1 *Let $p\in(0,1)$. How do we describe $\mathcal{Q}_p$ functions geometrically?*

More specifically, we may ask

Question 5.2 *How to describe $\mathcal{Q}_p$ functions using the logarithmic capacity?*

6.6 $\mathcal{Q}_K$ spaces

We know that for special choices of p, the space $\mathcal{Q}_p$ gives the following: $\mathcal{Q}_0=\mathcal{D}$, $\mathcal{Q}_1=BMOA$, and $\mathcal{Q}_p=\mathcal{B}$ for $1<p<\infty$. Is there a more general structure behind these facts? For example, a natural question is whether or not there exist Möbius invariant function spaces between BMOA and the Bloch space.

6.6.1 Definition

Let $K:[0,\infty)\to[0,\infty)$ be a right-continuous and nondecreasing function. We say that $f\in H(\mathbb{D})$ belongs to the space $\mathcal{Q}_K$ if

$$||f||^2_{\mathcal{Q}_K}=\sup_{a\in\mathbb{D}}\int_{\mathbb{D}}|f'(z)|^2K(g(z,a))\,dA(z)<\infty. \tag{6.25}$$

Modulo constants, $\mathcal{Q}_K$ is a Banach space under the norm defined in (6.25). Also, $\mathcal{Q}_K$ is Möbius invariant. For $0<p<\infty$, $K(t)=t^p$ gives the space $\mathcal{Q}_p$. Choosing $K(t)=1$, we get the Dirichlet space $\mathcal{D}$. Basic properties of $\mathcal{Q}_K$ can be found in [EW2000, EWX2006].

- The space $\mathcal{Q}_K$ is trivial (only constant functions) if and only if

$$\sup_{a\in\mathbb{D}}\int_{\mathbb{D}}\frac{(1-|a|^2)^2}{|1-z\bar{a}|^4}K\left(\log\frac{1}{|z|}\right)dA(z)=\infty. \tag{6.26}$$

- $\mathcal{D}\subset\mathcal{Q}_K\subset\mathcal{B}$ for all increasing functions K if the left side of (6.26) is finite.
- $\mathcal{Q}_K=\mathcal{B}$ if and only if

$$\int_0^1 K(\log(1/r))(1-r^2)^{-2}\,r\,dr<\infty. \tag{6.27}$$

Choosing $K(r)=r^p$, we obtain that (6.27) holds for $p>1$ and that (6.27) fails for $0<p\le 1$. Thus, $\mathcal{Q}_p=\mathcal{B}$ for all $p>1$.

- $\mathcal{Q}_K=\mathcal{D}$ if and only if $K(0)>0$.
- Let $K_1\le K_2$ in $(0,1)$ and assume furthermore that $K_1(r)/K_2(r)\to 0$ as $r\to 0$.

If the integral in (6.27) is divergent for $K=K_2$, then $\mathcal{Q}_{K_2}\subsetneqq\mathcal{Q}_{K_1}$. In fact, this result gives a general method to construct a space between two different spaces $\mathcal{Q}_{K_1}$ and $\mathcal{Q}_{K_2}$. For example, one may consider

$$K_0(t)=\begin{cases}\dfrac{t}{|\log t|}, & 0<t\le 1/e,\\ t, & t>1/e,\end{cases}$$

which satisfies

$$BMOA\subsetneqq\mathcal{Q}_{K_0}\subsetneqq\mathcal{B}.$$

- $\log\frac{1}{1-z}\in\mathcal{Q}_K$ if and only if

$$\int_0^1(1-r^2)^{-1}K(\log(1/r))r\,dr<\infty. \tag{6.28}$$

Question 6.1 *Find necessary and sufficient conditions on K such that* $\mathcal{Q}_K=BMOA$.

- A study on more general $\mathcal{Q}_K$ spaces can be found in [WZh2006].

6.6.2 Univalent functions

For analytic univalent function f in $\mathbb{D}$, Ch. Pommerenke [Po1977] proved that $f\in\mathcal{B}\Longrightarrow f\in BMOA$, which easily implies a result of A. Baernstein about univalent Bloch functions: *If* $g(z)\neq 0$ *is an analytic univalent function in* $\mathbb{D}$, *then* $\log g\in BMOA$; see [Ba1976]. Ch. Pommerenke's result mentioned above was generalized to $\mathcal{Q}_p$ spaces and $\mathcal{Q}_K$; see [ALXZ1997] and [Wul2004].

A function $f\in H(\mathbb{D})$ is said to be q-valent if the equation $f(z)=w$ never has more than q solutions. Let $n(w)=n(w,f)$ be the number of roots of this equation in $\mathbb{D}$, counted according to their multiplicities, and let

$$p(R)=(2\pi)^{-1}\int_0^{2\pi}n(Re^{i\theta})d\theta.$$

If there exists a positive number q such that

$$\int_0^R p(\rho)\, d(\rho^2) \le qR^2$$

or $p(R) \le q$ for all $R > 0$, we say that f is areally mean q-valent or circumferentially mean q-valent, respectively; see [Ha1994]. If f is univalent, then f is areally and circumferentially mean 1-valent. For an areally mean q-valent function f, if

$$\int_0^1 \left(\log \frac{1}{1-r}\right)(1-r)^{-1} K(\log(1/r))\, r dr < \infty, \tag{6.29}$$

then $f \in \mathcal{B}$ if and only if $f \in \mathcal{Q}_K$. Moreover, for a circumferentially mean q-valent function f, if (6.29) holds, then $\log f \in \mathcal{Q}_K$.

Question 6.2 *Does there exist a weaker assumption than (6.29) on K such that $f \in \mathbb{B}$ belongs to $\mathcal{Q}_K$ if we also assume that f is univalent or q-valent?*

6.6.3 Structure of the weight function K

The function theory of $\mathcal{Q}_K$ spaces obviously depends on the properties of K. To summarize, we make the following standing assumptions on the weight function K from this point on, unless otherwise noted.

(a) $K : [0,\infty) \to [0,\infty)$ is non-decreasing.

(b) $K(0) = 0$.

(c) K is right continuous everywhere.

(d) $K(t) > 0$ for all $t > 0$.

Consider an auxiliary function φ_K.

$$\varphi_K(s) = \sup_{0<t\le 1} \frac{K(st)}{K(t)}, \qquad 0 < s < \infty.$$

Note that $\varphi_K(s)$ is always well defined if we allow it to take infinite values. In many situations we will need to consider the following constraints on K:

$$\int_0^1 \varphi_K(s) \frac{ds}{s} < \infty, \tag{6.30}$$

and

$$\int_1^\infty \frac{\varphi_K(s)}{s^{1+\sigma}}\, ds < \infty, \qquad \sigma > 0. \tag{6.31}$$

It is clear that the weight function $K(t) = t^p$ satisfies (6.30) for all $0 < p < \infty$ and it satisfies condition (6.31) whenever $0 < p < \sigma$. We will see that many properties of

$\mathcal{Q}_K$ spaces depend on one or two of these conditions. For example, see [EWX2006].

- Suppose K satisfies (6.30) and the doubling condition on $(0,1)$. Then there exists a weight function K_1, still satisfying all standing assumptions on weights, such that K_1 is comparable with K on $(0,1)$, $\mathcal{Q}_K = \mathcal{Q}_{K_1}$ and the function $t^{-c}K_1(t)$ is increasing on $(0,1)$ for all sufficiently small positive constants c. Conversely, if the function $t^{-c}K(t)$ is increasing on $(0,1)$ for some $c > 0$, then K satisfies condition (6.30).
- Suppose K satisfies (6.31) for some $\sigma > 0$. Then there exists a weight function K_2, still satisfying all the standing assumptions on weights, such that K_2 is comparable with K on $(0,1)$, $\mathcal{Q}_K = \mathcal{Q}_{K_2}$ and the functions $K_2(t)/t^{\sigma}$ and $K_2(t)/t^{\sigma-c}$ are both decreasing on $(0,\infty)$ for all sufficiently small positive constants c. Conversely, if the function $K(t)/t^{\sigma-c}$ is decreasing for some $c > 0$, then K satisfies condition (6.31) for all $\sigma' > \sigma - c$.
- Suppose K satisfies (6.31) for some $0 < \sigma \leq 1$. Then $\mathcal{Q}_K \subset BMOA$.
- If K is convex, then $BMOA \subset \mathcal{Q}_K$; if K is concave, then $\mathcal{Q}_K \subset BMOA$.
- Some properties of functions in $\mathcal{Q}_K$ defined by lacunary series are obtained in [WZ2007-2] and [Zho2016].

6.7 K-Carleson measures

6.7.1 Definition

A very useful tool in the study of $\mathcal{Q}_K$ spaces is the notion of K-Carleson measures. For $|I| < 1$, define the Carleson box by

$$S_G(I) = \begin{cases} \{r\zeta \in G : 1-|I| < r < 1, \zeta \in I\}, & G = \mathbb{D}, \\ \{r\zeta \in G : 1 < r < 1+|I|, \zeta \in I\}, & G = \mathbb{C} \setminus \overline{\mathbb{D}}. \end{cases}$$

If $|I| \geq 1$, we simply set $S_{\mathbb{D}}(I) = \mathbb{D}$ and

$$S_{\mathbb{C}\setminus\overline{\mathbb{D}}}(\mathbb{T}) = \{z \in \mathbb{C} : \ 1 < |z| < 2\}.$$

Following [EWX2006] and [WY2015], a positive Borel measure μ on $G = \mathbb{D}$ or $G = \mathbb{C} \setminus \overline{\mathbb{D}}$ is said to be a K-Carleson measure if

$$\|\mu\|_K = \sup_{I \subset \mathbb{T}} \int_{S_G(I)} K\left(\frac{|1-|z||}{|I|}\right) d\mu(z) < \infty.$$

Clearly, if $K(t) = t^p$, then $d\mu$ is a K-Carleson measure on $\mathbb{D}$ if and only if

$$(1-|z|^2)^p \, d\mu(z)$$

is a p-Carleson measure on $\mathbb{D}$. Let K satisfy (6.30) and the doubling condition. Then

- μ is a K-Carleson measure on $\mathbb{D}$ if and only if

$$\sup_{a\in\mathbb{D}}\int_{\mathbb{D}} K(1-|\varphi_a(z)|^2)d\mu(z)<\infty.$$

- μ is a K-Carleson measure on $\mathbb{C}\setminus\overline{\mathbb{D}}$ if and only if

$$\sup_{a\in\mathbb{D}}\int_{1<|z|<2} K(|\varphi_a(z)|^2-1)d\mu(z)<\infty.$$

By [WZh2007] and [WZ2008] we know that $\mathcal{Q}_K$ spaces can be characterized in terms of higher order derivatives.

- $f\in\mathcal{Q}_K$ if and only if $|f^{(n)}(z)|^2(1-|z|^2)^{2n-2}dA(z)$ is a K-Carleson measure on $\mathbb{D}$.

Fix some $b>1$ and define the α-order derivative of $f\in H(\mathbb{D})$:

$$f^{(\alpha)}(z)=\frac{\Gamma(b+\alpha)}{\pi\Gamma(b)}\int_{\mathbb{D}}\frac{(1-|w|^2)^{b-1}}{(1-\overline{w}z)^{b+\alpha}}\overline{w}^{[\alpha-1]}f'(w)dA(w),\quad b+\alpha>0,$$

where Γ is the gamma function and $[\alpha]$ denotes the smallest integer which is larger than or equal to α. Characterizations of $\mathcal{Q}_K$ spaces can be given in terms of fractional derivatives.

- Let K satisfy conditions (6.30) and (6.31) for some $0<\sigma<2$. Then $f\in\mathcal{Q}_K$ if and only if

$$|f^{(\alpha)}(z)|^2(1-|z|^2)^{2(\alpha-1)}dA(z),\quad \alpha>1/2,$$

is a K-Carleson measure.

- A duality theory of $\mathcal{Q}_K$ spaces was developed in [Zho2013] and [ZC2014].
- Korenblum's inequality for $\mathcal{Q}_K$ spaces can be found in [AS2004].

6.7.2 Decomposition

For $z\in\mathbb{D}$, δ_z is the point measure (the Dirac measure at z) on $\mathbb{D}$ defined by $\delta_z(w)=1$ if $z=w$ and $\delta_z(w)=0$ if $z\neq w$. A positive measure $\sum_{j=1}^{\infty}\delta_{z_j}$ is a K-Carleson measure if

$$\|\sum_{j=1}^{\infty}\delta_{z_j}\|_K=\sup_{I\subset\partial\mathbb{D}}\sum_{z_j\in S(I)}K\left(\frac{1-|z_j|}{|I|}\right)<\infty.$$

H. Wulan and J. Zhou [WZh2014] established decomposition theorems for a general $\mathcal{Q}_K$ space. Let K satisfy conditions (6.30) and (6.31) for some $0<\sigma<2$. For $b>\max\{1,\sigma\}$, there exists a sequence $\{z_j\}_{j=1}^{\infty}$ in $\mathbb{D}$ such that the following are true:

(i) If $\sum_{j=1}^{\infty}|\lambda_j|^2\delta_{z_j}$ is a K-Carleson measure for a complex-valued sequence $\{\lambda_j\}_{j=1}^{\infty}$, then the function

$$f(z)=\sum_{j=1}^{\infty}\lambda_j\left(\frac{1-|z_j|^2}{1-\overline{z_j}z}\right)^b \tag{6.32}$$

belongs to $\mathcal{Q}_K$ with

$$\|f\|_{\mathcal{Q}_K} \lesssim \left\| \sum_{j=1}^{\infty} |\lambda_j|^2 \delta_{z_j} \right\|_K .$$

(ii) If $f \in \mathcal{Q}_K$, then there exists a sequence $\{\lambda_j\}_{j=1}^{\infty}$ such that f admits a decomposition as in (6.32) and the measure $\sum_{j=1}^{\infty} |\lambda_j|^2 \delta_{z_j}$ is a K-Carleson measure with

$$\left\| \sum_{j=1}^{\infty} |\lambda_j|^2 \delta_{z_j} \right\|_K \lesssim \|f\|_{\mathcal{Q}_K}.$$

For certain special choices, the above gives the corresponding results for the Bloch space by R. Rochberg [Ro1985], for BMOA by R. Rochberg and S. Semmes [RS1986], and for $\mathcal{Q}_p$ spaces by Z. Wu and C. Xie [WX2002].

Here we say that a sequence $\{z_k\}$ of distinct points in $\mathbb{D}$ is an interpolating sequence for $\mathcal{Q}_K \cap H^\infty$ if for every sequence $\{v_k\} \in l^\infty$ there exists a function $f \in \mathcal{Q}_K \cap H^\infty$ such that $f(z_k) = v_k$ for all $k \geq 1$. The following result can be found in [Zho2014]. Suppose K satisfies conditions (6.30) and (6.31) for $\sigma = 1$. Then a sequence $\{z_k\}$ of distinct points in $\mathbb{D}$ is an interpolating sequence for $\mathcal{Q}_K \cap H^\infty$ if and only if

$$\delta = \inf_i \prod \left\{ \left| \frac{z_i - z_j}{1 - z_i \bar{z}_j} \right| : j \geq 1, j \neq i \right\} > 0$$

and $\mu = \sum_{k=1}^{\infty} \delta_{z_k}$ is a K-Carleson measure.

An inner function I is an analytic on $\mathbb{D}$ having the following properties: $|I(z)| \leq 1$ for all $z \in \mathbb{D}$ and $|I(e^{i\theta})| = 1$ almost everywhere on $\partial\mathbb{D}$. Of course, any inner function belongs to BMOA. However, this fact is no longer true for $\mathcal{Q}_p$. In the general case, suppose K satisfies (6.30) and (6.31) for $\sigma = 1$. Then the following are equivalent for an inner function I (cf. [EWX2006]):

(i) $I \in \mathcal{Q}_K$.

(ii) I is a multiplier of $\mathcal{Q}_K \cap H^\infty$.

(iii) I is a Blaschke product with zeros $\{z_n\}$ and $\sum \delta_{z_n}$ is a K-Carleson measure.

An outer function has the form

$$O_\psi(z) = \exp\left(\int_{\partial\mathbb{D}} \frac{\zeta + z}{\zeta - z} \log \psi(\zeta) \frac{|d\zeta|}{2\pi} \right),$$

where $\psi \geq 0$ a.e. on $\partial\mathbb{D}$ and $\log \psi \in L^1(\partial\mathbb{D})$. An inner-outer decomposition of $\mathcal{Q}_K$ functions was given in [EWX2006] as follows: Let K satisfy (6.30) and (6.31) for $\sigma = 1$. Let $f \in H^2$ with $f \not\equiv 0$. Then $f \in \mathcal{Q}_K$ if and only if $f = IO$, where I is an inner function, O is an outer function in $\mathcal{Q}_K$, and

$$\sup_{a \in \mathbb{D}} \int_{\mathbb{D}} |O(z)|^2 (1 - |I(z)|^2) \frac{K(1 - |\varphi_a(z)|^2)}{(1 - |z|^2)^2} \, dA(z) < \infty.$$

This result can also be used to construct outer functions in $\mathcal{Q}_K$ and to represent every $\mathcal{Q}_K$-function as the ratio of two bounded functions in $\mathcal{Q}_K$. That is, suppose K satisfies conditions (6.30) and (6.31) for $\sigma = 1$. If $f \in \mathcal{Q}_K$, then there are functions f_1 and f_2 in $H^\infty \cap \mathcal{Q}_K$ such that $f = f_1/f_2$.

6.7.3 Derivative-free characterizations

Using the mean oscillation of a function in the Bergman metric, H. Wulan and K. Zhu [WZ2007-1] gave several characterizations of $\mathcal{Q}_K$ that are free of the use of derivatives.

- If K satisfies (6.31) for $\sigma = 2$, then $f \in H(\mathbb{D})$ belongs to $\mathcal{Q}_K$ if and only if

$$\sup_{a\in\mathbb{D}} \int_{\mathbb{D}} \int_{\mathbb{D}} \frac{|f(z) - f(w)|^2}{|1 - z\overline{w}|^4} K(1 - |\varphi_a(z)|^2)\, dA(z)dA(w) < \infty.$$

Fix a positive radius R and denote by

$$A_R(f)(z) = \frac{1}{|D(z,R)|} \int_{D(z,R)} f(w)\, dA(w)$$

the average of f over the Bergman metric ball $D(z,R)$. Define

$$MO_R(f)(z) = \left[\frac{1}{|D(z,R)|} \int_{D(z,R)} |f(w) - A_R(f)(z)|^2\, dA(w)\right]^{\frac{1}{2}}$$

the mean oscillation of f in the Bergman metric at the point z. It is easy to verify that

$$[MO_R(f)(z)]^2 = A_R(|f|^2)(z) - |A_R(f)(z)|^2.$$

- If K satisfies (6.31) for $\sigma = 2$, then $f \in H(\mathbb{D})$ belongs to $\mathcal{Q}_K$ if and only if

$$\sup_{a\in\mathbb{D}} \int_{\mathbb{D}} [MO_R(f)(z)]^2 K(1 - |\varphi_a(z)|^2) \frac{dA(z)}{(1 - |z|^2)^2} < \infty, \quad R > 0.$$

To state a result about membership in $\mathcal{Q}_K$ in terms of the absolute values of a function in $\mathbb{D}$, we use the following family of measures based on the Poisson kernel:

$$d\mu_z(\zeta) = \frac{1 - |z|^2}{2\pi|\zeta - z|^2}|d\zeta|, \quad z \in \mathbb{D}, \quad \zeta \in \mathbb{T}.$$

- If K satisfies (6.30) and (6.31) for $\sigma = 1$, then $f \in H(\mathbb{D})$ belongs to $\mathcal{Q}_K$ if and only if

$$\sup_{a\in\mathbb{D}} \int_{\mathbb{D}} \left(\int_{\mathbb{T}} |f|^2\, d\mu_z - |f(z)|^2\right) \frac{K(1 - |\varphi_a(z)|^2)}{(1 - |z|^2)^2}\, dA(z) < \infty.$$

It is clear that $f \in \mathcal{Q}_K(\mathbb{T}) \Rightarrow |f| \in \mathcal{Q}_K(\mathbb{T})$, but the converse is not always true. G. Bao, Z. Lou, R. Qian and H. Wulan [BLQW2016] investigated the effect of absolute values on the behavior of functions f in the spaces $\mathcal{Q}_K$. They found that a condition involving the modulus of the function f in the Hardy space H^2 together with $|f| \in \mathcal{Q}_K(\mathbb{T})$ is equivalent to $f \in \mathcal{Q}_K$. As an application, a new criterion for inner-outer factorization of $\mathcal{Q}_K$ spaces is given.

6.7.4 Multiplication operators

Let X be a Banach space of analytic functions in $\mathbb{D}$. A function $g \in H(\mathbb{D})$ is said to be a multiplier of X if the multiplication operator

$$M_g(f)(z) = g(z)f(z), \qquad f \in X,$$

is bounded on X. The space of all multipliers of X is denoted by $M(X)$. Multiplication operators are closely related to integration operators J_g and I_g, induced by symbols $g \in H(\mathbb{D})$:

$$J_g(f)(z) = \int_0^z f(w)g'(w)dw, \qquad z \in \mathbb{D},$$

and

$$I_g(f)(z) = \int_0^z f'(w)g(w)dw, \qquad z \in \mathbb{D}, f \in H(\mathbb{D}).$$

The relationship

$$J_g(f)(z) = M_g(f)(z) - f(0)g(0) - I_g(f)(z) \tag{6.33}$$

essentially says that if g is a symbol for which two of the operators J_g, I_g, M_g are bounded on X then so is the third. It also says that it is possible for two of the operators to be unbounded but the third is bounded due to cancellation.

6.7.5 The space of multipliers

The space of multipliers is known for several of the classical spaces such as Hardy and Bergman spaces. In particular, for $H^2 = \mathscr{D}_1$ the space of multipliers $M(H^2) = H^\infty$, the algebra of bounded analytic functions. For other Dirichlet spaces $\mathscr{D}_p, p \in (0,1)$, the situation is more complicated.

It is convenient at this point to use the space $\mathscr{W}_p$ of functions $g \in H(\mathbb{D})$ such that the measure

$$d\mu_g(z) = |g'(z)|^2(1-|z|^2)^p\, dA(z)$$

is a $\mathscr{D}_p$-Carleson measure. This space were introduced by R. Rochberg and Z. Wu [RW1993]. The multipliers of $\mathscr{D}_p$ were described in [St1980-1] as follows.

$$M(\mathscr{D}_p) = H^\infty \cap \mathscr{W}_p.$$

On the other hand, the multipliers of $\mathscr{Q}_p$ are completely described in [PP2009-2] and [XX2014] as follows.

$$M(\mathscr{Q}_p) = H^\infty \cap \mathscr{Q}_{p,\log}, \qquad 0 < p < 1,$$

where $\mathscr{Q}_{p,\log}$ denotes the space of functions that satisfy

$$\sup_{I \subset \mathbb{T}} \frac{(\log \frac{1}{|I|})^2}{|I|^p} \int_{S(I)} |g'(z)|^2(1-|z|^2)^p\, dA(z) < \infty.$$

It is not difficult to check that

$$\mathscr{Q}_{p,\log} \subset \mathscr{W}_p \subset \mathscr{Q}_p, \qquad 0 < p < 1.$$

Question 7.1 *What is $M(\mathcal{Q}_K)$?*

6.7.6 $\mathcal{Q}_K$-Teichmüller spaces

The Schwarzian derivative of a locally univalent function f is

$$S_f = \left(\frac{f''}{f'}\right)' - \frac{1}{2}\left(\frac{f''}{f'}\right)^2.$$

The universal Teichmüller space, denoted by $T(1)$, is defined as the set of all functions $g = \log f'$ in the unit disk $\mathbb{D}$, where f is conformal in $\mathbb{D}$ and has a quasiconformal extension to the complex plane $\mathbb{C}$. It is well known that the universal Teichmüller space $T(1)$ is the interior of S in $\mathscr{B}$. As a bridge between the space of univalent functions and general Teichmüller spaces, $T(1)$ is the simplest Teichmüller space.

The BMO-Teichmüller space, denoted by $\mathscr{T}$, is the subset of $T(1)$ consisting of all functions $g = \log f' \in T(1)$ such that

$$|\mu_f(z)|^2(|z|^2-1)^{-1}\,dA(z)$$

is a Carleson measure on $\mathbb{C}\setminus\overline{\mathbb{D}}$. Note that the BMO-Teichmüller space $\mathscr{T}$ is also a subset of BMOA.

The theory of Teichmüller spaces including $T(1)$ was developed by Ahlfors and Bers [Ah1966, AB1960], Gehring [Ge1977], Astala-Gehring [AG1986], and Lehto [Le1987]. The BMO-Teichmüller space was introduced by Astala and Zinsmeister [AZ1991]. Some results on the BMO-Teichmüller space can be found in [SW1999] and [WS2014]. The more general $\mathcal{Q}_K$-Teichmüller spaces were introduced and investigated in [WY2014] and [Zho2012].

For a given weight function K, we consider the subset $\mathscr{T}_K$ of $T(1)$ consisting of all functions $g = \log f' \in T(1)$ such that

$$|\mu_f(z)|^2(|z|^2-1)^{-2}\,dA(z)$$

is a K-Carleson measure on $\mathbb{C}\setminus\overline{\mathbb{D}}$.

- Suppose K satisfies (6.30) and the doubling condition. Then $\mathscr{T}_K = T(1)$ if and only if

$$\int_0^1 \frac{K(t)}{t^2}\,dt < \infty.$$

- Suppose K satisfies (6.30) and the doubling condition. If f is univalent on $\mathbb{D}$, then $\log f' \in \mathcal{Q}_K$ if and only if

$$|S_f(z)|^2(1-|z|^2)^2\,dA(z)$$

is a K-Carleson measure on $\mathbb{D}$.

- Suppose K satisfies (6.30) and (6.31) for some $0 < \sigma < 2$. Then $\mathscr{T}_K \subset \mathcal{Q}_K$.

6.7.7 Hadamard products

This subsection gives a characterization of functions in $\mathcal{Q}_K$ spaces in terms of the Hadamard products. J. Anderson, J. Clunie and Ch. Pommerenke [ACP1974] proved that if $f, g \in \mathcal{B}$, then $f * g \in \mathcal{B}$ and

$$\|f * g\|_{\mathcal{B}} \leq C\|f\|_{\mathcal{B}}\|g\|_{\mathcal{B}}.$$

Using a result in [MP1990] and Fefferman's duality theorem $\left(H^1\right)^* = BMOA$, we have that if $f \in BMOA$ and $g \in \mathcal{B}$ then $f * g \in BMOA$ and

$$\|f * g\|_{BMOA} \leq C\|f\|_{BMOA}\|g\|_{\mathcal{B}}.$$

For the space $\mathcal{Q}_p, 0 < p < 1$, R. Aulaskari, D. Girela and H. Wulan [AGW2001] obtained that if

$$f(z) = \sum_{n=0}^{\infty} a_n z^n \in \mathcal{Q}_p$$

with $a_n \geq 0$ for all n and $g \in \mathcal{B}$, then $f * g \in \mathcal{Q}_p$ and

$$\|f * g\|_{\mathcal{Q}_p} \leq C\|f\|_{\mathcal{Q}_p}\|g\|_{\mathcal{B}}.$$

M. Pavlović [Pa2005] proved the preceding result without any more assumption. Moreover, H. Wulan and Y. Zhang [WZha2008] showed a similar result: Let K satisfy (6.31) for $\sigma = 1$. If $f \in \mathcal{Q}_K$ and $g \in \mathcal{B}$, then $f * g \in \mathcal{Q}_K$ and

$$\|f * g\|_{\mathcal{Q}_K} \leq C\|f\|_{\mathcal{Q}_K}\|g\|_{\mathcal{B}}.$$

6.7.8 Morrey K-spaces

The $\mathcal{Q}_K$ spaces represent a natural generalization of the classical function space BMO. Now we consider another class of function spaces which are also natural generalizations of BMO. Thus for any weight function K we let H_K^2 (called Morrey K-spaces) denote the space of all functions f in the Hardy space H^2 such that

$$\|f\|_{H_K^2} = \left(\sup_{I \subset \mathbb{T}} \frac{1}{K(|I|)} \int_I |f(\zeta) - f_I|^2 |d\zeta|\right)^{\frac{1}{2}} < \infty, f_I = \frac{1}{|I|}\int_I f(\zeta)\,|d\zeta|.$$

When $K(t) = t^p$, the resulting Morrey K-spaces H_K^2 were characterized in [WX2002] in terms of p-Carleson measures. H. Wulan and J. Zhou [WZh2013] gave a simple relationship between $\mathcal{Q}_K$ and H_K^2. Let K satisfy conditions (6.30) and (6.31) for some $0 < \sigma < 2$. If $q \in (0, \sigma]$ is sufficiently close to σ, then $f \in \mathcal{Q}_K$ implies that the fractional derivative $f^{(\frac{1-q}{2})}$ belongs to H_K^2. Conversely, there exists some $q \in (0, \sigma)$ such that $f \in H_K^2$ implies $f^{(\frac{q-1}{2})} \in \mathcal{Q}_K$.

6.8 Boundary $\mathcal{Q}_K$ spaces

Corresponding to $\mathcal{Q}_K$ spaces on $\mathbb{D}$, $f \in L^2$ belongs to $\mathcal{Q}_K(\partial\mathbb{D})$, the boundary $\mathcal{Q}_K$ spaces, if

$$||f||^2_{\mathcal{Q}_K(\partial\mathbb{D})} = \sup_{I\subset\partial\mathbb{D}} \int_I\int_I \frac{|f(\zeta)-f(\eta)|^2}{|\zeta-\eta|^2} K\left(\frac{|\zeta-\eta|}{|I|}\right) |d\zeta||d\eta| < \infty. \tag{6.34}$$

Let $\hat{f}$ denote the Poisson extension of f; that is

$$\hat{f}(z) = \frac{1}{2\pi}\int_0^{2\pi} f(e^{i\theta})P_z(\theta)d\theta,\, P_z(\theta) = \frac{1-|z|^2}{|e^{i\theta}-z|^2}.$$

The following result in [EWX2006] shows that $f \in H^2$, under some assumptions for K, belongs to $\mathcal{Q}_K$ if and only if its boundary function belongs to the boundary $\mathcal{Q}_K$ space. Meanwhile, $f \in L^2(\mathbb{T})$ belongs to $\mathcal{Q}_K(\mathbb{T})$ if and only if $|\nabla \hat{f}(z)|^2 dA(z)$ is a K-Carleson measure.

Using P. Jones's constructive solution of the $\bar{\partial}$-equation [Jo1983], A. Nicolau and J. Xiao [NX1997] proved the following result: for $p \in (0,1)$ if $|g(z)|^2(1-|z|^2)^p dA(z)$ is a p-Carleson measure, then there exists a function f defined on $\bar{\mathbb{D}} = \mathbb{D}\cup\mathbb{T}$ such that $\bar{\partial}f = g$ on $\mathbb{D}$ and $f \in \mathcal{Q}_p(\mathbb{T})\cap L^\infty(\mathbb{T})$.

6.8.1 Fefferman-Stein decomposition

The well-known Fefferman-Stein decomposition theorem [FS1972] states that an arbitrary BMO function can be represented as the sum of an L^∞ function and the conjugate of another such function. An analogue of this theorem in the context of $\mathcal{Q}_p$ spaces, due to Nicolau and Xiao [NX1997], can be stated as follows: for $p \in (0,1)$ $f \in H^2$ belongs to $\mathcal{Q}_p$ if and only if f can be written as f_1+f_2, where $f_1, f_2 \in H(\mathbb{D})$ and $\Re f_1, \Re f_2 \in L^\infty(\mathbb{T})\cap\mathcal{Q}_p(\mathbb{T})$. A similar result (cf. [Pa2008]) for the boundary $\mathcal{Q}_K$ spaces can be written as follows. Suppose K satisfies (6.30) and (6.31) for $\sigma = 1$. Then a function $f \in L^2(\mathbb{T})$ belongs to $\mathcal{Q}_K(\mathbb{T})$ if and only if $f = u+\tilde{v}$ where $u, v \in \mathcal{Q}_K(\mathbb{T})\cap L^\infty(\mathbb{T})$.

6.8.2 The corona theorem

I would like to mention here a little more about the paper [NX1997] because it provides an important contribution to the recently initiated study of $\mathcal{Q}_p$ spaces. The celebrated corona theorem of L. Carleson [Ca1962] states that the maximal ideals formed by functions which vanish at a prescribed point in the disc are dense in the maximal ideal space of H^∞. This can also be reformulated as follows: given functions $f_1, f_2, \cdots, f_n$ in H^∞ such that

$$|f_1(z)|+|f_2(z)|+\cdots+|f_n(z)| \geq \delta$$

for some $\delta > 0$, there exist functions $g_1, g_2, \cdots, g_n$ in H^∞ such that

$$f_1 g_1 + f_2 g_2 + \cdots + f_n g_n \equiv 1.$$

The authors of [NX1997] proved an analogous theorem which is obtained by replacing H^∞ by $\mathcal{Q}_p \cap H^\infty$ everywhere in the above formulation.

We show that the corona theorem holds for the algebras $\mathcal{Q}_K \cap H^\infty$. It means that the unit disk $\mathbb{D}$ is dense in the maximal ideal spaces of $\mathcal{Q}_K \cap H^\infty$.

- Suppose K satisfies (6.30) and (6.31) for $\sigma = 1$. Let $f_1, \cdots, f_n \in \mathcal{Q}_K \cap H^\infty$ with

$$\inf_{z \in \mathbb{D}} \sum_{k=1}^{n} |f_k(z)| > 0.$$

Then there exist $g_1, \cdots, g_n \in \mathcal{Q}_K \cap H^\infty$ such that

$$f_1 g_1 + \cdots + f_n g_n = 1$$

on $\mathbb{D}$.

6.8.3 $\bar{\partial}$-problem

For a classical Carleson measure μ on $\mathbb{D}$, it is well known (see [Ga1981]) that the $\bar{\partial}$-problem $\bar{\partial} F = \mu$ has a solution F, in the sense of distributions, satisfying $\|F\|_{L^\infty(\mathbb{T})} \le C\|\mu\|_1$. Furthermore, it was shown in [Jo1983] that such a solution F can be given by the following simple formula,

$$F(z) = \int_{\mathbb{D}} K_\mu(z, \zeta)\, d\mu(\zeta),$$

where

$$K_\mu(z, \zeta) = \frac{1 - |\zeta|^2}{\pi(1 - \bar{\zeta} z)(z - \zeta)} \exp\left\{ \int_{|w| \ge |\zeta|} \left(\frac{1 + \bar{w}\zeta}{1 - \bar{w}\zeta} - \frac{1 + \bar{w}z}{1 - \bar{w}z} \right) \frac{d\mu(w)}{\|\mu\|_1} \right\}$$

and

$$\int_{\mathbb{D}} |K_\mu(e^{i\theta}, \zeta)|\, d\mu(\zeta) \le C_1 \|\mu\|_1.$$

In particular, if $|g(z)|\, dA(z)$ is a 1-Carleson measure, then the equation $\bar{\partial} F = g$ has a solution $F \in L^\infty(\mathbb{T})$. A similar result for the spaces $\mathcal{Q}_K(\mathbb{T}) \bigcap L^\infty(\mathbb{T})$ can be found in [Pa2008]. Suppose K satisfies (6.30) and (6.31) for $\sigma = 1$. If $d\lambda(z) = |g(z)|^2\, dA(z)$ is a K-Carleson measure, then there is a function f on $\overline{\mathbb{D}}$ with boundary values in $\mathcal{Q}_K(\mathbb{T}) \cap L^\infty(\mathbb{T})$ such that $\bar{\partial} f(z) = g(z)$ for $z \in \mathbb{D}$. Furthermore,

$$\|f\|_{\mathcal{Q}_K(\mathbb{T})} + \|f\|_{L^\infty(\mathbb{T})} \lesssim \|\lambda\|_K^{1/2}.$$

6.8.4 $\mathcal{Q}_K$ spaces of several real variables

Concerning the dyadic BMO [GJ1982], the author of [Ja1999] considers the space $\mathcal{Q}_p^d(\mathbb{T})$, the dyadic counterpart of $\mathcal{Q}_p(\mathbb{T})$. Motivated by the well-known work of C. Fefferman, the three authors Essén, Janson, Peng and Xiao [EJPX2000] defined an analogue of $\mathcal{Q}_p(\mathbb{T})$ on $\mathbb{R}^n$, $n \geq 1$, with extension to half spaces $\mathbb{R} \times \mathbb{R}_+^n$. They extended the standard Carleson characterization of BMO in terms of wavelet coefficients to the full range of $\mathcal{Q}_p(\mathbb{R}^n)$ spaces and gave a number of properties of $\mathcal{Q}_p(\mathbb{R}^n)$. G. Bao and H. Wulan [BW2015] gave a series of results about $\mathcal{Q}_K(\mathbb{R}^n)$ of several real variables.

6.9 Composition operators on $\mathcal{Q}_p$ and $\mathcal{Q}_K$ spaces

6.9.1 Compactness of composition operators on BMOA

Let φ be an analytic self-map of the unit disk $\mathbb{D}$. The equation $C_\varphi(f) = f \circ \varphi$ defines a composition operator C_φ on the space of holomorphic functions in $\mathbb{D}$. A fundamental problem in the study of composition operators is to determine when C_φ maps boundedly or compactly between various Banach spaces in $H(\mathbb{D})$ and the study of such operators has attracted a lot of attention (cf. [Sh1993]). Related spaces include the Hardy space, the Bergman space, the Bloch space, the Dirichlet space, the Besov space, BMOA and so on. Composition operators played a key role in De Branges's renowned proof of the Bieberbach conjectures [Br1985, Br1987]. The first paper considering composition operator C_φ on $\mathcal{Q}_p$ spaces is [SZ1997] in which the boundedness criteria of composition operators from $\mathcal{B}$ to $\mathcal{Q}_p$ are shown. J. Xiao [Xi2000-2] provided a function-theoretic characterization of when the operator $C_\varphi : \mathcal{B} \to \mathcal{Q}_p$ is compact. It should be mentioned that the notion of Carleson measures and its various generalizations are important tools in the study of composition operators.

It is not difficult to see that every composition operator C_φ is bounded on BMOA, as a consequence of Littlewood's Inequality. Significant progress on the compactness of composition operators C_φ on BMOA were made by P. Bourdon, J. Cima and A. Matheson [BCM1999] and W. Smith [Smi1999]. The first result characterized compactness of the composition operators C_φ on BMOA via a Carleson measure condition: for every $\varepsilon > 0$ there is an $r, 0 < r < 1$, such that

$$\int_{R(I)} 1_{\Omega_r}(z)(1-|z|^2)|f'(\varphi(z))|^2|\varphi'(z)|^2 dm(z) \leq \varepsilon |I| \tag{6.35}$$

for every arc $I \subset \partial\mathbb{D}$ and every $f \in BMOA$ with $\|f\|_{BMOA} \leq 1$, where $R(I)$ is a Carleson box.

Z. Lou [Lo2001] extends the work of Bourdon, Cima and Matheson [BCM1999] to more general spaces such as $\mathcal{Q}_p$ spaces. However, these compactness conditions not only depend on the map φ but also contain a requirement that all functions in these spaces satisfy an integral condition.

W. Smith considered in [Smi1999] the same problem by studying the Nevanlinna counting function of an analytic self-map φ of $\mathbb{D}$. In his conditions below Smith drops the requirement for all functions in BMOA corresponding to (6.35) and his compact characterization involves only the symbol φ of the operator as the following:

(i)

$$\lim_{|\varphi(a)|\to 1} \sup_{0<|w|<1} |w|^2 N(\sigma_{\varphi(a)} \circ \varphi \circ \sigma_a, w) = 0; \tag{6.36}$$

(ii) For all $0 < R < 1$,

$$\lim_{t\to 1} \sup_{\{a:|\varphi(a)|\le R\}} m(\sigma_a(E(\varphi,t))) = 0. \tag{6.37}$$

Here

$$E(\varphi,t) = \left\{e^{i\theta} : |\varphi(e^{i\theta})| > t\right\}, \quad 0 < t < 1,$$

and σ_a is a Möbius transformation of $\mathbb{D}$.

We know that compactness of C_φ on BMOA implies compactness on H^2. Thus, conditions (i) and (ii) above must imply that

$$\lim_{|w|\to 1} \frac{N(\varphi,w)}{\log(1/|w|)} = 0,$$

which is a compactness criterion for C_φ on H^2; see [Sh1987]. However, it is not easy to see this fact from Smith's conditions just as W. Smith mentioned in his paper [Smi1999] that he does not know a direct proof of this.

The following result [Wul2007] provides a new and simple characterization depending only on φ for compactness of C_φ on BMOA and from which a direct proof of the fact that compactness of C_φ on BMOA implies its compactness on H^2 is easily obtained. This is, conditions (6.36) and (6.37) can be replaced by the following (i) and (ii), or (i) and (iii).

(i)

$$\lim_{n\to\infty} \|\varphi^n\|_* = 0.$$

(ii)

$$\lim_{r\to 1} \sup_{\lambda\in\mathbb{D}} \int_{|\varphi(z)|>r} \frac{|\varphi'(z)|^2}{|1-\bar{a}\varphi(z)|^2} (1-|\sigma_\lambda(z)|^2)\,dA(z) = 0$$

uniformly for $a \in \mathbb{D}$.

(iii)

$$\lim_{|a|\to 1} \|C_\varphi \sigma_a\|_* = 0.$$

H. Wulan, D. Zheng and K. Zhu [WZZ2009] showed that condition (i) above alone is enough to be a necessary and sufficient condition for C_φ to be compact on BMOA. J. Laitila, E. Nieminen, E. Saksman and O. Tylli [LNST2013] proved that (ii) or (iii) is also a necessary and sufficient condition for C_φ to be compact on BMOA.

6.9.2 Composition operators on $\mathcal{Q}_p$ and $\mathcal{Q}_K$

We know nothing about boundedness or compactness of composition operators on $\mathcal{Q}_p$ spaces for $0 < p < 1$, despite the effort by several mathematicians in recent years.

Question 9.1 *Give a function-theoretic characterization of φ such that the composition operator C_φ on $\mathcal{Q}_p$, $0 < p < 1$, is bounded or compact.*

Question 9.2 *Give a function-theoretic characterization of φ such that the composition operator C_φ on $\mathcal{Q}_K$ is bounded or compact.*

The final result [LiW2007, Wul2005] stated in this paper is a characterization of compact composition operators $C_\varphi : \mathcal{Q}_{K_1} \to \mathcal{Q}_{K_2}$, which is an analogy of the result by P. Bourdon, J. Cima and A. Matheson [BCM1999] for composition operator $C_\varphi : \mathcal{Q}_{K_1} \to \mathcal{Q}_{K_2}$ as follows. Let φ be an analytic self-map of $\mathbb{D}$ and $C_\varphi(\mathcal{Q}_{K_1}) \subset \mathcal{Q}_{K_2}$. Then $C_\varphi : \mathcal{Q}_{K_1} \to \mathcal{Q}_{K_2}$ is compact if and only if

$$\lim_{t\to 1} \sup_{a\in\mathbb{D}, f\in\mathbb{B}_{\mathcal{Q}_{K_1}}} \int_{|\varphi(z)|>t} |f'(\varphi(z))\varphi'(z)|^2 K_2(g(z,a))dA(z) = 0.$$

Question 9.3 *Give a function-theoretic characterization of φ such that the composition operator C_φ from $\mathcal{Q}_{K_1}$ to $\mathcal{Q}_{K_2}$ is bounded or compact.*

Acknowledgment

The author's research is supported by the National Natural Science Foundation of China (Grant Number 11720101003).

References

[Ah1966] L. Ahlfors, *Lectures on Quasi-conformal Mappings*, Princeton: Van Nostrand, 1966.

[AB1960] L. Ahlfors and L. Bers, Riemann's mapping theorem for variable metrics, *Ann. Math.* **72** (1960), 385-404.

[AG1986] K. Astala and F. Gehring, Injectivity, the BMO norm, and the universal Teichmüller space, *J. Anal. Math.* **46** (1986), 16-57.

[AZ1991] K. Astala and M. Zinsmeister, Teichmüller spaces and BMOA, *Math. Ann.* **289** (1991), 613-625.

[AlCP2007-1] A. Aleman, M. Carleson, and A. Persson, Preduals of $\mathcal{Q}_p$ spaces, *Complex Var. Elliptic Equ.* **52** (2007), 605-628.

[AlCP2007-2] A. Aleman, M. Carleson, and A. Persson, Preduals of $\mathcal{Q}_p$ spaces II, Carleson embeddings and atomic decompositions, *Complex Var. Elliptic Equ.* **52** (2007), 629-653.

[AS2004] A. Aleman and A. Simbotin, Estimates in Möbius invariant spaces of analytic functions, *Complex Variables* **49** (2004), 487-510.

[ACP1974] J. Anderson, J. Clunie, and Ch. Pommerenke, On Bloch functions and normal functions, *J. Reine Angew. Math.* **270** (1974), 12-37.

[AFP1985] J. Arazy, S. Fisher, and J. Peetre, Möbius invariant function spaces, *J. Reine Angew. Math.* **363** (1985), 110-145.

[ARS2000] N. Arcozzi, R. Rochberg and E. Sawyer, Carleson measures for analytic Besov spaces, *Rev. Mat. Iberoamericana* **18** (2002), 443-510.

[AGW2001] R. Aulaskari, D. Girela, and H. Wulan, Taylor coefficients and mean growth of the derivative of $\mathcal{Q}_p$ functions, *J. Math. Anal. Appl.* **258** (2001), 415-428.

[AGW2000] R. Aulaskari, D. Girela and H. Wulan, $\mathcal{Q}_p$ spaces, Hadamand products and Carleson measures, *Math. Reports*, **52** (2000), 421-430.

[AL1994] R. Aulaskari and P. Lappan, Criteria for an analytic function to be Bloch and a harmonic or meromorphic function to be normal, in *Complex analysis and its applications*, Pitman Research Notes in Mathematics **305**, Longman Scientific & Technical, Harlow, 1994, 136-146.

[ALXZ1997] R. Aulaskari, P. Lappan, J. Xiao and R. Zhao, On α-Bloch spaces and multipliers of Dirichlet spaces, *J. Math. Anal. Appl.* **209** (1997), 103-121.

[ANZ1998] R. Aulaskari, M. Novake, and R. Zhao, The *n*th derivative characterization of the Möbius bounded Dirichlet space, *Bull. Austral. Math. Soc.* **58** (1998), 43-56.

[ASX1996] R. Aulaskari, D. Stegenga, and J. Xiao, Some subclasses of BMOA and their characterization in terms of Carleson measures, *Rocky Mountain J. Math.* **26** (1996), 485-506.

[AXZ1995] R. Aulaskari, J. Xiao, and R. Zhao, On subspaces and subsets of BMOA and UBC, *Analysis* **15** (1995), 101-121.

[AWZ2000] R. Aulaskari, H. Wulan and R. Zhao, Carleson measure and some classes of meromorphic functions, *Proc. Amer. Math. Soc.* **128** (2000), 2329-2335.

[Ax1986] S. Axler, The Bergman spaces, the Bloch space, and commutators of multiplication operators, *Duke Math. J.* **53** (1986), 315-332.

[Ba1976] A. Baernstein II, Univalence and bounded mean oscillation, *Michigan Math. J.* **23** (1976), 217-223.

[Ba1980] A. Baernstein II, Analytic functions of bounded mean oscillation, in *Aspects of Contemporary Complex Analysis*, Academic Press, London, 1980, 3-36.

[Br1985] L. de Branges, A proof of the Bieberbach conjecture, *Acta Math.* **154**(1985), 137-152.

[Br1987] L. de Branges, Underlying concepts in the proof of the Bieberbach conjecture, *Proceedings of the International Congress of Mathematicians (Berkeley, 1986), Amer. Math. Soc., Providence*, (1987), 25-42.

[BLQW2016] G. Bao, Z. Lou, R. Qian, and H. Wulan, On absolute values of $\mathcal{Q}_K$ functions, *Bull. Korean Math. Soc.* **53** (2016), 561-568.

[BP2016] G. Bao and J. Pau, Boundary multipliers of a family of Möbius invariant function spaces, *Ann. Acad. Sci. Fenn. Math.* **41** (2016), 199-220.

[BW2015] G. Bao and H. Wulan, John-Nirenberg type inequality and wavelet characterization for $\mathcal{Q}_K(\mathcal{R}^n)$ spaces, *Sci. China* **45** (2015), 1833-1846.

[BCM1999] P. Bourdon, J. Cima, and A. Matheson, Compact composition operators on BMOA, *Trans. Amer. Math. Soc.* **351** (1999), 2183-2196.

[Ca1958] L. Carleson, An interpolation problem for bounded analytic functions, *Amer. J. Math.* **80** (1958), 921-930.

[Ca1962] L. Carleson, Interpolations by bounded analytic functions and the corona problem, *Ann. Math.* **76** (1962), 547-559.

[Ca1980] L. Carleson, BMO-10 years' development, 18th Scandinavian Congress of Mathematicians (Aarhus, 1980), Progr. Math., 11, Birkhäuser, 3-21.

[Du1970] P. Duren, *Theory of H^p Spaces*, Academic Press, New York and London, 1970.

[Du1969] P. Duren, Extension of a theorem of Carleson's, *Bull. Amer. Math.* **75** (1969), 143-146.

[EW2000] M. Essén and H. Wulan, On analytic and meromorphic functions and spaces of $\mathcal{Q}_K$ type, *Uppsala University, Department of Mathematics Report* **32** (2000), 1-26. *Illinois J. Math.* **46** (2002), 1233-1258.

[EJPX2000] M. Essén, S. Janson, L. Peng and J. Xiao, *Q*-spaces of several real variables, *Indiana Univ. Math. J.*, **49** (2000), 575-615.

[EWX2006] M. Essén, H. Wulan, and J. Xiao, Function-theoretic aspects of Möbius invariant $\mathcal{Q}_K$ spaces, *J. Funct. Anal.* **230** (2006), 78-115.

[EX1997] M. Essén and J. Xiao, Some results on $\mathcal{Q}_p$ spaces, $0 < p < 1$, *J. Reine Angew. Math.* **485** (1997), 173-195.

[Fe1971] C. Fefferman, Characterizations of bounded mean oscillations, *Bull. Amer. Math. Soc.* **77** (1971), 587-588.

[FS1972] C. Fefferman and E. Stein , H^p spaces of several variables, *Acta Math.* **129** (1972), 137-193.

[Ga1981] J. Garnett, *Bounded Analytic Functions*, Academic Press, New York, 1981.

[GJ1982] J. Garnett and P. Jones, BMO from dyadic BMO, *Pacific J. Math.* **99** (1982), 351-371.

[Ge1977] F. Gehring, Univalent functions and the Schwarzian derivative, *Comment. Math. Helv.* **52** (1977), 561-572.

[Ha1994] W. Hayman, *Multivalent Functions (second edition)*, Cambridge Tracts in Mathematics **110**, Cambridge University Press, London, 1994.

[HP1978] W. Hayman and Ch. Pommerenke, On analytic functions of bounded mean oscillation, *Bull. London Math. Soc.* **10**(1978), 219-224.

[Ja1999] S. Janson, On the space $\mathcal{Q}_p$ and its dyadic counterpart, *Complex analysis and differential equations* (Uppsala, 1997), 194-205, Acta Univ. Upsaliensis Skr. Uppsala Univ. C Organ. Hist., 64, Uppsala Univ., Uppsala, 1999.

[JN1961] F. John and L. Nirenberg, On functions of bounded mean oscillation, *Comm. Pure Appl. Math.* **14** (1961), 415-426.

[Jo1983] P. Jones, L^∞ estimates for the $\bar{\partial}$-problem in a half plane, *Acta Math.* **150** (1983), 137-152.

[LNST2013] J. Laitila, E. Nieminen, E. Saksman and O. Tylli, Compact and weakly compact composition operators on BMOA, *Complex Analysis and Operator Theory* **7** (2013), 163-181.

[Le1987] O. Lehto, *Univalent Functions and Teichmüller Spaces*, Springer-Verlag, 1987.

[LiW2007] S. Li and H. Wulan, Composition operators on $\mathcal{Q}_K$ spaces, *J. Math. Anal. Appl.* **327** (2007), 948-958.

[Lo2001] Z. Lou, Composition operators on $\mathcal{Q}_p$ spaces, *J. Austral. Math. Soc.* **70** (2001), 161-188.

[Lu1985] H. Luecking, Forward and reverse Carleson inequalities for functions in Bergman spaces and their derivatives, *Amer. J. Math.* **107** (1985), 85-111.

[MP1990] M. Mateljević and M. Pavlović, Multipliers of H^p and BMOA, *Pacific J. Math.* **146** (1990), 71-84.

[NX1997] A. Nicolau and J. Xiao, Bounded functions in Möbius invariant Dirichlet spaces, *J. Funct. Anal.* **150** (1997), 383-425.

[PX2006] M. Pavlović and J. Xiao, Splitting planar isoperimetric inequality through preduality of $\mathscr{Q}_p, 0 < p < 1$, *J. Funct. Anal.* **233** (2006), 40-59.

[Pa2008] J. Pau, Bounded Möbius invariant $\mathscr{Q}_K$ spaces, *J. Math. Anal. Appl.* **338** (2008), 1029-1042.

[PP2009-2] J. Pau and J. Peláez, Logarithms of the derivative of univalent functions in $\mathscr{Q}_p$ spaces, *J. Math. Anal. Appl.* **350** (2009), 184-194.

[Pa2005] M. Pavlović, Hadamard product in $\mathscr{Q}_p$ spaces, *J. Math. Anal. Appl.* **305** (2005), 589-598.

[Po1977] Ch. Pommerenke, Schlichte funktionen und analytische funktionen von beschränkten mittlerer oszillation, *Comm. Math. Helv.* **52**(1977), 591-602.

[Po1980] S. Power, Vanishing Carleson measures, *Bull. London Math. Soc.* **12** (1980), 207-210.

[Rä2001] J. Rättyä, On some complex function spaces and classes, Ann. Acad. Sci. Fenn. Math. Dissertationes **124** (2001), 73 pages.

[Ro1985] R. Rochberg, Decomposition theorems for Bergman space and their applications in operators and function theory, in *Operator and Function Theory* (S. Power, editor), NATO ASI Series C, *Math. and Physical Sci.* **153** (1985), 225-277.

[RS1986] R. Rochberg and S. Semmes, A decomposition theorem for BMO and applications, *J. Funct. Anal.* **37** (1986), 228-263.

[RW1993] R. Rochberg and Z. Wu, A new characterization of Dirichlet type spaces and applications, *Illinois J. Math.* **37** (1993), 101-122.

[RT1979] L. Rubel and R. Timoney, An extremal property of the Bloch space, *Proc. of Amer. Math. Soc.* **75** (1979), 45-49.

[Ru1974] W. Rudin, *Real and Complex Analysis*, 2nd ed. , McGraw-Hill, New York, 1974.

[Sh1987] J. Shapiro, The essential norm of a composition operator, *Ann. Math.* **125** (1987), 375-404.

[Sh1993] J. Shapiro, *Composition Operators and Classical Function Theory*, Springer-Verlag, New York, 1993.

[Smi1999] W. Smith, Compactness of composition operators on BMOA, *Proc. Amer. Math. Soc.* **127** (1999), 2715-2725.

[SW1999] Y. Shen and H. Wei, Universal Teichmüller space and BMO, *Adv. in Math.* **234** (1999), 129-148.

[SZ1997] W. Smith and R. Zhao, Composition operators mapping into the $\mathcal{Q}_p$ spaces, *Analysis* **17** (1997), 239-263.

[St1980-1] D. Stegenga, Multipliers of the Dirichlet space, *Illinois J. Math.* **24** (1980), 113-139.

[St1980-2] D. Stegenga, A geometric condition which implies BMOA, *Michigan Math. J.* **27**(1980), 247-252.

[SS1981] D. Stegenga and K. Stephenson, A geometric characterization of analytic functions with bounded mean oscillation, *J. London Math. Soc.* **24**(1981), 243-254.

[WS2014] H. Wei and Y. Shen, On the tangent space to the BMO-Teichmüller space, *J. Math. Anal. Appl.* **419**(2014), 715-726.

[WX2002] Z. Wu and C. Xie, Decomposition theorems for $\mathcal{Q}_p$ spaces, *Ark. Mat.* **40** (2002), 383-401.

[WX2003] Z. Wu and C. Xie, $\mathcal{Q}_p$ spaces and Morrey spaces, *J. Funct. Anal.* **201** (2003), 282-297.

[Wul1993] H. Wulan, Carleson measure and the derivatives of BMO functions (Chinese), *J. Inner Mongolia Normal Univ.* **2** (1993), 1-9.

[Wul1998] H. Wulan, On some classes of meromorphic functions, *Ann. Acad. Sci. Fenn. Math. Diss.* **116** (1998), 1-57.

[Wul2005] H. Wulan, Compactness of the composition operators from the Bloch space $\mathcal{B}$ to $\mathcal{Q}_K$ spaces, *Acta. Math. Sin.* **21** (2005), 1415-1424.

[Wul2004] H. Wulan, Multivalent functions and $\mathcal{Q}_K$ spaces, *Internat. J. Math. Math. Sci.* **48** (2004), 2537-2546.

[Wul2007] H. Wulan, Compactness of composition operators on BMOA and VMOA, *Sci. China* **50** (2007), 997-1004.

[WW2001] P. Wu and H. Wulan, Characterizations of $\mathcal{Q}_T$ spaces, *J. Math. Anal. Appl.* **254** (2001), 484-497.

[WY2015] H. Wulan and F. Ye, Some results in Möbius invariant $\mathcal{Q}_K$ spaces, *Complex Var. Elliptic Equ.* **60** (2015), 1602-1611.

[WY2014] H. Wulan and F. Ye, Universal Teichmüller space and $\mathcal{Q}_K$ spaces, *Ann. Acad. Sci. Fenn. Math.* **39** (2014), 691-709

[WZZ2009] H. Wulan, D. Zheng and K. Zhu, Compact composition operators on the Bloch and BMOA spaces, *Proc. Amer. Math. Soc.* **137** (2009), 3861-3868.

[WZh2006] H. Wulan and J. Zhou, $\mathcal{Q}_K$ type spaces of analytic functions, *J. Function Spaces and Appl.* **4** (2006), 73-84.

[WZh2007] H. Wulan and J. Zhou, The higher order derivatives of $\mathcal{Q}_K$ spaces, *J. Math. Anal. Appl.* **332** (2007), 1216-1228.

[WZh2013] H. Wulan and J. Zhou, $\mathcal{Q}_K$ and Morrey type spaces, *Ann. Acad. Sci. Fenn. Math.* **38** (2013), 193-207.

[WZh2014] H. Wulan and J. Zhou, Decomposition theorems for $\mathcal{Q}_K$ spaces and applications, *Forum Math.* **26** (2014), 467-495.

[WZha2008] H. Wulan and Y. Zhang, Hadamard products and $\mathcal{Q}_K$ spaces, *J. Math. Anal. Appl.* **337** (2008), 11142-1150.

[WZ2008] H. Wulan and K. Zhu, $\mathcal{Q}_K$ spaces via higher order derivatives, *Rocky Mountain J. Math.* **38** (2008), 329-350.

[WZ2007-1] H. Wulan and K. Zhu, Derivative free characterizations of $\mathcal{Q}_K$ spaces, *J. Australian Math. Soc.* **82** (2007), 283-295.

[WZ2007-2] H. Wulan and K. Zhu, Lacunary series in $\mathcal{Q}_K$ spaces, *Studia Math.* **173** (2007), 217-230.

[WZ2017] H. Wulan and K. Zhu, Möbius invariant $\mathcal{Q}_K$ spaces, Springer-Verlag, Berlin, 2017.

[Xi2001] J. Xiao, *Holomorphic $\mathcal{Q}$ classes*, Lecture Notes in Mathematics **1767**, Springer-Verlag, Berlin, 2001.

[Xi2006] J. Xiao, *Geometric $\mathcal{Q}_p$ Functions*, Birkhäuser Verlag, Basel-Boston-Berlin, 2006.

[Xi2008] J. Xiao, The $\mathcal{Q}_p$ Carleson measure problem, *Adv. Math.* **217** (2008), 2075-2088.

[Xi2000-1] J. Xiao, Some essential properties of $\mathcal{Q}_p(\mathbb{T})$-spaces, *J. Fourier Anal. Appl.* **6** (2000), 311-323.

[Xi2000-2] J. Xiao, Composition operators: N_α to the Bloch space to Q_β, *Studia Math.* **139** (2000), 245-260.

[XX2014] J. Xiao and W. Xu, Composition operators between analytic Campanato spaces, *J. Geom. Anal.* **24** (2014), 649-666.

[ZC2014] M. Zhan and G. Cao, Duality of $\mathcal{Q}_K$-type spaces, *Bull. Korean Math. Soc.* **51** (2014), 1411-1423.

[Zha1996] R. Zhao, On a general family of function spaces, *Fenn. Ann. Acad. Sci. Fenn. Math.* **105** (1996).

[Zho2016] J. Zhou, A note on lacunary series in $\mathcal{Q}_K$ spaces, *Bull. Iranian Math. Soc.* **42** (2016), 195-200.

[Zho2012] J. Zhou, Schwarzian derivative, geometric conditions, and $\mathscr{Q}_K$ spaces, *Sci. Sin. Math.* (Chinese) **42** (2012), 939-950.

[Zho2013] J. Zhou, Predual of $\mathscr{Q}_K$ spaces, *J. Funct. Spaces Appl.* **2013** (2013), Article ID 252735, 6 pages.

[Zho2014] J. Zhou, Interpolating sequences in $\mathscr{Q}_K$ spaces, preprint.

[ZB2015] J. Zhou and G. Bao, Analytic version of $\mathscr{Q}_1(\mathbb{T})$ space, *J. Math. Anal. Appl.* **422** (2015), 1091-1192.

[Zhu2007] K. Zhu, *Operator Theory in Function Spaces*, Second Edition, American Mathematical Society, 2007.

[Zo1986] C. Zorko, Morrey space, *Proc. Amer. Math. Soc.* **98** (1986), 586-592.

Chapter 7

Analytical Aspects of the Drury-Arveson Space

Quanlei Fang

Department of Mathematics and Computer Science
CUNY-BCC, Bronx, NY 10453, USA
quanlei.fang@bcc.cuny.edu

Jingbo Xia

Department of Mathematics, SUNY at Buffalo, Buffalo, NY 14260, USA
jxia@buffalo.edu

CONTENTS

7.1 Introduction

The space that is now simply denoted H_n^2, first appeared in [24, 40, 41]. In [41], Lubin used this space to produce the first example of a tuple of commuting subnormal operators that does not admit a joint normal extension. Drury's motivation in [24] was to find the correct multi-operator analogue of the von Neumann inequality for contractions.

But it was Arveson's seminal paper [5] published twenty years ago that really brought H_n^2 to the attention of the operator-theory community. Although it initially went under various appellations, the community seems to have now settled on the name "Drury-Arveson space" for H_n^2. Therefore we will use the term "Drury-Arveson space" in this survey, as we have always done in our previous writings.

Let $\mathbf{B}$ be the open unit ball in $\mathbf{C}^n$. Throughout the article, the complex dimension n is always assumed to be greater than or equal to 2. Recall that the Drury-Arveson space H_n^2 is the Hilbert space of analytic functions on $\mathbf{B}$ that has the function

$$K_z(\zeta) = \frac{1}{1 - \langle \zeta, z \rangle} \tag{7.1}$$

as its reproducing kernel [5]. Equivalently, H_n^2 can be described as the Hilbert space of analytic functions on $\mathbf{B}$ where the inner product is given by

$$\langle h, g \rangle = \sum_{\alpha \in \mathbf{Z}_+^n} \frac{\alpha!}{|\alpha|!} a_\alpha \overline{b_\alpha}$$

for

$$h(\zeta) = \sum_{\alpha \in \mathbf{Z}_+^n} a_\alpha \zeta^\alpha \quad \text{and} \quad g(\zeta) = \sum_{\alpha \in \mathbf{Z}_+^n} b_\alpha \zeta^\alpha.$$

Here and throughout, we follow the standard multi-index notation [46, page 3].

Arveson derived the space H_n^2 from the viewpoint of dilation theory, and he was the first to recognize that H_n^2 is a reproducing-kernel Hilbert space with (7.1) as its reproducing kernel [5]. Based on this Arveson regarded H_n^2 as a generalization of the Hardy space. At the same time, Arveson recognized that H_n^2 is fundamentally different from the traditional Hardy space and Bergman space in that it cannot be defined in terms of a measure.

Today, we view the Drury-Arveson space, the Hardy space and the Bergman space all as members of a continuum family of reproducing-kernel Hilbert spaces parametrized by their "weight", $t \in [-n, \infty)$.

However one views the Drury-Arveson space, it is a place where one can do a lot of exciting operator theory, function theory and analysis. In the last twenty years, the collective effort of the operator-theory community has produced a huge body of literature on H_n^2. In this article we will offer some of our own perspectives on the theory of Drury-Arveson space. We will also take a look at some of the recent developments. This review is not intended to be comprehensive in any way. On the contrary, we will limit the article to the aspects of the Drury-Arveson space that are most familiar to us.

7.2 von Neumann inequality for row contractions

Both Drury and Arveson arrived at the space H_n^2 by considering the proper generalization of the von Neumann inequality to commuting tuples $(A_1, \ldots, A_n)$ of operators. After initial experimentations [44, 48] with $\|\cdot\|_\infty$ as the "right-hand side" of such an inequality, it was quickly realized that a different norm is needed, and that the condition that each A_i individually be a contraction is not enough. The proper setting for such a generalization is the *row contractions*.

A commuting tuple of operators $(A_1, \ldots, A_n)$ on a Hilberts space $\mathscr{H}$ is said to be a row contraction if the operator inequality

$$A_1 A_1^* + \cdots + A_n A_n^* \le 1$$

holds on $\mathscr{H}$. Equivalently, $(A_1, \ldots, A_n)$ is a row contraction if and only if

$$\|A_1 x_1 + \cdots + A_n x_n\|^2 \le \|x_1\|^2 + \cdots + \|x_n\|^2$$

for all $x_1, \ldots, x_n \in \mathscr{H}$. For such a tuple, the von Neumann inequality reads

$$\|p(A_1, \ldots, A_n)\| \le \|p\|_{\mathscr{M}} \tag{7.2}$$

for every $p \in \mathbf{C}[z_1, \ldots, z_n]$ [5, 24]. Here, $\|p\|_{\mathscr{M}}$ is the *multiplier norm* of p, the norm of the operator of multiplication by the polynomial p on H_n^2. Both Drury and Arveson showed that in general $\|p\|_{\mathscr{M}}$ is not dominated by the supremum norm of p on $\mathbf{B}$.

Drury's proof of (7.2) in [24] is particularly simple and can be easily explained here. First of all, Drury observed that in order to prove (7.2), it suffices to consider commuting tuples $A = (A_1, \ldots, A_n)$ on $\mathscr{H}$ satisfying the condition

$$A_1 A_1^* + \cdots + A_n A_n^* \le r \tag{7.3}$$

for some $0 < r < 1$. For such a tuple, one easily verifies that the combinatorial identity

$$\sum_{\alpha \in \mathbf{Z}_+^n} \frac{|\alpha|!}{\alpha!} A^\alpha (1 - A_1 A_1^* - \cdots - A_n A_n^*) A^{*\alpha} = 1 \tag{7.4}$$

holds. Now define the operator $Z : \mathscr{H} \to H_n^2 \otimes \mathscr{H}$ by the formula

$$(Zx)(\zeta) = \sum_{\alpha \in \mathbf{Z}_+^n} \frac{|\alpha|!}{\alpha!} (1 - A_1 A_1^* - \cdots - A_n A_n^*)^{1/2} A^{*\alpha} x \zeta^\alpha, \quad x \in \mathscr{H}. \tag{7.5}$$

Then (7.4) ensures that Z is an *isometry*. It is straightforward to verify that

$$Z p(A_1^*, \ldots, A_n^*) = (p(M_{\zeta_1}^*, \ldots, M_{\zeta_n}^*) \otimes 1) Z \tag{7.6}$$

for every $p \in \mathbf{C}[z_1, \ldots, z_n]$. Since Z is an isometry, this implies (7.2).

We would like to make two comments at this point. First, if one is aware of the combinatorial identity (7.4), one can go backwards to figure out the proper definition for H_n^2 based on the requirement that the operator Z defined by (7.5) be an isometry. Second, (7.4) is what one usually calls a *resolution* of the identity operator. Such resolutions can be exploited in various ways, and we will come back to this point later.

By Drury's combinatorial argument, for a commuting row contraction $(A_1,\ldots,A_n)=A$, the resolution of identity (7.4) holds whenever

$$\lim_{k\to\infty}\sum_{|\alpha|=k}\frac{|\alpha|!}{\alpha!}A^{\alpha}A^{*\alpha}=0 \tag{7.7}$$

in the strong operator topology. Obviously, (7.7) is a much weaker condition than (7.3). Thus (7.4) holds for many more commuting row contractions than those satisfying condition (7.3). A commuting row contraction satisfying (7.7) is said to be *pure* [5, 6].

7.3 The multipliers

A fundamental contribution that Arveson made in [5] was that he took the idea of $\|\cdot\|_{\mathcal{M}}$ one step further to introduce general *multipliers* for H_n^2, which turn out to be a constant source of fascination for the Drury-Arveson space community today. A function $f\in H_n^2$ is said to be a multiplier of the Drury-Arveson space if $fh\in H_n^2$ for every $h\in H_n^2$ [5]. We will write $\mathcal{M}$ for the collection of the multipliers of H_n^2. If $f\in\mathcal{M}$, then the multiplication operator M_f is bounded on H_n^2 [5], and the multiplier norm $\|f\|_{\mathcal{M}}$ is defined to be the operator norm $\|M_f\|$ on H_n^2.

There are many more questions about the multipliers than there are answers. We begin with the few things that we do know about $\mathcal{M}$. The most significant piece of knowledge about $\mathcal{M}$ is the corona theorem due to Costea, Sawyer and Wick.

Theorem 3.1 ([16]) *For $g_1,\ldots,g_k\in\mathcal{M}$, if there is a $c>0$ such that*

$$|g_1(\zeta)|+\cdots+|g_k(\zeta)|\ge c$$

for every $\zeta\in\mathbf{B}$, then there exist $f_1,\ldots,f_k\in\mathcal{M}$ such that

$$f_1g_1+\cdots+f_kg_k=1.$$

This is so far the only success story with regard to the corona theorem in the multi-variable setting. A special case of the corona theorem is the one-function corona theorem:

Corollary 3.2 *For $f\in\mathcal{M}$, if there is a $c>0$ such that $|f(\zeta)|\ge c$ for every $\zeta\in\mathbf{B}$, then $1/f\in\mathcal{M}$.*

While the proof of the full version of the corona theorem involves difficult analysis [16] and is very long, the one-function corona theorem can be directly proved with a short, soft and elementary argument [30]. An essential ingredient in the direct proof of Corollary 3.2 in [30] was the von Neumann inequality (7.2). But it is worth mentioning that other than the use of (7.2), the rest of the proof in [30] was completely self-contained.

Since the publication of [30], there has been a proof of Corollary 3.2 in [45] that claims to be "significantly shorter". But the proof in [45] relies heavily on facts established in [1, 12, 43]. To begin with, the proof in [45] required the fact that $\mathscr{M}$ is contained in the collection $\mathscr{C}H_n^2$ defined in that paper, and for this inclusion [45] simply cited [43]. Thus if one follows the argument in [45], then much of the work for proving Corollary 3.2 was actually done in [1, 12, 43]. This differs significantly from our purpose in [30], namely to give a proof of Corollary 3.2 that does not require references.

From Corollary 3.2 we immediately obtain

Corollary 3.3 *Let $f \in \mathscr{M}$. Then the spectrum of the multiplication operator M_f on H_n^2 is contained in the closure of $\{f(z) : z \in \mathbf{B}\}$.*

In the case $f \in \mathbf{C}[z_1, \ldots, z_n]$, this fact was already known to Arveson in [5], where he proved it using a Banach-algebra argument. The converse of Corollary 3.3, namely for $f \in \mathscr{M}$ the spectrum of M_f contains $\{f(z) : z \in \mathbf{B}\}$, is trivial. In fact, using the reproducing kernel, for every $f \in \mathscr{M}$ we have $M_f^* K_z = \overline{f(z)} K_z$, $z \in \mathbf{B}$. Thus for each $z \in \mathbf{B}$, $\overline{f(z)}$ is an eigenvalue for M_f^*.

Nowadays, one commonly views the Drury-Arveson space H_n^2 as a Hilbert module over the polynomial ring $\mathbf{C}[z_1, \ldots, z_n]$. In fact, one often refers to H_n^2 as *the Drury-Arveson module*, in various contexts. But where there are modules, there are submodules. A *submodule* of the Drury-Arveson module is a closed linear subspace $\mathscr{S}$ of H_n^2 that is invariant under the multiplication by $\mathbf{C}[z_1, \ldots, z_n]$. In other words, a submodule is what one would otherwise call an invariant subspace of H_n^2. The next proposition sets H_n^2 apart from other reproducing-kernel Hilbert spaces on $\mathbf{B}$.

Proposition 3.4 ([6]) *Let $\mathscr{S}$ be a submodule of the Drury-Arveson module. If $\mathscr{S} \neq \{0\}$, then $\mathscr{S} \cap \mathscr{M} \neq \{0\}$.*

One obvious consequence of Proposition 3.4 is the following:

Corollary 3.5 *Let $\mathscr{S}_1$, $\mathscr{S}_2$ be submodules of the Drury-Arveson module. If $\mathscr{S}_1 \neq \{0\}$ and $\mathscr{S}_2 \neq \{0\}$, then $\mathscr{S}_1 \cap \mathscr{S}_2 \neq \{0\}$.*

In the jargon of invariant-subspace theory, Corollary 3.5 says that there are no non-trivial invariant subspaces of H_n^2 that are *disjoint*. If $n = 1$, then this is a consequence of Beurling's theorem. So one might interpret Corollary 3.5 as saying that this one particular aspect of Beurling's theorem is retained by H_n^2 for all $n \geq 2$.

This is definitely not the case for some of the other reproducing-kernel Hilbert spaces. For example, one can easily construct invariant subspaces N_1, N_2 of the Bergman space $L_a^2(\mathbf{B}, dv)$ such that $N_1 \cap N_2 = \{0\}$ while $N_1 \neq \{0\}$ and $N_2 \neq \{0\}$. In particular, this implies that $N_1 \cap H^\infty(\mathbf{B}) = \{0\}$ and $N_2 \cap H^\infty(\mathbf{B}) = \{0\}$.

Incidentally, Proposition 3.4 can also be proved using Drury's ideas. In fact, define

$$\mathscr{Z}_{\mathscr{S},i} = M_{\zeta_i}|\mathscr{S}, \quad i = 1, \ldots, n. \tag{7.8}$$

It is easy to verify that $(\mathscr{Z}_{\mathscr{S},1}, \ldots, \mathscr{Z}_{\mathscr{S},n})$ is a row contraction on $\mathscr{S}$. Moreover, it is easy to show that (7.7) holds for the tuple $(A_1, \ldots, A_n) = (\mathscr{Z}_{\mathscr{S},1}, \ldots, \mathscr{Z}_{\mathscr{S},n})$; consequently so does the resolution of identity (7.4). Then, taking adjoints on both sides of (7.6), we find that

$$p(\mathscr{Z}_{\mathscr{S},1}, \ldots, \mathscr{Z}_{\mathscr{S},n})Z^* = Z^*(p(M_{\zeta_1}, \ldots, M_{\zeta_n}) \otimes 1), \tag{7.9}$$

$p \in \mathbf{C}[z_1, \ldots, z_n]$. For each $y \in \mathscr{S}$, set $\varphi_y = Z^*(1 \otimes y) \in \mathscr{S}$. The above gives us

$$\|p\varphi_y\|_{H_n^2} = \|p\varphi_y\|_{\mathscr{S}} \leq \|p \otimes y\|_{H_n^2 \otimes \mathscr{S}} = \|p\|_{H_n^2}\|y\|_{\mathscr{S}}$$

for every $p \in \mathbf{C}[z_1, \ldots, z_n]$. This means that $\varphi_y \in \mathscr{M}$. Finally, since the linear span of all $p\varphi_y = Z^*(p \otimes y)$ is dense in $\mathscr{S}$, the condition $\mathscr{S} \neq \{0\}$ implies that $\varphi_y \neq 0$ for some $y \in \mathscr{S}$.

The argument above has the virtue that it is based on (7.6), the same thing that establishes the von Neumann inequality for commuting row contractions. This connection clearly makes Proposition 3.4 a "close cousin" of the von Neumann inequality (7.2).

The above argument can be pushed even further. Let $\{u_j : j \in J\}$ be an orthonormal basis in $\mathscr{S}$. For each $j \in J$, write $\varphi_j = Z^*(1 \otimes u_j)$, which we now know is a multiplier for H_n^2. Let $\{f_j : j \in J\} \subset H_n^2$ be such that $\sum_{j \in J} \|f_j\|^2 < \infty$. Then it follows from (7.9) that

$$Z^* \sum_{j \in J} f_j \otimes u_j = \sum_{j \in J} f_j \varphi_j = \sum_{j \in J} M_{\varphi_j} f_j.$$

In other words, there is a unitary operator U such that Z^*U is the row operator

$$[M_{\varphi_1}, M_{\varphi_2}, \ldots] : H_n^2 \oplus H_n^2 \oplus \cdots \to \mathscr{S}.$$

Since $Z^*Z = 1$ on $\mathscr{S}$, this leads to the representation

$$P_{\mathscr{S}} = \sum_{j \in J} M_{\varphi_j} M_{\varphi_j}^*$$

for the orthogonal projection $P_{\mathscr{S}} : H_n^2 \to \mathscr{S}$, which first appeared in Arveson's paper [6].

Theorem 3.6 ([27]) *Let $\mathscr{S}$ be a submodule of the Drury-Arveson module H_n^2 and define the corresponding defect operator*

$$D_{\mathscr{S}} = [\mathscr{Z}_{\mathscr{S},1}^*, \mathscr{Z}_{\mathscr{S},1}] + \cdots + [\mathscr{Z}_{\mathscr{S},n}^*, \mathscr{Z}_{\mathscr{S},n}].$$

Suppose that $\mathscr{S} \neq \{0\}$. Then there is an $\varepsilon = \varepsilon(\mathscr{S}) > 0$ such that

$$s_1(D_{\mathscr{S}}) + \cdots + s_k(D_{\mathscr{S}}) \geq \varepsilon k^{(n-1)/n}$$

for every $k \in \mathbf{N}$. Consequently, $D_{\mathscr{S}}$ does not belong to the Schatten class $\mathscr{C}_n$.

An analogue of Theorem 3.6 also holds for submodules of the Hardy module [27]. But, thanks to Proposition 3.4, the proof of Theorem 3.6 is much easier than the proof of its analogue on the Hardy space. This is a good example of applications of multipliers. Incidentally, the analogue of Theorem 3.6 is not known to be true or false in the Bergman-space case, although one would certainly expect this analogue to be true. In fact, the possible Bergman-space analogue of Theorem 3.6 seems to be a very hard problem.

One reason the Bergman space and the Hardy space attract attention is that there are Toeplitz operators defined there, and every aspect of these operators is thoroughly studied. For the Drury-Arveson space H_n^2, $n \geq 2$, there is no L^2 space associated with it. Thus the only kind of "Toeplitz operators" on H_n^2 are the multipliers and their adjoints. Nevertheless, these operators retain certain familiar properties of Toeplitz operators. One noticeable such property is the essential commutativity. Arveson showed in [5] that the commutators $[M_{\zeta_i}^*, M_{\zeta_j}]$ on H_n^2 all belong to the Schatten class $\mathscr{C}_p$ for $p > n$. Such essential commutativity was later generalized to include multipliers:

Theorem 3.7 ([28]) *Let $f \in \mathscr{M}$ and $j \in \{1, \ldots, n\}$. Then on H_n^2, the commutator $[M_f^*, M_{\zeta_j}]$ belongs to the Schatten class $\mathscr{C}_p$ for $p > 2n$. Moreover, for each $2n < p < \infty$, there is a C such that*

$$\|[M_f^*, M_{\zeta_j}]\|_p \leq C\|f\|_{\mathscr{M}}$$

for all $f \in \mathscr{M}$ and $j \in \{1, \ldots, n\}$.

An enduring challenge in the theory of the Drury-Arveson space, since its very inception, has been the quest for a *good* characterization of the membership in $\mathscr{M}$. In other words, here we ask a very instinctive question: what does a general $f \in \mathscr{M}$ look like?

Let $k \in \mathbf{N}$ be such that $2k \geq n$. Then given any $f \in H_n^2 \cap H^\infty(\mathbf{B})$, one can define the measure $d\mu_f$ on $\mathbf{B}$ by the formula

$$d\mu_f(\zeta) = |(R^k f)(\zeta)|^2 (1 - |\zeta|^2)^{2k-n} dv(\zeta), \tag{7.10}$$

where dv is the normalized volume measure on $\mathbf{B}$ and R denotes the radial derivative $z_1\partial_1 + \cdots + z_n\partial_n$. Ortega and Fàbrega showed that f is a multiplier of the Drury-Arveson space if and only if $d\mu_f$ is an H_n^2-Carleson measure [43]. That is, $f \in \mathscr{M}$ if and only if there is a C such that

$$\int |h(\zeta)|^2 d\mu_f(\zeta) \leq C\|h\|^2$$

for every $h \in H_n^2$. In [4], Arcozzi, Rochberg and Sawyer gave a characterization for all the H_n^2-Carleson measures on $\mathbf{B}$.

For a given Borel measure on $\mathbf{B}$, the conditions in [4, Theorem 34] are not the easiest to verify. More to the point, [4, Theorem 34] deals with all Borel measures on $\mathbf{B}$, not just the class of measures $d\mu_f$ of the form (7.10). Thus it is natural to ask, is there a simpler, or a more direct, characterization of the membership $f \in \mathscr{M}$?

Since the Drury-Arveson space is a reproducing-kernel Hilbert space, it is natural to turn to the reproducing kernel for possible answers. Recall that the normalized reproducing kernel for H_n^2 is given by the formula

$$k_z(\zeta) = \frac{(1-|z|^2)^{1/2}}{1-\langle \zeta, z\rangle},$$

$z, \zeta \in \mathbf{B}$. For any $f \in H_n^2$, its Berezin transform

$$\langle f k_z, k_z \rangle$$

is none other than $f(z)$ itself. Given what is known about H_n^2 (see the related discussions in Section 4 below), one does not expect the boundedness of the Berezin transform on $\mathbf{B}$ to be enough to guarantee the membership $f \in \mathcal{M}$. But what about something stronger than the Berezin transform of f? For example, anyone who gives any thought about multipliers is likely to come up with the natural question, for $f \in H_n^2$, does the condition

$$\sup_{|z|<1} \|f k_z\| < \infty$$

imply the membership $f \in \mathcal{M}$? Conditions of this type are now called "reproducing-kernel thesis" [42] and are among the first things that one would check when it comes to boundedness. One can rephrase the question thus: is it enough to determine the boundedness of M_f by testing it on the special subset $\{k_z : |z| < 1\}$ of the unit ball in H_n^2?

It is very tempting to think that the answer to the above question might be affirmative, and that was what we thought for quite a while. After all, an affirmative answer would provide a very simple characterization of the membership $f \in \mathcal{M}$. But that would be too simplistic, as it turns out. The answer is just the opposite:

Theorem 3.8 ([32]) *There exists an $f \in H_n^2$ satisfying the conditions $f \notin \mathcal{M}$ and*

$$\sup_{|z|<1} \|f k_z\| < \infty.$$

Thus, unfortunately, we are back where we started, looking for a non-trivial characterization of the membership $f \in \mathcal{M}$ that is reasonably simple and straightforward.

7.4 A family of reproducing-kernel Hilbert spaces

For each real number $-n \leq t < \infty$, consider the kernel

$$\frac{1}{(1-\langle \zeta, z\rangle)^{n+1+t}}, \quad \zeta, z \in \mathbf{B}.$$

Let $\mathcal{H}^{(t)}$ be the corresponding reproducing kernel Hilbert space of analytic functions on $\mathbf{B}$. Alternately, one can describe $\mathcal{H}^{(t)}$ as the completion of $\mathbf{C}[z_1, \ldots, z_n]$ with respect to the norm $\|\cdot\|_t$ arising from the inner product $\langle\cdot,\cdot\rangle_t$ defined according to the following rules: $\langle z^\alpha, z^\beta\rangle_t = 0$ whenever $\alpha \neq \beta$,

$$\langle z^\alpha, z^\alpha\rangle_t = \frac{\alpha!}{\prod_{j=1}^{|\alpha|}(n+t+j)}$$

if $\alpha \in \mathbf{Z}_+^n\backslash\{0\}$, and $\langle 1,1\rangle_t = 1$. Clearly, we have

$$\mathscr{H}^{(s)} \subset \mathscr{H}^{(t)}$$

for all $-n \leq s < t < \infty$.

Obviously, $\mathscr{H}^{(0)}$ is the Bergman space $L_a^2(\mathbf{B}, dv)$, $\mathscr{H}^{(-1)}$ is the Hardy space $H^2(S)$, and $\mathscr{H}^{(-n)}$ is none other than the Drury-Arveson space H_n^2. Moreover, for each $-1 < t < \infty$, $\mathscr{H}^{(t)}$ is a weighted Bergman space. One can view the Bergman space $\mathscr{H}^{(0)} = L_a^2(\mathbf{B}, dv)$ as a benchmark, against which the other spaces in the family should be compared.

This suggests that for each $-n \leq t < \infty$, we should think of t as the weight of the space $\mathscr{H}^{(t)}$. In particular, the Drury-Arveson space H_n^2 has weight $-n$. This approach tells us that the Drury-Arveson space is but one member of a continuum family of reproducing kernel Hilbert spaces. Thus, for example, if one obtains a result on one space of the family, then one should take it as a hint for things to come: does its analogue also hold for other spaces in the family? There is no simple answer to this general question, and often different techniques are required for different spaces.

Using the radial derivative $R = z_1\partial_1 + \cdots + z_n\partial_n$, the norm $\|\cdot\|_t$ can be expressed in an equivalent form. Indeed for a given $-n \leq t < \infty$, let m be any non-negative integer such that $2m + t > -1$. Then it is easy to verify that

$$\|f\|_t^2 \approx |f(0)|^2 + \int |(R^m f)(\zeta)|^2 (1 - |\zeta|^2)^{2m+t} dv(\zeta)$$

for $f \in \mathscr{H}^{(t)}$. For computations and estimates, the right-hand side is often more convenient.

For each $-n \leq t < \infty$, consider $(M_{\zeta_1}^{(t)}, \ldots, M_{\zeta_n}^{(t)})$, the tuple of multiplication by the coordinate functions on $\mathscr{H}^{(t)}$. It was Lubin who first proved that, when $n \geq 2$, the tuple $(M_{\zeta_1}^{(-n)}, \ldots, M_{\zeta_n}^{(-n)})$ is not jointly subnormal [41]. Arveson arrived at the same conclusion in [5] by showing that there is some $p \in \mathbf{C}[z_1, \ldots, z_n]$ with the property that the operator of multiplication by p on H_n^2 is not hyponormal. Later, Arazy and Zhang showed that for each $-n < t < -1$, the tuple $(M_{\zeta_1}^{(t)}, \ldots, M_{\zeta_n}^{(t)})$ on $\mathscr{H}^{(t)}$ is also not jointly subnormal [3]. This means that when $-n < t < -1$, $\mathscr{H}^{(t)}$ has more in common with the Drury-Arveson space than with the Hardy space or the Bergman space.

In practical terms, the lack of joint subnormality for $(M_{\zeta_1}^{(t)}, \ldots, M_{\zeta_n}^{(t)})$ means that it is more difficult to do estimates on $\mathscr{H}^{(t)}$, and analytical results are hard to come by. This problem is particularly acute with H_n^2, $n \geq 2$. For example, the proof of Theorem

3.8 in [32] practically represents all the analytical techniques we have on H_n^2 at the moment.

One can even define $\mathscr{H}^{(t)}$ and $(M_{\zeta_1}^{(t)}, \ldots, M_{\zeta_n}^{(t)})$ in the range $-n-1 < t < -n$. But when $t < -n$, the tuple $(M_{\zeta_1}^{(t)}, \ldots, M_{\zeta_n}^{(t)})$ is no longer a row contraction. That is why we only consider the weight range $-n \le t < \infty$.

Arveson was the first to notice that, when $n \geq 2$, H_n^2 does not contain $H^\infty(\mathbf{B})$, the collection of bounded analytic function on $\mathbf{B}$. In [5], he explicitly constructed an $f \in H^\infty(\mathbf{B})$ that does not belong to H_n^2. This construction was based on the function

$$\theta(\zeta_1, \ldots, \zeta_n) = \zeta_1 \cdots \zeta_n$$

on $\mathbf{B}$. Arveson observed that

$$\|\theta^k\|_\infty = \frac{1}{n^{kn/2}} \quad \text{while} \quad \|\theta^k\|_{H_n^2} = \left(\frac{(k!)^n}{(nk)!}\right)^{1/2} \approx \frac{k^{(n-1)/4}}{n^{kn/2}}.$$

Once this is seen, for $n \geq 2$ it is easy to come up with coefficients $a_0, a_1, \ldots a_k, \ldots$ such that $f = \sum_{k=0}^\infty a_k \theta^k$ is in $H^\infty(\mathbf{B})$ but not in H_n^2. In fact, one can even require f to be continuous on the closure of the unit ball $\mathbf{B}$.

It should be mentioned that examples of $f \in H^\infty(\mathbf{B})$, $f \notin H_n^2$ actually existed in plain sight. From the last chapter of Rudin's famous book [46] we know that when $n \geq 2$, if u is a non-constant inner function on $\mathbf{B}$, then $|\nabla u|$ is not square-integrable with respect to the volume measure on $\mathbf{B}$. Using the spaces introduced in this section, we can rephrase this result as saying that if u is a non-constant inner function on $\mathbf{B}$, then $u \notin \mathscr{H}^{(-2)}$. In particular, $\mathscr{H}^{(-n)} = H_n^2$ does not contain any non-constant inner function.

But when Rudin's book was published in 1980, it was not yet known whether non-constant inner functions existed in the case $n \geq 2$. In fact, Rudin offered the gradient result in his book as evidence against the existence of non-constant inner functions. But shortly thereafter, non-constant inner functions were successfully constructed by Løw [39] and Aleksandrov [2]. Amazingly, H_n^2 somehow misses all these functions!

We know that the Hardy space $H^2(S) = \mathscr{H}^{(-1)}$ contains all the inner functions, whereas $\mathscr{H}^{(-2)}$ contains none, other than the constants. This comparison raises an interesting open question: what about the weights $-2 < t < -1$? That is, if $-2 < t < -1$, does the space $\mathscr{H}^{(t)}$ contain any non-constant inner function? To us, this question is extremely interesting. But unfortunately we have no clue to offer, one way or the other.

7.5 Essential normality

If $\mathscr{S}$ is a submodule of H_n^2 as defined in Section 3, then $\mathscr{Q} = H_n^2 \ominus \mathscr{S}$ is a quotient module of the Drury-Arveson module. The term "quotient module" is justified by the fact that for $f, g \in \mathbf{C}[z_1, \ldots, z_n]$ and $h \in H_n^2$, we have

$$P_{\mathscr{Q}} fgh = P_{\mathscr{Q}} f P_{\mathscr{Q}} gh,$$

where $P_{\mathscr{Q}} : H_n^2 \to \mathscr{Q}$ is the orthogonal projection. For a submodule $\mathscr{S}$, recall that the module operators $\mathscr{Z}_{\mathscr{S},1}, \ldots, \mathscr{Z}_{\mathscr{S},n}$ are given by (7.8). For the corresponding quotient module $\mathscr{Q}$, we also have the module operators

$$\mathscr{Z}_{\mathscr{Q},i} = P_{\mathscr{Q}} M_{\zeta_i} | \mathscr{Q},$$

$i = 1, \ldots, n$. Suppose that $\mathscr{F}$ is either a submodule or a quotient module. Then for $1 \le p < \infty$, the module $\mathscr{F}$ is said to be *p-essentially normal* if the commutators

$$[\mathscr{Z}_{\mathscr{F},i}^*, \mathscr{Z}_{\mathscr{F},j}], \quad i, j \in \{1, \ldots, n\},$$

all belong to the Schatten class $\mathscr{C}_p$. In this regard, we have the famous

Arveson Conjecture. [8, 9] Every graded submodule $\mathscr{G}$ of $H_n^2 \otimes \mathbf{C}^m$ is p-essentially normal for every $p > n$.

Here, the term "graded" means that $\mathscr{G}$ admits an orthogonal decomposition in terms of homogeneous polynomials. Arveson's original thinking was that the study of the module operators on such a $\mathscr{G}$ is really an operator-theoretic version of algebraic geometry, which has broad implications. There has been a lot of work on the Arveson Conjecture [8, 18, 26, 34, 35, 36, 37, 38, 47], and the best results to date are due to Guo and K. Wang [36].

Douglas made a more refined conjecture in [19] that also covers more modules. Together these essential-normality problems are now called the Arveson-Douglas Conjecture, which is quite all encompassing. We state a somewhat specialized version of it below:

Arveson-Douglas Conjecture. Suppose that I is an ideal in $\mathbf{C}[z_1, \ldots, z_n]$, and let V_I denote the zero variety of I. Then the quotient module

$$\mathscr{Q} = \{h \in H_n^2 : h \perp I\}$$

is p-essentially normal for all $p > \dim_{\mathbf{C}} V_I$.

This is a more refined conjecture in the respect that it asserts $p > \dim_{\mathbf{C}} V_I$, which is more than just $p > n$. In practice, though, it is a serious challenge to reach the lower limit $p > \dim_{\mathbf{C}} V_I$, and this is true even on the Bergman space. However, as is shown in [49], reaching the lower limit $p > \dim_{\mathbf{C}} V_I$ leads to real applications.

If one only considers graded submodules, the work is inherently algebraic in nature. In 2011, Douglas and K. Wang made a breakthrough in which analysis became predominant:

Theorem 5.1 ([22]) *For every $q \in \mathbf{C}[z_1, \ldots, z_n]$, the submodule $[q]$ of the Bergman module is p-essentially normal for every $p > n$.*

What is remarkable about this result is that it is *unconditional* in the sense that it makes no assumptions about the polynomial q. This sets a very high standard for all the essential-normality results to come. In [29], we were able to show that the analogue of Theorem 5.1 holds for the Hardy module $H^2(S)$.

As it turns out, for essential-normality problems on the unit ball, there is a very simple parameter that measures both progress and the level of difficulty: it is the weight t introduced in Section 3. Theorem 5.1 covers the weight $t = 0$, whereas our Hardy-space analogue [29] covers $t = -1$. It is the value of t that actually determines the difficulty of the problem: the more negative the value of t, the harder it is to solve the corresponding essential-normality problem. For the Drury-Arveson space analogue of Theorem 5.1, the level of difficulty is set at $t = -n$. The following is the best that we can do at the moment:

Theorem 5.2 ([33]) *Let q be an arbitrary polynomial in $\mathbf{C}[z_1, \ldots, z_n]$. Then for every $-3 < t < \infty$, the submodule $[q]^{(t)}$ of $\mathscr{H}^{(t)}$ is p-essentially normal for every $p > n$.*

Specializing this to the weight $t = -2$ gives us the only unconditional essential normality that we have at the moment in a Drury-Arveson space case:

Corollary 5.3 *For every $q \in \mathbf{C}[z_1, z_2]$, the submodule $[q]$ of the two-variable Drury-Arveson module H_2^2 is p-essentially normal for every $p > 2$.*

For three variables, we must deal with the weight $t = -3$, which represents substantially more difficulty. At the moment, we only have partial results for the weight $t = -3$ [33], and even that requires non-trivial work. Although we are not able to solve it at the moment, the stumbling block for the case $t = -3$ can be stated as a very explicit estimate:

Problem 5.4 *Denote $R = z_1\partial_1 + \cdots + z_n\partial_n$, the radial derivative on $\mathbf{B}$. Given an arbitrary $q \in \mathbf{C}[z_1, \ldots, z_n]$, does there exist a constant $C = C(q)$ such that*

$$\int |(Rq)(\zeta)f(\zeta)|^2 dv(\zeta) \leq C \int |R(qf)(\zeta)|^2 dv(\zeta)$$

for every $f \in \mathbf{C}[z_1, \ldots, z_n]$ satisfying the condition $f(0) = 0$?

For weights $t < -3$, the main difficulty can also be stated in terms of explicit estimates of this type [33, Definition 1.7]. In fact, we think that the value of these essential-normality problems lies precisely in the fact that they embody such non-trivial analysis.

7.6 Expanding on Drury's idea

A reasonable way to interpret the von Neumann inequality (7.2) is to say that the tuple $(M_{\zeta_1}, \ldots, M_{\zeta_n})$ on H_n^2 "dominates" every other row contraction. In other words, the row contraction $(M_{\zeta_1}, \ldots, M_{\zeta_n})$ on H_n^2 is the "master" among all row contractions.

This interpretation of (7.2) inspires us to consider the following question. Suppose that we have two row contractions, $(A_1,\ldots,A_n)$ and $(B_1,\ldots,B_n)$. It seems fair to say that $(B_1,\ldots,B_n)$ dominates $(A_1,\ldots,A_n)$ if the inequality

$$\|p(A_1,\ldots,A_n)\| \le \|p(B_1,\ldots,B_n)\|$$

holds for every polynomial $p \in \mathbf{C}[z_1,\ldots,z_n]$. Or, perhaps one can relax this condition somewhat: if there is a constant $0 < C < \infty$ such that

$$\|p(A_1,\ldots,A_n)\| \le C\|p(B_1,\ldots,B_n)\|$$

for every polynomial $p \in \mathbf{C}[z_1,\ldots,z_n]$, one might still say that the tuple $(B_1,\ldots,B_n)$ dominates the tuple $(A_1,\ldots,A_n)$.

The main point is this: we can also ask the rather restricted question whether a given tuple $(B_1,\ldots,B_n)$ dominates (whatever the word means) a particular $(A_1,\ldots,A_n)$, not just the question whether it dominates a general class of $(A_1,\ldots,A_n)$'s. In other words, the tuple $(B_1,\ldots,B_n)$ may not be as dominating as the tuple $(M_{\zeta_1},\ldots,M_{\zeta_n})$ on H_n^2, but does it dominate a particular $(A_1,\ldots,A_n)$ nonetheless?

The first hint of a possible hierarchical structure among commuting tuples comes from the fact that the Drury-Arveson space H_n^2 is really "the head" of the family of reproducing-kernel Hilbert spaces $\{\mathcal{H}^{(t)} : -n \le t < \infty\}$ introduced in Section 4. An obvious question is: what about the "lesser" tuples $(M_{\zeta_1}^{(t)},\ldots,M_{\zeta_n}^{(t)})$, $-n < t < \infty$? What do they dominate?

In other words, here we are asking whether there is some sort of hierarchy, albeit partial, among commuting tuples of operators. Obviously, such a general question represents a monumental undertaking, one that perhaps requires the efforts of many researchers over many years.

But one thing encouraging is that there are plenty of interesting examples of such a hierarchy. One way to construct such examples is to consider reproducing-kernel Hilbert spaces $\mathcal{H}^{(v)}$ that are even more general than the $\mathcal{H}^{(t)}$ introduced in Section 4.

Suppose that $v = \{v_\alpha : \alpha \in \mathbf{Z}_+^n\}$ is a set of positive numbers satisfying the condition

$$\sum_{\alpha \in \mathbf{Z}_+^n} v_\alpha |w^\alpha|^2 < \infty$$

for every $w \in \mathbf{B}$. We define an inner product $\langle \cdot,\cdot \rangle_v$ on $\mathbf{C}[z_1,\ldots,z_n]$ according to the following rules: $\langle z^\alpha, z^\beta \rangle_v = 0$ whenever $\alpha \ne \beta$, and

$$\langle z^\alpha, z^\alpha \rangle_v = 1/v_\alpha$$

for $\alpha \in \mathbf{Z}_+^n$. Let $\|\cdot\|_v$ be the norm induced by the inner product $\langle \cdot,\cdot \rangle_v$, and let $\mathcal{H}^{(v)}$ be the Hilbert space obtained as the completion of $\mathbf{C}[z_1,\ldots,z_n]$ with respect to $\|\cdot\|_v$. Let $(M_{\zeta_1}^{(v)},\ldots,M_{\zeta_n}^{(v)})$ be the tuple of multiplication by the coordinate functions on $\mathcal{H}^{(v)}$. Each $\mathcal{H}^{(v)}$ has its own collection of multipliers, and for each multiplier f we write $M_f^{(v)}$ for the operator of multiplication by f on $\mathcal{H}^{(v)}$.

For each $j \in \{1, \dots, n\}$, let ε_j denote the element in $\mathbf{Z}_+^n$ whose j-th component is 1 and whose other components are 0. If $\alpha \in \mathbf{Z}_+^n$ and if the j-th component of α is not 0, then, of course, $\alpha - \varepsilon_j \in \mathbf{Z}_+^n$. Given a set of positive numbers $v = \{v_\alpha : \alpha \in \mathbf{Z}_+^n\}$, for $\alpha \in \mathbf{Z}_+^n$ we define $v_{\alpha-\varepsilon_j} = 0$ if the j-th component of α is 0. Suppose that $v = \{v_\alpha : \alpha \in \mathbf{Z}_+^n\}$ satisfies the condition

$$\sum_{j=1}^{n} \frac{v_{\alpha-\varepsilon_j}}{v_\alpha} \leq 1 \quad \text{for every } \alpha \in \mathbf{Z}_+^n.$$

Then it is easy to verify that $(M_{\zeta_1}^{(v)}, \dots, M_{\zeta_n}^{(v)})$ is a row contraction on $\mathscr{H}^{(v)}$.

In our view, identity (7.4) is the heart and soul of the theory of Drury-Arveson space. Therefore one way to uncover a possible hierarchy described above is to tinker with (7.4) and see what happens. For example, we can try to replace the coefficients $|\alpha|!/\alpha!$ in (7.4) with general v_α. If $|\alpha|!/\alpha!$ is replaced by v_α, then obviously the defect operator

$$D = 1 - A_1A_1^* - \cdots - A_nA_n^*$$

in (7.4) also needs to be replaced accordingly. But what replaces D?

Theorem 6.1 ([31]) *Let $A = (A_1, \dots, A_n)$ be a commuting row contraction on a Hilbert space H. Suppose that there is a positive operator W on H such that the sum*

$$Y = \sum_{\alpha \in \mathbf{Z}_+^n} v_\alpha A^\alpha W A^{*\alpha}$$

converges in the weak operator topology. Furthermore, suppose that the sum Y satisfies the operator inequality $c \leq Y \leq C$ on H for some scalars $0 < c \leq C < \infty$. Then the inequality

$$\|f(A)\| \leq (C/c)\|M_f^{(v)}\|$$

holds for every multiplier f of the space $\mathscr{H}^{(v)}$.

Let B be a bounded operator on a Hilbert space H. Then its *essential norm* is

$$\|B\|_{\mathscr{Q}} = \inf\{\|B + K\| : K \in \mathscr{K}(H)\},$$

where $\mathscr{K}(H)$ is the collection of compact operators on H.

Theorem 6.2 ([31]) *Let $A = (A_1, \dots, A_n)$ be a commuting row contraction on a separable Hilbert space H. Suppose that there is a positive, compact operator W on H such that the sum*

$$Y = \sum_{\alpha \in \mathbf{Z}_+^n} v_\alpha A^\alpha W A^{*\alpha}$$

converges in the weak operator topology. Furthermore, suppose that the operator Y has the following two properties:

(a) There are scalars $0 < c \leq C < \infty$ *such that the operator inequality* $c \leq Y \leq C$ *holds on* H*;*

(b) $Y = 1 + K$*, where* K *is a compact operator on* H*.*

Then the inequality

$$\|f(A)\|_{\mathscr{Q}} \leq \|M_f^{(v)}\|_{\mathscr{Q}}$$

holds for every multiplier f *of the space* $\mathscr{H}^{(v)}$*.*

Families of non-trivial (albeit quite technical) examples of A, v, W were given in [31]. In particular, if $-n < s < t < \infty$, then the tuple $(M_{\zeta_1}^{(s)}, \ldots, M_{\zeta_n}^{(s)})$ dominates the tuple $(M_{\zeta_1}^{(t)}, \ldots, M_{\zeta_n}^{(t)})$.

7.7 Closure of the polynomials

Let $\mathscr{A}$ be the closure of $\mathbf{C}[z_1, \ldots, z_n]$ in $\mathscr{M}$ with respect to the multiplier norm. It is easy to understand that $\mathscr{A}$ is a special set of multipliers. Obviously, $\mathscr{A}$ is contained in $A(\mathbf{B})$, the ball algebra. Thus all multipliers in $\mathscr{A}$ are continuous on $\overline{\mathbf{B}}$, but not all continuous multipliers are in $\mathscr{A}$, although this latter statement is not completely trivial. Arveson showed in [5] that the maximal ideal space of $\mathscr{A}$ is homeomorphic to $\overline{\mathbf{B}}$.

In [13], Clouâtre and Davidson identified the first and second dual of $\mathscr{A}$: there is a commutative von Neumann algebra W such that

$$\mathscr{A}^* \simeq \mathscr{M}_* \oplus_1 W_* \quad \text{and} \quad \mathscr{A}^{**} \simeq \mathscr{M} \oplus_\infty W.$$

They established analogues of several classical results concerning the dual space of the ball algebra. These developments are deeply intertwined with the problem of peak interpolation for multipliers. It is also worth mentioning that these results shed light on the nature of the extreme points of the unit ball of $\mathscr{A}^*$.

Theorem 7.1 ([13]) *Let* $f \in \mathscr{A}$ *with* $\|f\|_\infty < \|f\|_{\mathscr{M}} = 1$*. The set*

$$\mathscr{F} = \{\Psi \in \mathscr{M}_* : \|\Psi\|_{\mathscr{A}^*} = 1 = \Psi(f)\}$$

has extreme points, which are also extreme points of the closed unit ball of $\mathscr{M}_*$*.*

Building on the work in [13], Clouâtre and Davidson further gave a complete characterization of absolutely continuous commuting row contractions in measure theoretic terms [14]. They also showed that completely non-unitary row contractions are necessarily absolutely continuous, which is a direct analogue of the single-operator case.

Acknowledgment

We are grateful to Ronald G. Douglas for bringing Lubin's work [40, 41] to our attention.

References

[1] J. Agler and J. McCarthy, *Pick interpolation and Hilbert function spaces*, Graduate Studies in Mathematics **44**, American Mathematical Society, Providence, RI, 2002.

[2] A. Aleksandrov, The existence of inner functions in a ball, (Russian) *Mat. Sb. (N.S.)* **118(160)** (1982), 147-163, 287; (English) *Math. USSR Sbornik* **46** (1983), 143-159.

[3] J. Arazy and G. Zhang, Homogeneous multiplication operators on bounded symmetric domains, *J. Funct. Anal.* **202** (2003), 44-66.

[4] N. Arcozzi, R. Rochberg and E. Sawyer, Carleson Measures for the Drury-Arveson Hardy space and other Besov-Sobolev spaces on complex balls, *Adv. Math.* **218** (2008), 1107-1180.

[5] W. Arveson, Subalgebras of C^*-algebras. III. Multivariable operator theory, *Acta Math.* **181** (1998), 159-228.

[6] W. Arveson, The curvature invariant of a Hilbert module over $\mathbf{C}[z_1, \ldots, z_d]$, *J. Reine Angew. Math.* **522** (2000), 173-236.

[7] W. Arveson, The Dirac operator of a commuting d-tuple, *J. Funct. Anal.* **189** (2002), 53-79.

[8] W. Arveson, p-summable commutators in dimension d, *J. Operator Theory* **54** (2005), 101-117.

[9] W. Arveson, Myhill Lectures, SUNY Buffalo, April 2006.

[10] W. Arveson, Quotients of standard Hilbert modules, *Trans. Amer. Math. Soc.* **359** (2007), 6027-6055.

[11] J. Ball, T. Trent and V. Vinnikov, Interpolation and commutant lifting for multipliers on reproducing kernel Hilbert spaces, *Op. Th. Adv. and App.* **122** (2001), 89-138.

[12] C. Cascante, J. Fàbrega, and J. Ortega, On weighted Toeplitz, big Hankel operators and Carleson measures, *Integral Equations Operator Theory* **66** (2010), 495-528.

[13] R. Clouâtre and K. Davidson, Duality, convexity and peak interpolation in the Drury-Arveson space, *Adv. Math.* **295** (2016), 90-149.

[14] R. Clouâtre and K. Davidson, Absolute continuity for commuting row contractions, *J. Funct. Anal.* **271** (2016), 620-641.

[15] R. Clouâtre and K. Davidson, Ideals in a multiplier algebra on the ball, *Trans. Amer. Math. Soc.* **370** (2018), 1509-1527.

[16] S. Costea, E. Sawyer and B. Wick, The corona theorem for the Drury-Arveson Hardy space and other holomorphic Besov-Sobolev spaces on the unit ball in $\mathbf{C}^n$, *Anal. PDE* **4** (2011), 499-550.

[17] K. Davidson, The mathematical legacy of William Arveson, *J. Operator Theory* **68** (2012), 307-334.

[18] R. Douglas, Essentially reductive Hilbert modules, *J. Operator Theory* **55** (2006), 117-133.

[19] R. Douglas, A new kind of index theorem, *Analysis, geometry and topology of elliptic operators*, 369-382, World Sci. Publ., Hackensack, NJ, 2006.

[20] R. Douglas, K. Guo and Y. Wang, On the p-essential normality of principal submodules of the Bergman module on strongly pseudoconvex domains, preprint 2017, arXiv:1708.04949

[21] R. Douglas, X. Tang and G. Yu, An analytic Grothendieck Riemann Roch theorem, *Adv. Math.* **294** (2016), 307-331.

[22] R. Douglas and K. Wang, Essential normality of cyclic submodule generated by any polynomial, *J. Funct. Anal.* **261** (2011), 3155-3180.

[23] R. Douglas and Y. Wang, Geometric Arveson-Douglas conjecture and holomorphic extension, *Indiana Univ. Math. J.* **66** (2017), 1499-1535.

[24] S. Drury, A generalization of von Neumann's inequality to the complex ball, *Proc. Amer. Math. Soc.* **68** (1978), 300-304.

[25] M. Engliš and J. Eschmeier, Geometric Arveson-Douglas conjecture, *Adv. Math.* **274** (2015), 606-630.

[26] J. Eschmeier, Essential normality of homogeneous submodules, *Integr. Equ. Oper. Theory* **69** (2011), 171-182.

[27] Q. Fang and J. Xia, Defect operators associated with submodules of the Hardy module, *Indiana Univ. Math. J.* **60** (2011), 729-749.

[28] Q. Fang and J. Xia, Commutators and localization on the Drury-Arveson space, *J. Funct. Anal.* **260** (2011), 639-673.

[29] Q. Fang and J. Xia, Essential normality of polynomial-generated submodules: Hardy space and beyond, *J. Funct. Anal.* **265** (2013), 2991-3008.

[30] Q. Fang and J. Xia, Corrigendum to “Multipliers and essential norm on the Drury-Arveson space”, *Proc. Amer. Math. Soc.* **141** (2013), 363-368.

[31] Q. Fang and J. Xia, A hierarchy of von Neumann inequalities?, *J. Operator Theory* **72** (2014), 219-239.

[32] Q. Fang and J. Xia, On the problem of characterizing multipliers for the Drury-Arveson space, *Indiana Univ. Math. J.* **64** (2015), 663-696.

[33] Q. Fang and J. Xia, On the essential normality of principal submodules of the Drury-Arveson module, *Indiana Univ. Math. J.* **67** (2018), 1439-1498.

[34] K. Guo, Defect operators for submodules of H_d^2, *J. Reine Angew. Math.* **573** (2004), 181-209.

[35] K. Guo and K. Wang, Essentially normal Hilbert modules and K-homology. II. Quasi-homogeneous Hilbert modules over the two dimensional unit ball, *J. Ramanujan Math. Soc.*, **22** (2007), 3, 259-281.

[36] K. Guo and K. Wang, Essentially normal Hilbert modules and K-homology, *Math. Ann.* **340** (2008), 907-934.

[37] K. Guo and Ch. Zhao, p-essential normality of quasi-homogeneous Drury-Arveson submodules, *J. Lond. Math. Soc.* (2) **87** (2013), 899-916.

[38] M. Kennedy and O. Shalit, Essential normality and the decomposability of algebraic varieties, *New York J. Math.* **18** (2012), 877-890.

[39] E. Løw, A construction of inner functions on the unit ball in $\mathbf{C}^p$, *Invent. Math.* **67** (1982), 223-229.

[40] A. Lubin, Models for commuting contractions, *Michigan Math. J.* **23** (1976), 161-165.

[41] A. Lubin, Weighted shifts and products of subnormal operators, *Indiana Univ. Math. J.* **26** (1977), 839-845.

[42] M. Mitkovski and B. Wick, A reproducing kernel thesis for operators on Bergman-type function spaces, arXiv:1212.0507v4

[43] J. Ortega and J. Fàbrega, Pointwise multipliers and decomposition theorems in analytic Besov spaces, *Math. Z.* **235** (2000), 53-81.

[44] S. Parrott, Unitary dilations for commuting contractions, *Pacific J. Math.* **34** (1970), 481-490.

[45] S. Richter and J. Sunkes, Hankel operators, invariant subspaces, and cyclic vectors in the Drury-Arveson space, *Proc. Amer. Math. Soc.* **144** (2016), 2575-2586.

[46] W. Rudin, *Function Theory in the Unit Ball of* $\mathbf{C}^n$, Springer-Verlag, New York, 1980.

[47] O. Shalit, Stable polynomial division and essential normality of graded Hilbert modules, *J. Lond. Math. Soc.* (2) **83** (2011), 273-289.

[48] N. Varopoulos, On an inequality of von Neumann and an application of the metric theory of tensor products to operators theory, *J. Funct. Anal.* **16** (1974), 83-100.

[49] Y. Wang and J. Xia, Essential normality for quotient modules and complex dimensions, *J. Funct. Anal.* **276** (2019), 1061-1096.

Chapter 8

A Brief Survey of Operator Theory in $H^2(\mathbb{D}^2)$

Rongwei Yang

Department of Mathematics and Statistics, SUNY at Albany, Albany, NY 12222, USA
ryang@albany.edu

In memory of Keiji Izuchi

CONTENTS

8.1 Introduction

This survey aims to give a brief introduction to operator theory in the Hardy space over the bidisc $H^2(\mathbb{D}^2)$. As an important component of multivariable operator theory, the theory in $H^2(\mathbb{D}^2)$ focuses primarily on two pairs of commuting operators that are naturally associated with invariant subspaces (or submodules) in $H^2(\mathbb{D}^2)$. The connection between operator-theoretic properties of the pairs and the structure of the invariant subspaces is the main subject. The theory in $H^2(\mathbb{D}^2)$ is motivated by and still tightly related to several other influential theories, namely, the Nagy-Foias theory on operator models, Ando's dilation theorem of commuting operator pairs, Rudin's function theory on $H^2(\mathbb{D}^n)$, and Douglas-Paulsen's framework of Hilbert modules. Due to the simplicity of the setting, a great supply of examples in particular, the operator theory in $H^2(\mathbb{D}^2)$ has seen remarkable growth in the past two decades. This survey is far from a full account of this development but rather a glimpse from the author's perspective. Its goal is to show an organized structure of this theory, to bring together some results and references and to inspire curiosity in new researchers.

8.2 Background

8.2.1 The unilateral shift

A bounded linear operator T on a complex separable Hilbert space $\mathscr{H}$ is said to be normal if $T^*T = TT^*$. A milestone in operator theory is the functional calculus

$$f(T) = \int_{\sigma(T)} f(\lambda)dE(\lambda),$$

which identifies a continuous function f on the spectrum $\sigma(T)$ with an operator $f(T)$ in the commutative C^*-algebra $C^*(T)$ generated by T and the identity I. However, many important operators are not normal. A classical example is the unilateral shift S defined by $Se_n = e_{n+1},\ \ n \geq 0$, where $\{e_n \mid n \geq 0\}$ is an orthonormal basis for a complex Hilbert space $\mathscr{H}$. In this case $S^*S - SS^* = e_0 \otimes e_0$ which is the orthogonal projection from $\mathscr{H}$ to the one-dimensional subspace $\mathbb{C}e_0$. Since S is fairly simple, one naturally wonders whether its invariant subspaces can be fully described. About 70 years ago A. Beurling solved the problem using a representation of S on the Hardy space $H^2(\mathbb{D})$. Let $\{z^n : n \geq 0\}$ be the standard orthonormal basis for $H^2(\mathbb{D})$ and let $U : \mathscr{H} \to H^2(\mathbb{D})$ be the unitary defined by $Ue_n = z^n$, $n \geq 0$. Then USU^* is the multiplication by z on functions of $H^2(\mathbb{D})$; i.e., it is the Toeplitz operator T_z. Hence the invariant subspace problem for S is equivalent to the invariant subspace problem for T_z.

Theorem 2.1 (Beurling [9]) *A closed subspace $M \subset H^2(\mathbb{D})$ is invariant for T_z if and only if $M = \theta H^2(\mathbb{D})$ for some inner function θ.*

Here an inner function θ is a function in $H^2(\mathbb{D})$ such that $|\theta(z)| = 1$ almost everywhere on the unit circle $\mathbb{T}$. Observe that the statement of Beurling's theorem uses two properties of $H^2(\mathbb{D})$ which are absent in the abstract Hilbert space $\mathscr{H}$, namely boundary value and multiplication of functions in $H^2(\mathbb{D})$. The study of the unilateral shift S, and Beurling's theorem in particular, has made a wide and far reaching impact on analytic function theory, operator theory and operator algebras, etc. It is thus an appealing question whether Beurling's theorem has multivariable generalizations.

8.2.2 Rudin's examples and a theorem of Ahern and Clark

The pair (T_{z_1}, T_{z_2}) (or simply (T_1, T_2)) of Toeplitz operators on the Hardy space over the bidisc $H^2(\mathbb{D}^2)$ is a natural generalization of the unilateral shift T_z on $H^2(\mathbb{D})$. A closed subspace $M \subset H^2(\mathbb{D}^2)$ is said to be invariant if it is invariant under multiplications by both T_1 and T_2, i.e., $z_1 M \subset M$ and $z_2 M \subset M$. Examples of such subspaces are rich. For instance, if J is an ideal in the polynomial ring $\mathscr{R} := \mathbb{C}[z_1, z_2]$ then its closure $[J]$ in $H^2(\mathbb{D}^2)$ is an invariant subspace. More generally, if $f_1, f_2, \cdots, f_n$ are functions in $H^2(\mathbb{D}^2)$ then the closure

$$[f_1, f_2, \cdots, f_n] := cls\{f_1\mathscr{R} + f_2\mathscr{R} + \cdots + f_n\mathscr{R}\}$$

is an invariant subspace generated by the set $\{f_1, f_2, \cdots, f_n\}$. The minimal cardinality of such generating sets for an invariant subspace M is called the rank of M and shall be denoted by $\text{rank}(M)$. It was observed in [78] that the invariant subspace $[z_1 - z_2]$ is not of the form $\theta H^2(\mathbb{D}^2)$ for any two variable inner function $\theta \in H^2(\mathbb{D}^2)$. In fact, more exotic examples of invariant subspaces were constructed by Rudin ([78, 79]).

Example 2.2 *Let M be the set of all functions $f \in H^2(\mathbb{D}^2)$ which have a zero of order greater than or equal to n at $(\alpha_n, 0)$, where $\alpha_n = 1 - n^{-3}, n = 1, 2, \cdots$. Then M is not finitely generated; i.e., there exists no finite set $\{f_1, f_2, \cdots, f_n\} \subset H^2(\mathbb{D}^2)$ such that $M = [f_1, f_2, \cdots, f_n]$.*

Example 2.3 *Fix a number $2 > R > 1$ and let*

$$f(z_1, z_2) = \prod_{j=1}^{\infty}\left(1 - R\left(\frac{z_1 + \overline{\alpha_j} z_2}{2}\right)^{n_j}\right),$$

where $|\alpha_j| = 1$ such that the value of each α_j repeats infinite many times in the sequence (α_j), and (n_j) is an increasing sequence of natural numbers which are chosen such that $f \in H^2(\mathbb{D}^2)$. Then the singly-generated invariant subspace $[f]$ contains no bounded functions other than 0.

These two somewhat pathological examples, which stand as extreme contrasts to Beurling's theorem, manifest the complexity of invariant subspaces in $H^2(\mathbb{D}^2)$.

Nevertheless, progress was made by Ahern and Clark in 1970 regarding invariant subspaces of finite codimension. Here, as well as in many other places in this survey, we state the result in the two variable case.

Theorem 2.4 (Ahern and Clark [1]) *Suppose M is an invariant subspace in $H^2(\mathbb{D}^2)$ of codimension $k<\infty$ and let*

$$Z(J)=\{z\in\mathbb{C}^2 \mid p(z)=0,\ \forall p\in J\},$$

where J is an ideal in $\mathscr{R}$. Then $\mathscr{R}\cap M$ is an ideal in $\mathscr{R}$ such that

(a) $\mathscr{R}\cap M$ *is dense in* M,

(b) $\dim(\mathscr{R}/\mathscr{R}\cap M)=k$,

(c) $Z(\mathscr{R}\cap M)$ *is a finite subset of* $\mathbb{D}^2$.

Conversely if $J\subset\mathscr{R}$ is an ideal with $Z(J)$ being a finite subset of $\mathbb{D}^2$ then $[J]$ is an invariant subspace of $H^2(\mathbb{D}^2)$ with $\dim[J]^\perp=\dim(\mathscr{R}/J)$ and $[J]\cap\mathscr{R}=J$.

8.2.3 Module formulation and the rigidity phenomena

In view of the complexity of invariant subspaces in $H^2(\mathbb{D}^2)$, Douglas and Paulsen in [31] proposed a natural algebraic approach. Since an invariant subspace M is invariant under multiplications by the coordinates z_1 and z_2, it is invariant under multiplications by functions in $\mathscr{R}$, and more generally it is invariant under multiplications by functions in the bidisc algebra $A(\mathbb{D}^2)$ which is the closure of $\mathscr{R}$ in $C(\overline{\mathbb{D}}^2)$. In other words, an invariant subspace can be viewed as a submodule of $H^2(\mathbb{D}^2)$ over the bidisc algebra $A(\mathbb{D}^2)$. This point of view gives rise to the following two canonical equivalence relations for submodules.

Definition 2.5 *Two submodules M_1 and M_2 in $H^2(\mathbb{D}^2)$ are said to be similar if there is an invertible module map $T: M_1\to M_2$ such that*

$$fTh=T(fh),\quad f\in A(\mathbb{D}^2),\ h\in M_1.$$

Further, M_1 and M_2 are said to be unitarily equivalent if T is unitary.

Unitary equivalence of submodules is a subject in many studies. For a good reference we refer readers to [16] and the many references therein.

Theorem 2.6 (Agrawal, Clark and Douglas [2], Douglas and Yan [34]) *Two submodules M_1 and M_2 are unitarily equivalent if and only if there exists a $\varphi\in L^\infty(\mathbb{T}^2)$ with $|\varphi(z_1,z_2)|=1$ a.e. such that $M_2=\varphi M_1$.*

The following result describes all submodules that are equivalent to $H^2(\mathbb{D}^2)$.

Corollary 2.7 *A submodule M is unitarily equivalent to $H^2(\mathbb{D}^2)$ if and only if $M = \theta H^2(\mathbb{D}^2)$ for some inner function $\theta \in H^2(\mathbb{D}^2)$.*

Proof If M is unitarily equivalent to $H^2(\mathbb{D}^2)$ then by Theorem 2.6 there exists a $\varphi \in L^\infty(\mathbb{T}^2)$ with $|\varphi| = 1$ a.e. such that $M = \varphi H^2(\mathbb{D}^2)$. Since $1 \in H^2(\mathbb{D}^2)$, we have $\varphi \in M \subset H^2(\mathbb{D}^2)$ (which means φ has an analytic extension to $\mathbb{D}^2$). Hence φ is inner. The other direction is clear because multiplication by φ is a unitary map from $H^2(\mathbb{D}^2)$ to M.

A bit of function theory from [78] is needed to proceed. The Nevanlinna class $N(\mathbb{D}^2)$ consists of holomorphic functions f on $\mathbb{D}^2$ such that

$$\sup_{0 \le r < 1} \int_{\mathbb{T}^2} \log^+ |f(rz)| dm(z) < \infty,$$

where $\log^+ x = \log x$ for $x \ge 1$ and $\log^+ x = 0$ for $0 \le x < 1$. If $f \in N(\mathbb{D}^2)$ then the radial limit f^* exists a.e. on $\mathbb{T}^2$ with $\log|f^*| \in L^1(\mathbb{T}^2)$, and there is a real singular measure $d\sigma_f$ on $\mathbb{T}^2$ such that the least harmonic majorant $u(\log|f|)$ of $\log|f|$ is given by

$$u(\log|f|)(z) = P_z(\log|f^*| + d\sigma_f), \quad z \in \mathbb{D}^2, \tag{8.1}$$

where P_z is the Poisson integration. Every function $f \in H^p(\mathbb{D}^2), 0 < p < \infty$, is in $N(\mathbb{D}^2)$ with $d\sigma_f \le 0$. For a submodule M, the following two sets are defined in [34]:

$$Z(M) = \{z \in \mathbb{D}^2 \mid f(z) = 0, \forall f \in M\}, \; Z_\partial(M) = \inf\{-d\sigma_f \mid f \in M, \; f \ne 0\}.$$

A submodule is said to satisfy condition (*) if $Z_\partial(M) = 0$ and the real 2-dimensional Hausdorff measure of $Z(M)$ is 0.

Theorem 2.8 (Douglas and Yan [34]) *If M_1 and M_2 are submodules of $H^2(\mathbb{D}^2)$ which satisfy (*), then the following are equivalent:*

(a) M_1 and M_2 are unitarily equivalent.

(b) M_1 and M_2 are similar.

(c) $M_1 = M_2$.

To see why condition (*) matters, one observes that $H^2(\mathbb{D}^2)$ and $z_1 H^2(\mathbb{D}^2)$ are unitarily equivalent but not identical. Further study which involves algebraic geometry and commutative algebras was made in [32]. Theorem 2.8 reveals the so-called rigidity phenomenon of submodules. The following example provides the simplest illustration.

Example 2.9 *Fix any $\alpha \in \mathbb{D}^2$ and set*

$$H_\alpha = \{f \in H^2(\mathbb{D}^2) \mid f(\alpha) = 0\}.$$

Clearly, H_α satisfies (*). Then it follows from Theorem 2.8 that for two points $\alpha, \beta \in \mathbb{D}^2$, the submodules H_α and H_β are similar if and only if $\alpha = \beta$.

Given a submodule $M \subset H^2(\mathbb{D}^2)$, its orthogonal complement $N := H^2(\mathbb{D}^2) \ominus M$ is a module over $A(\mathbb{D}^2)$ with module action defined by

$$f \cdot h = P_N(fh), \quad h \in N, \ f \in A(\mathbb{D}^2),$$

where P_N stands for the orthogonal projection from $H^2(\mathbb{D}^2)$ onto N. Such N are called quotient modules, and they exhibit a stronger rigidity phenomenon.

Theorem 2.10 (Douglas and Foias [30]) *Two quotient modules N_1 and N_2 in $H^2(\mathbb{D}^2)$ are unitarily equivalent if and only if $N_1 = N_2$.*

We shall give a simple proof to this theorem in Section 8.5.

The singular measure $d\sigma_f$ and $Z_\partial(M)$ turned out to be useful in several other places. A connection with joint invariant subspaces of (S_1, S_2) will be mentioned in Section 8.5. An application to the study of multipliers of submodule M can be found in Nakazi [72]. His other pioneering work related to the topics of this survey can be found in [70, 71]. For some ideals $J \subset \mathscr{R}$ (for instance $J = (z_1 - z_2)$), the associated quotient module $H^2(\mathbb{D}^2) \ominus [J]$ naturally gives rise to a Cowen-Douglas operator. Hence a particular Hermitian bundle (kernel bundle) exists over the variety $Z(J)$. Study of the curvature of the bundle and its Chern class is an exciting interplay between operator theory and complex geometry. Since the focus of this survey is operator theory, we refer the readers to [14, 15, 24, 25, 26, 31] for research along that line.

8.3 Nagy-Foias theory in $H^2(\mathbb{D}^2)$

Let E be a complex separable infinite dimensional Hilbert space. A contraction T on E is a bounded linear operator such that $\|T\| \le 1$. It is said to be of $C_{\cdot 0}$-class if $(T^*)^n$ converges to 0 in strong operator topology. It is said to be of C_0-class if there exists a nontrivial $\varphi \in H^\infty(\mathbb{D})$ such that $\varphi(T) = 0$. In functional model theory the vector-valued Hardy space $H^2(\mathbb{D}) \otimes E$ is used to construct universal models for the $C_{\cdot 0}$-class and C_0-class contractions. For references on this important part of operator theory we refer readers to [8, 10, 13, 27, 64, 74, 77]. In particular, [10] includes many updates of the theory. Here we only mention a few key ingredients in the Nagy-Foias theory ([10]). Let the defect operators of a contraction T be defined by

$$D_T = (I - T^*T)^{1/2}, \quad D_{T^*} = (I - TT^*)^{1/2},$$

and let the closure of their ranges be denoted by $\mathscr{D}_T$ and $\mathscr{D}_{T^*}$, respectively. The characteristic operator function of T is defined as

$$\theta_T(\lambda) = [-T + \lambda D_{T^*}(I - \lambda T^*)^{-1} D_T] \,|_{\mathscr{D}_T}, \qquad \lambda \in \mathbb{D}. \tag{8.2}$$

8.3.1 A formulation in $H^2(\mathbb{D}^2)$

Since one can identify E with $H^2(\mathbb{D})$, the Nagy-Foias theory can be formulated in $H^2(\mathbb{D})\otimes E \cong H^2(\mathbb{D}^2)$. A closed subspace $M \subset H^2(\mathbb{D}^2)$ is said to be z_1-invariant if it is invariant under multiplication by z_1. Then by Beurling-Lax-Halmos theorem ([63, 74]) every z_1-invariant subspace $M \in H^2(\mathbb{D}^2)$ is of the form $\Theta H^2(E')$, where E' is some Hilbert space and Θ is an analytic function on $\mathbb{D}$ such that $\Theta(z)$ is a bounded linear operator from E' to $E = H^2(\mathbb{D})$ for each z, and the radial limit $\lim_{r\to 1}\Theta(re^{i\theta})$ exists and is an isometry for almost every $\theta \in [0, 2\pi)$. Up to multiplication by constant unitary on the right such Θ is unique, and it is called the left-inner function for the z_1-invariant subspace M. As before, we set $N = H^2(\mathbb{D}^2)\ominus M$. Then the compression S_1 of multiplication by z_1 to N is defined by $S_1 h = P_N(z_1 h),\quad h \in N$. The contraction S_1 is a universal model in the following sense.

Theorem 3.1 *Let T be a $C_{\cdot 0}$-class contraction. Then there is a z_1-invariant subspace $M \subset H^2(\mathbb{D}^2)$ such that T is unitarily equivalent to S_1.*

The left-inner function Θ for M and the characteristic function θ_{S_1} for S_1 are related by the identity

$$\Theta(\lambda) = U\theta_{S_1}(\lambda)V \oplus W, \qquad \lambda \in \mathbb{D}, \tag{8.3}$$

where U and V are constant unitaries and W is either a constant unitary or 0. We will say more about W a bit later.

Let $H^2_{z_i}$ denote the Hardy space over the unit disc with coordinate functions $z_i, i = 1, 2$. Note that $H^2_{z_1}$ and $H^2_{z_2}$ are different subspaces of $H^2(\mathbb{D}^2)$. For $\lambda \in \mathbb{D}$, the left and right evaluation operators $L(\lambda): H^2(\mathbb{D}^2) \to H^2_{z_2}$ and respectively $R(\lambda): H^2(\mathbb{D}^2) \to H^2_{z_1}$ are defined by

$$L(\lambda)h(z_2) = h(\lambda, z_2),\;\; R(\lambda)h = h(z_1, \lambda),\;\; h \in H^2(\mathbb{D}^2).$$

The restriction of $L(\lambda)$ and $R(\lambda)$ to a closed subspace $K \subset H^2(\mathbb{D}^2)$ shall be denoted by $L_K(\lambda)$ and $R_K(\lambda)$, respectively. For a z_1-invariant subspace M we let $M_1 = M \ominus z_1 M$. Then another natural operator $D: M_1 \to N$ is defined by

$$Dh = \frac{h(z_1, z_2) - h(0, z_2)}{z_1}, \quad h \in M_1.$$

Clearly, D is the restriction of the backward shift $T^*_{z_1}$ to M_1. The following proposition describes more explicitly the defect operators of S_1.

Proposition 3.2 ([95]) *Let M be a z_1-invariant subspace. Then on $N = H^2(\mathbb{D}^2)\ominus M$ we have*

(a) $S_1^* S_1 + DD^* = I,$

(b) $S_1 S_1^* + L_N^*(0)L_N(0) = I.$

The restriction $L_{M_1}(\lambda)$ turns out to coincide with the left inner function $\Theta(\lambda)$ for M. In this case, the W in (8.3) is exactly the identity operator on the intersection $M_1 \cap H^2_{z_2}$. Hence $W = 0$ if and only if $M_1 \cap H^2_{z_2}$ is trivial. Moreover, it can be shown that

$$L_{M_1}(\lambda) = L_{M_1}(0) + \lambda L_N(0)(I - \lambda S_1^*)^{-1} D, \qquad \lambda \in \mathbb{D}. \tag{8.4}$$

The following spectral connections hold (cf. [10, 95]).

Proposition 3.3 *Let M be a z_1-invariant subspace and $\lambda \in \mathbb{D}$. Then*

(a) *$S_1 - \lambda$ is invertible if and only if $L_{M_1}(\lambda) : M_1 \to H^2(\mathbb{D})$ is invertible.*

(b) *$S_1 - \lambda$ is Fredholm if and only if $L_{M_1}(\lambda) : M_1 \to H^2(\mathbb{D})$ is Fredholm, and in this case,*

$$\operatorname{ind}(S_1 - \lambda) = \operatorname{ind} L_{M_1}(\lambda).$$

The following fact, which was not observed in the Nagy-Foias theory, is very important to the study of submodules.

Lemma 3.4 ([94]) *Let M be a z_1-invariant subspace. Then $R_{M_1}(\lambda)$ is Hilbert-Schmidt for every $\lambda \in \mathbb{D}$.*

Submodules with dimension

$$dim(M \ominus (z_1 M + z_2 M)) = \infty$$

constitute a formidable class. The following corollary is thus a fine application of Proposition 3.3 and Lemma 3.4.

Corollary 3.5 ([95]) *If M is a submodule with*

$$dim(M \ominus (z_1 M + z_2 M)) = \infty,$$

then

$$\sigma_e(S_1) = \sigma_e(S_2) = \overline{\mathbb{D}}.$$

Proof Let $\{f_n \mid n = 1, 2, \cdots\}$ be an orthonormal basis for

$$M \ominus (z_1 M + z_2 M) = M_1 \cap M_2.$$

Since $R_{M_1}(\lambda)$ is Hilbert-Schmidt by Lemma 3.4, we have

$$\sum_{n=1}^{\infty} \|R_{M_1}(\lambda) f_n\|^2 \le \|R_{M_1}(\lambda)\|_{HS}^2,$$

where $\|\cdot\|_{HS}$ stands for the Hilbert-Schmidt norm. This implies that

$$\lim_{n\to\infty} \|R_{M_2}(\lambda) f_n\| = \lim_{n\to\infty} \|R_{M_1}(\lambda) f_n\| = 0.$$

This shows that $R_{M_2}(\lambda)$ is not Fredholm, and therefore $\lambda \in \sigma_e(S_2)$ by the parallel statement of Proposition 3.3 for S_2. Proof for S_1 is similar.

The space $M \ominus (z_1 M + z_2 M)$ is sometimes called the defect space for the submodule M and its importance will become evident later. Its dimension is less than or equal to the rank of M ([36]). For Rudin's submodule M in Example 2.2, it is shown that the dimension of $M \ominus (z_1 M + z_2 M)$ is 2, though $\text{rank}(M) = \infty$ ([96]). Hence the defect space is in general not a generating set for M, unlike the situation for the Bergman space $L_a^2(\mathbb{D})$ ([7]) or the classical Hardy space $H^2(\mathbb{D})$. It was a question due to Nakazi ([69]) whether every rank one submodule $M = [f]$ is generated by the defect spae $M \ominus (z_1 M + z_2 M)$. The question is solved only recently.

Theorem 3.6 (Izuchi [59]) *There exists a nontrivial function $f \in H^2(\mathbb{D}^2)$ such that $[f] \ominus (z_1[f] + z_2[f])$ does not generate $[f]$.*

Since S_1 on a quotient module N is much less general than the S_1 in Theorem 3.1, the reformulation of Nagy-Foias theory in $H^2(\mathbb{D}^2)$ thus gives rise to the following

Problem 1. Characterize the $C_{\cdot 0}$-class contraction T which is unitarily equivalent to S_1 on a quotient module $N \subset H^2(\mathbb{D}^2)$.

8.3.2 Important examples

One well-known example of Problem 1 is the Bergman shift B, i.e., multiplication by the coordinate function w on the classical Bergman space $L_a^2(\mathbb{D})$.

Example 3.7 *Let $e'_n = \sqrt{n+1}w^n, n \geq 0$ be the standard orthonormal basis for $L_a^2(\mathbb{D})$. Then it is easy to check that*

$$Be'_n = \sqrt{\frac{n+1}{n+2}}e'_{n+1}, \quad n \geq 0.$$

If $M = [z_1 - z_2]$ then $N = M^\perp$ has the orthonormal basis

$$e_n(z) = \frac{1}{\sqrt{n+1}} \frac{z_1^{n+1} - z_2^{n+1}}{z_1 - z_2}, \quad n = 0, 1, \cdots.$$

One checks that $S_1 e_n = \sqrt{\frac{n+1}{n+2}} e_{n+1}$, which verifies that S_1 is unitarily equivalent to B.

A remarkable application of Example 3.7 is that it leads to alternative approaches to some important problems on the Bergman shift B. For instance, in [84] Sun and Zheng reproved the Aleman-Richter-Sundberg theorem ([7]) that every B-invariant subspace $\mathscr{M} \subset L_a^2(\mathbb{D})$ is generated by its defect space $\mathscr{M} \ominus w\mathscr{M}$, or in other words, $\mathscr{M} = [\mathscr{M} \ominus w\mathscr{M}]$. Reducing subspaces of $\varphi(B)$, where φ is a finite Blaschke product, can also be studied via $H^2(\mathbb{D}^2)$ ([40, 85]).

Another example of Problem 1 is the C_0-class operators. For a single variable inner function $\theta \in H^2(\mathbb{D})$, the quotient $K(\theta) = H^2(\mathbb{D}) \ominus \theta H^2(\mathbb{D})$ is often called the model space. The operator $S(\theta)$ defined by

$$S(\theta)h = P_{K(\theta)}(zh), \quad h \in K(\theta),$$

is often called the associated Jordan block and it has been very well-studied ([8]). Clearly, $S(\theta)$ is in C_0-class. A sequence of inner functions $(q_j)_{j=0}^{\infty}$ in $H^2_{z_1}$ is called an inner sequence if q_{j+1} divides q_j for each j. An inner-sequence-based submodule is of the form

$$M = \bigoplus_{j=0}^{\infty} q_j H^2_{z_1} z_2^j,$$

where (q_j) is an inner sequence. In this case,

$$N = H^2(\mathbb{D}^2) \ominus M = \bigoplus_{j=0}^{\infty} (H^2_{z_1} \ominus q_j H^2_{z_1}) z_2^j,$$

and hence S_1 is unitarily equivalent to $\oplus_{j=0}^{\infty} S(q_j)$. The following fact follows from the observation above and a classical result about C_0-class operators ([8, 10, 83]).

Theorem 3.8 *Every C_0-class contraction is quasi-similar to S_1 for some inner-sequence-based submodule $N^{\perp}$.*

Here, two operators T_i on respective Banach spaces $H_i, i = 1,2$ are said to be quasi-similar if there are bounded operators $A : H_1 \to H_2$ and $B : H_2 \to H_1$, both injective with dense range, such that $AT_1 = T_2A$ on H_1 and $T_1B = BT_2$ on H_2.

The rank of inner-sequence-based submodules is carefully studied in the case q_0 is a Blaschke product ([50]). Setting $\xi_n = q_n/q_{n+1}, n = 0, 1, \cdots$, then each ξ_n is also a Blaschke product. Then for every $\alpha \in \mathbb{D}$, we define

$$N_\alpha = \{n \in \mathbb{N} \mid \xi_{n-1}(\alpha) = 0\},$$

and denote its cardinality by $|N_\alpha|$. The following theorem has a rather difficult proof.

Theorem 3.9 (K. J. Izuchi, K. H. Izuchi and Y. Izuchi [50]) *Let M be an inner-sequence-based submodule with q_0 being a Blaschke product. Then*

$$\operatorname{rank} M = \sup_{\alpha \in \mathbb{D}} |N_\alpha| + 1.$$

It is interesting to observe that Rudin's submodule in Example 2.2 is in fact inner-sequence-based with inner sequence defined by

$$q_0(z_1) = \prod_{n=1}^{\infty} b_n^n(z_1), \text{ and } q_j = q_{j-1} / \prod_{n=j}^{\infty} b_n^n(z_1), \quad j \geq 1,$$

where

$$b_n(z_1) = \frac{z_1 - \alpha_n}{1 - \overline{\alpha_n} z_1}, \qquad n \geq 0.$$

Hence

$$\xi_j(z_1) = \prod_{n=j+1}^{\infty} b_n^n(z_1), \qquad j \geq 0.$$

Therefore, by Theorem 3.9 we have

$$\operatorname{rank} M \geq |N_{\alpha_n}| + 1 = n + 1, \qquad \forall n \in \mathbb{N}.$$

This verifies that Rudin's submodule in Example 2.2 has infinite rank.

A distinguished property of inner-sequence-based submodules is that the characteristic functions for S_1 and S_2 are both simple and elegant ([75]). The readers shall find great fun computing for themselves. Due to their simple structure, inner-sequence-based submodules are useful for many purposes. We refer the readers to [82, 103, 105] for some of the applications.

8.4 Commutators

It is not hard to check that T_1 and T_2 doubly commute in the sense that $[T_1, T_2] = [T_1^*, T_2] = 0$. Given a submodule M, we denote by R_1^M and R_2^M the restrictions of T_1 and T_2 to M, and by S_1^N and S_2^N the compression of T_1 and T_2 to the quotient module $N = H^2(\mathbb{D}^2) \ominus M$, respectively, i.e.,

$$S_1^N f = P_N(z_1 f), \ S_2^N f = P_N z_2 f, \qquad f \in N.$$

It is easy to see that two submodules M and M' are unitarily equivalent if and only if the pairs (R_1^M, R_2^M) and $(R_1^{M'}, R_2^{M'})$ are unitarily equivalent in the sense that there exists a unitary $U: M \to M'$ such that $UR_i^M = R_i^{M'}U, \ i = 1, 2$. This fact, together with Theorem 2.10, indicates that all information of submodule M is contained in the pairs (R_1^M, R_2^M) and (S_1^N, R_2^N). Moreover, the pairs (R_1^M, R_2^M) and (S_1^N, S_2^N) must also be intimately connected because they are both faithful representations of M. When a submodule M is fixed we shall write the two pairs simply as $R = (R_1, R_2)$ and $S = (S_1, S_2)$. They are the primary subject of study in the operator theory in $H^2(\mathbb{D}^2)$.

8.4.1 Double commutativity

The pairs (R_1, R_2) and (S_1, S_2) are both commuting pairs, but they doubly commute only for very special submodules M.

Theorem 4.1 (Gatage-Mandrekar [39]; Mandrekar [68]) *Let M be a submodule. Then the commutator $[R_1^*, R_2] = 0$ if and only if $M = \theta H^2(\mathbb{D}^2)$ for some inner function $\theta \in H^2(\mathbb{D}^2)$.*

In view of Corollary 2.7, this theorem implies that a submodule M is unitarily equivalent to $H^2(\mathbb{D}^2)$ if and only if $[R_1^*, R_2] = 0$. The case $[S_1^*, S_2] = 0$ is considered in [36], and a complete solution is obtained in [56].

Theorem 4.2 (Izuchi, Nakazi and Seto [56]) *Let M be a submodule. Then $[S_1^*, S_2] = 0$ if and only if M is of the form*

$$M = q_1(z_1)H^2(\mathbb{D}^2) + q_2(z_2)H^2(\mathbb{D}^2),$$

where q_1 and q_2 are either 0 or one variable inner functions.

This theorem provides another important example of submodules, and we shall come back to it several times later. For convenience we shall call them the Izuchi-Nakazi-Seto type submodule (or INS-submodule for short). It is worth noting that both Theorem 4.1 and 4.2 have clean generalizations to $H^2(\mathbb{D}^n)$ ([22, 66, 80, 81]). A question raised in [36] is whether INS-submodule M is necessarily of rank 2 when q_1 and q_2 are nonconstant inner functions. This problem is harder than it looks and is settled only recently.

Theorem 4.3 (Chattopadhyay, Dias and Sarkar [17]) *Let q_1 and q_2 be nonconstant inner functions in $H^2(\mathbb{D})$. Then the INS-submodule has rank 2.*

An interesting but only partially solved problem about this submodule is raised in [56].

Problem 2. Is $\left(q_1H^2(\mathbb{D}^2) + q_2H^2(\mathbb{D}^2)\right) \cap H^\infty(\mathbb{D}^2) = q_1H^\infty(\mathbb{D}^2) + q_2H^\infty(\mathbb{D}^2)$?

Theorem 4.4 (Nakazi and Seto [73]) *If either q_1 or q_2 is a finite Blaschke product then the answer to Problem 2 is positive.*

INS-submodules seem to appear first in [60] where a special case of Theorem 4.3 was proved. Submodules that have structure similar to INS or inner-sequenced-based submodules are constructed in [62] through the so-called generalized inner functions, i.e., functions $f \in H^\infty(\mathbb{D}^2)$ such that $|f| \geq c > 0$ a.e. on $\mathbb{T}^2$ for some constant c. It is observed in [61] that $fH^2(\mathbb{D}^2)$ is a closed subspace in $H^2(\mathbb{D}^2)$ (and hence is a submodule) if and only if f is a generalized inner function.

8.4.2 Hilbert-Schmidtness

For general operators A_1 and A_2 we call $[A_1^*, A_2]$ their cross commutator. Results in the last subsection make one wonder if the cross commutators $[R_1^*, R_2]$ and $[S_1^*, S_2]$ are of finite rank or compact for more general submodules. In [35], it was shown that if J is an ideal in $\mathscr{R}$ with the zero set $Z(J)$ of codimension ≥ 2 then both S_1 and S_2 are essentially normal, which implies that $[S_1^*, S_2]$ is compact. Then in [19] it was shown that $[R_1^*, R_2]$ is Hilbert-Schmidt on $[J]$ for every homogeneous ideal J. Using Proposition 3.3 and Lemma 3.4, stronger results were obtained in [94] where it showed that $[R_1^*, R_2]$ and $[R_1^*, R_1][R_2^*, R_2]$ are both Hilbert-Schmidt on $[J]$ for every ideal J. A further generalization which went beyond polynomials was made in [96].

Theorem 4.5 *Let M be a submodule such that $\sigma_e(S_1)\cap\sigma_e(S_2)\neq\overline{\mathbb{D}}$. Then*

(a) $[R_1^*,R_2]$ *is Hilbert-Schmidt,*

(b) $[R_2^*,R_2][R_1^*,R_1]$ *is Hilbert-Schmidt,*

(c) $[S_1^*,S_2]$ *is Hilbert-Schmidt.*

Proof Here we just give a proof to (b) with the assumption that S_2 is Fredholm. The other cases follow similarly with technical modifications. First note that $[R_i^*,R_i]=I-R_iR_i^*$ is the orthogonal projection from M onto $M_i=M\ominus z_iM, i=1,2$. Check that

$$R(0)\left([R_2^*,R_2][R_1^*,R_1]\right)=R(0)[R_1^*,R_1]-R(0)R_2R_2^*[R_1^*,R_1]=R(0)[R_1^*,R_1].$$

By Lemma 3.4, the right-hand side is Hilbert-Schmidt. Since S_2 is Fredholm, the evaluation $R(0)$ on $M\ominus z_2M$ is Fredholm by Proposition 3.3. Hence the Hilbert-Schmidtness of $R(0)\left([R_2^*,R_2][R_1^*,R_1]\right)$ implies that $[R_2^*,R_2][R_1^*,R_1]$ is Hilbert-Schmidt.

A submodule for which $[R_1^*,R_2]$ and $[R_2^*,R_2][R_1^*,R_1]$ are both Hilbert-Schmidt is called a Hilbert-Schmidt submodule. Except for the submodules considered in Corollary 3.5, it seems all submodules we have encountered so far satisfy the condition in Theorem 4.5. and hence are Hilbert-Schmidt. For Rudin's Example 2.2 this fact was proved in [96]. For Rudin's Example 2.3 this was proved in [76] with a condition. For the so-called splitting submodules, the fact is proved in [49].

For simplicity, we set

$$\Sigma_0(M)=\|[R_2^*,R_2][R_1^*,R_1]\|_{HS}^2,\qquad \Sigma_1(M)=\|[R_1^*,R_2]\|_{HS}^2.$$

It is not hard to see that $\Sigma_0(M)$ and $\Sigma_1(M)$ are invariants with respect to the unitary equivalence of submodules. If $\{\varphi_n\mid n\geq 0\}$ and $\{\psi_m\mid m\geq 0\}$ are orthonormal bases for M_1 and M_2, respectively, then

$$\Sigma_0(M)=\sum_{m,n=0}^{\infty}|\langle\varphi_n,\ \psi_m\rangle|^2,\qquad \Sigma_1(M)=\sum_{m,n=0}^{\infty}|\langle z_2\varphi_n,\ z_1\psi_m\rangle|^2.$$

The following numerical relation is shown in [96, 100].

Theorem 4.6 *Let M be a Hilbert-Schmidt submodule. Then*

(a) $\Sigma_0(M)-\Sigma_1(M)=1$,

(b) $[S_1^*,S_2]$ *is Hilbert-Schmidt with* $\|[S_1^*,S_2]\|_{HS}^2+\|P_N1\|^2\leq\Sigma_0(M)$.

In fact, a sequence of numerical invariants can be defined as follows:

$$\Sigma_k(M)=\sum_{m,n=0}^{\infty}|\langle z_2^k\varphi_n,\ z_1^k\psi_m\rangle|^2,\qquad k\geq 0.$$

It is an interesting exercise to compute this sequence for the INS-submodule. In view of Theorem 4.1 the following problem seems puzzling.

Problem 3. For what submodule M is $\Sigma_2(M) = 0$?

The most difficult problem along this line, in the author's view, is the following conjecture ([94]).

Conjecture 4. If a submodule has finite rank then it is Hilbert-Schmidt.

Recently, the following preliminary result is obtained.

Theorem 4.7 (Luo, Izuchi and Yang [65]) *If a submodule M contains the function $z_1 - z_2$ then it is Hilbert-Schmidt if and only if it is finitely generated.*

Problem 5. Does Theorem 4.7 hold if $z_1 - z_2$ is replaced by other polynomials?

Many studies were made about the ranks of $[R_1^*, R_2]$ and $[S_1^*, S_2]$ (cf. [48, 53, 54, 55]). Here we just mention two results.

Theorem 4.8 (K. J. Izuchi and K. H. Izuchi [54]) *Let M be a Hilbert-Schmidt submodule. Then*

$$\operatorname{rank}[R_1^*, R_2] - 1 \leq \operatorname{rank}[S_1^*, S_2] \leq \operatorname{rank}[R_1^*, R_2] + 1.$$

It is also a good exercise to verify that for the INS-submodules with nonconstant q_1 and q_2 we have $\operatorname{rank}[R_1^*, R_2] = 1$. An interesting generalization is the following.

Example 4.9 (K. J. Izuchi and K. H. Izuchi [55]) *Let q_i, $i = 1,2,3,4$ be nonconstant one variable inner functions and define*

$$M = q_1(z_1)q_2(z_1)H^2(\mathbb{D}^2) + q_2(z_1)q_3(z_2)H^2(\mathbb{D}^2) + q_3(z_2)q_4(z_2)H^2(\mathbb{D}^2).$$

Then M is a submodule and $\operatorname{rank}[S_1^*, S_2] = 1$. *In fact it is shown that every submodule for which* $\operatorname{rank}[S_1^*, S_2] = 1$ *is either a variation of M or of the form $\theta H^2(\mathbb{D}^2)$ for some genuine two variable inner function.*

8.5 Two-variable Jordan block

For an INS-submodule M the quotient $N = H^2(\mathbb{D}^2) \ominus M$ is of the form

$$N = (H^2(\mathbb{D}) \ominus q_1 H^2(\mathbb{D})) \otimes (H^2(\mathbb{D}) \ominus q_2 H^2(\mathbb{D})),$$

with $S_1 = S(q_1) \otimes I$ and $S_2 = I \otimes S(q_2)$. In view of this connection, the pair (S_1, S_2) on a general quotient module is sometimes called a two-variable Jordan block.

8.5.1 Defect operators

Using the so-called hereditary functional calculus ([5]), for a pair of commuting contractions $A=(A_1,A_2)$ we define the defect operator

$$\Delta_A = I - A_1^*A_1 - A_2^*A_2 + A_1^*A_2^*A_1A_2.$$

It is computed in [101] that for the adjoint pair $S^*=(S_1^*,S_2^*)$ of the two-variable Jordan block, we have

$$\Delta_{S^*} = \varphi \otimes \varphi, \tag{8.5}$$

where $\varphi = P_N 1$. Since φ is nonzero for every nontrivial quotient module, this fact leads to a simple proof of the following:

Proposition 5.1 *For every quotient module N, the pair (S_1,S_2) has no nontrivial joint reducing subspace.*

Proof Suppose the pair $S=(S_1,S_2)$ has a nontrivial joint reducing subspace, say N'. Then both N' and $N''=N\ominus N'$ are nontrivial quotient modules. We let S' and S'' be the restriction of S to N' and N'', respectively. Then since $S=S'\oplus S''$, we have

$$\Delta_{S^*} = \Delta_{S'^*} \oplus \Delta_{S''^*} = (P_{N'}1 \otimes P_{N'}1) \oplus (P_{N''}1 \otimes P_{N''}1),$$

which contradicts the fact that Δ_{S^*} is of rank 1.

Observe that (8.5) implies that the C^*-algebra $C^*(S_1,S_2)$ generated by I,S_1 and S_2 contains a rank 1 operator. Then it follows from Proposition 5.1 that $C^*(S_1,S_2)$ contains all compact operators on N ([28]). Interestingly, Formula (8.5) also led to the following concise proof of Theorem 2.10.

Proof First, we see that the operator norm $\|\Delta_{S^*}\| = \|\varphi\|^2$. Then for every $\lambda \in \mathbb{D}^2$, we have

$$\begin{aligned}(1-\lambda_1S_1)^{-1}(1-\lambda_2S_2)^{-1}\Delta_{S^*}\frac{\varphi}{\|\varphi\|} &= \|\varphi\|(1-\lambda_1S_1)^{-1}(1-\lambda_2S_2)^{-1}P_N1\\ &= \|\varphi\|P_NK(\lambda,\cdot)\\ &= \|\varphi\|K^N(\lambda,\cdot),\end{aligned}$$

where K and K^N are the reproducing kernels for $H^2(\mathbb{D}^2)$ and N, respectively. It follows that the Hilbert-Schmidt norm

$$\|(1-\lambda_1S_1)^{-1}(1-\lambda_2S_2)^{-1}\Delta_{S^*}\|_{HS}^2 = \|\varphi\|^2K^N(\lambda,\lambda).$$

If $S=(S_1,S_2)$ is unitarily equivalent to the two-variable Jordan block $S'=(S_1',S_2')$ on a quotient module N', then by (8.5) we must have $\|\varphi\|=\|\varphi'\|$ and hence

$$K^N(\lambda,\lambda) = K^{N'}(\lambda,\lambda), \qquad \forall \lambda \in \mathbb{D}^2.$$

This implies that $N = N'$ ([37, 97]).

The defect operator Δ_S is more complicated and can be of infinite rank. The original form of the following theorem is shown in [101].

Theorem 5.2 *If M is a Hilbert-Schmidt submodule then Δ_S is Hilbert-Schmidt on $N = M^{\perp}$ with $\|\Delta_S\|_{HS}^2 \leq 2\left(\|P_N 1\|^2 + \Sigma_1(M)\right)$.*

8.5.2 Joint invariant subspace

Given a submodule M, if a submodule M' sits properly between M and $H^2(\mathbb{D}^2)$, i.e. $M \subsetneq M' \subsetneq H^2(\mathbb{D}^2)$ then it is called an intermediate submodule between M and $H^2(\mathbb{D}^2)$. It is not hard to verify that the pair (S_1, S_2) on $N = M^{\perp}$ has a nontrivial joint invariant subspace $N' \subset N$ if and only if $M' := N' \oplus M$ is an intermediate submodule between M and $H^2(\mathbb{D}^2)$. Clearly, if $2 \leq \dim N < \infty$ then (S_1, S_2) has a nontrivial joint invariant subspace. Indeed, by Theorem 2.4 there exists an $\alpha \in \mathbb{D}^2$ such that H_α is an intermediate submodule. If (S_1, S_2) doubly commutes then it follows from Theorem 4.2 that (S_1, S_2) has a nontrivial joint invariant subspace. The following is a simple observation.

Proposition 5.3 *Let $\theta \in H^2(\mathbb{D}^2)$ be a nontrivial inner function and*

$$N = H^2(\mathbb{D}^2) \ominus \theta H^2(\mathbb{D}^2).$$

Then (S_1, S_2) has a nontrivial joint invariant subspace.

Proof If θ vanishes at some point $\alpha \in \mathbb{D}^2$, then $H_\alpha \ominus \theta H^2(\mathbb{D}^2)$ is a nontrivial joint invariant subspace for (S_1, S_2), where H_α is defined as in Example 2.9. If θ has no zero in $\mathbb{D}^2$, then $\sqrt{\theta}$ is well-defined and is in $H^2(\mathbb{D}^2)$. To show that $\sqrt{\theta}H^2(\mathbb{D}^2)$ sits properly between $\theta H^2(\mathbb{D}^2)$ and $H^2(\mathbb{D}^2)$ is to check $\sqrt{\theta}H^2(\mathbb{D}^2) \neq H^2(\mathbb{D}^2)$. Suppose the equality holds. Then we would have $1/\sqrt{\theta} \in H^2(\mathbb{D}^2)$. The fact that $1/|\sqrt{\theta(z)}| = 1$ a.e. on $\mathbb{T}^2$ would then imply that $1/\sqrt{\theta}$ is inner, which is impossible since $1/\sqrt{\theta(0,0)} > 1$.

Inner functions θ not vanishing on $\mathbb{D}^2$ have a notable feature in terms of the singular measure defined in (8.1). Since in this case $\log|\theta(z)|$ is well-defined and harmonic, by (8.1) we have

$$\begin{aligned} 0 > \log|\theta(z)| &= u(\log|\theta(z)|) = P_z(\log|\theta^*| + d\sigma_\theta) \\ &= P_z(d\sigma_\theta), \quad \forall z \in \mathbb{D}^2, \end{aligned} \tag{8.6}$$

which means $d\sigma_\theta < 0$. About the singular measure $d\sigma_f$, the following two properties are worth mentioning here ([34, 78]).

1) For every $f \in H^p(\mathbb{D}^2)$, $0 < p < \infty$ we have $d\sigma_f \leq 0$.

2) For $f,g \in H^2(\mathbb{D}^2)$ we have $d\sigma_{gf} = d\sigma_g + d\sigma_f$.

These two properties and the preceding observation in fact give another proof to Proposition 5.3. Two other partial results, which are unrelated to inner functions, are as follows ([95, 97]).

Theorem 5.4 *Let N be a quotient module with $dim(N) \geq 2$. If either $\|S_1\| < 1$ or $\|S_2\| < 1$ then (S_1,S_2) has a nontrivial joint invariant subspace.*

Theorem 5.5 *If M is a submodule with*

$$dim(M \ominus (z_1M + z_2M)) \geq 2,$$

then (S_1,S_2) has a nontrivial joint invariant subspace.

However, to the best of the author's knowledge, the following general problem is open.

Problem 6. For an infinite dimensional quotient module N does the pair (S_1,S_2) necessarily have a nontrivial joint invariant subspace?

8.6 Fredholmness of the pairs (R_1,R_2) and (S_1,S_2)

For every pair (A_1,A_2) of commuting operators on a Hilbert space $\mathscr{H}$ there is an associated Koszul complex

$$K(A_1,A_2): \quad 0 \to \mathscr{H} \xrightarrow{d_1} \mathscr{H} \oplus \mathscr{H} \xrightarrow{d_2} \mathscr{H} \to 0,$$

where $d_1x = (-A_2x, A_1x)$ and $d_2(x,y) = A_1x + A_2y, x,y \in \mathscr{H}$. It is easy to check that $d_2d_1 = 0$. The sequence $K(A_1,A_2)$ is said to be exact if the kernel of d_2 coincides with the range of d_1. It is said to be Fredholm if d_1 and d_2 both have closed range and

$$\dim\ker(d_1) + \dim(\ker(d_2) \ominus d_1(\mathscr{H})) + \dim(\mathscr{H} \ominus d_2(\mathscr{H} \oplus \mathscr{H})) < +\infty.$$

In this case its index of (A_1,A_2) is defined as

$$\begin{aligned}\mathrm{ind}(A_1,A_2) := &\dim\ker(d_1) - \dim(\ker(d_2) \ominus d_1(\mathscr{H})) \\ &+ \dim(\mathscr{H} \ominus d_2(\mathscr{H} \oplus \mathscr{H})).\end{aligned}$$

The set

$$\sigma(A_1,A_2) = \{(\lambda_1,\lambda_2) \in \mathbb{C}^2 : K(A_1-\lambda_1, A_2-\lambda_2) \text{ is not exact}\}$$

is called the Taylor spectrum of (A_1,A_2), and the set

$$\sigma_e(A_1,A_2) = \{(\lambda_1,\lambda_2) \in \mathbb{C}^2 : K(A_1-\lambda_1,A_2-\lambda_2) \text{ is not Fredholm}\}$$

is called the essential Taylor spectrum of (A_1,A_2). The Taylor spectrum can be defined similarly for any tuple of commuting operators, and it is a pillar in multivariable operator theory. We refer readers to [20, 86, 87, 88] for its orginal definition and related functional calculus. Back to submodules, the following are known ([96, 101]).

Theorem 6.1 *Let M be a Hilbert-Schmidt submodule. Then (R_1,R_2) is Fredholm with* $\mathrm{ind}(R_1,R_2)=1$.

Theorem 6.2 *Let M be a Hilbert-Schmidt submodule. Then $\sigma_e(S_1,S_2) \subset \partial\mathbb{D}^2$.*

It was a question whether for every Hilbert-Schmidt submodule M the essential Taylor spectrum $\sigma_e(S_1,S_2)$ is a proper subset of $\partial\mathbb{D}^2$ (cf. [101]). Recall that $Z(M)$ is the set of common zeros of functions in M. Paper [44] defines

$$Z'_\partial(M) = \{\lambda \in \partial\mathbb{D}^2 \mid \text{there exist sequence } \lambda_n \in Z(M) \text{ such that } \lim_{n\to\infty}\lambda_n = \lambda\}.$$

Observe that $Z'_\partial(M)$ is a subset of $\partial\mathbb{D}^2$ but $Z_\partial(M)$ (defined in Section 8.2) is a singular measure on $\mathbb{T}^2$. The support of $Z_\partial(M)$ is not equal to $Z'_\partial(M)$ either. For example, if θ is an inner function that has no zero in $\mathbb{D}^2$ and $M=\theta H^2(\mathbb{D}^2)$, then $Z'_\partial(M)=\emptyset$ but $Z_\partial(M) = -d\sigma_\theta$ has nonempty support (see (8.6)).

Theorem 6.3 (Guo and P. Wang [44]) *For every submodule M, the set $Z'_\partial(M) \subset \sigma_e(S_1,S_2)$.*

This theorem gives rise to the following interesting example.

Example 6.4 ([44]) *Let*

$$\varphi(w) = \prod_{n=1}^{\infty} \frac{\alpha_n - w}{1-\overline{\alpha_n}w}$$

be an infinite Blaschke product such that the unit circle $\mathbb{T}$ is contained in the closure of

$$Z(\varphi) = \{\alpha_n \mid n \geq 1\}.$$

Set $\Phi(z_1,z_2) = \varphi(z_1)\varphi(z_2)$ and consider the Beurling-type submodule $M = \Phi H^2(\mathbb{D}^2)$. Then it is not hard to see that $Z'_\partial(M) = \partial\mathbb{D}^2$. Hence, in view of Theorem 6.2 and 6.3, we have $\sigma_e(S_1,S_2) = \partial\mathbb{D}^2$.

There is a result without the assumption of Hilbert-Schmidtness.

Theorem 6.5 (Lu, R. Yang and Y. Yang [67]) *Let $N = M^\perp$ be a quotient module. If (S_1,S_2) is Fredholm, then both $M \ominus (z_1M+z_2M)$ and $\ker S_1 \cap \ker S_2$ are finite dimensional and*

$$\mathrm{ind}(S_1,S_2) = \dim(M\ominus(z_1M+z_2M)) - \dim(\ker S_1 \cap \ker S_2) - 1.$$

Interestingly, it follows from an observation about core operator (cf. (8.8)) that whenever M is Hilbert-Schmidt we have $\mathrm{ind}(S_1,S_2)=0$.

8.7 Essential normality of quotient module

A bounded linear operator T on a Hilbert space is said to be essentially normal if the commutator $[T^*, T]$ is compact. A quotient module $N \subset H^2(\mathbb{D}^2)$ is said to be essentially normal (or essentially reductive) if both S_1 and S_2 are essentially normal. We have observed in Section 8.5 that the C^*-algebra $C^*(S_1, S_2)$ contains the ideal $\mathscr{K}$ of all compact operators on N. Therefore, if N is essentially normal then the quotient algebra $C^*(S_1, S_2)/\mathscr{K}$ is commutative and it is isomorphic to the C^*-algebra of continuous functions on the essential Taylor spectrum $\sigma_e(S_1, S_2)$. This fact is neatly expressed in the following short exact sequence:

$$0 \to \mathscr{K} \xrightarrow{i} C^*(S_1, S_2) \xrightarrow{\pi} C(\sigma_e(S_1, S_2)) \to 0,$$

where i is the inclusion and π is the quotient map. Since the Bergman shift is essentially normal, the quotient module $N = H^2(\mathbb{D}^2) \ominus [z_1 - z_2]$ in Example 3.7 is essentially normal. The following generalization holds.

Theorem 7.1 (Clark [18]; P. Wang [89]) *Let $q_1(z_1)$ and $q_2(z_2)$ be two one-variable inner functions. Then the quotient module*

$$H^2(\mathbb{D}^2) \ominus [q_1(z_1) - q_2(z_2)]$$

is essentially normal if and only if both q_1 and q_2 are finite Blaschke products.

This theorem's sufficiency part was proved by D. Clark in [18], while the necessity is proved only recently by P. Wang in [89]. A generalization of this theorem to the Hardy space over the polydisc $H^2(\mathbb{D}^n)$ also holds ([18, 89]).

If M is the INS-submodule then

$$[S_1^*, S_1] = [S^*(q_1), S(q_1)] \otimes I_{K(q_2)}$$

and

$$[S_2^*, S_2] = I_{K(q_1)} \otimes [S^*(q_2), S(q_2)].$$

It is well-known that the commutator $[S^*(\theta), S(\theta)]$ is of at most rank 2 for every one-variable inner function θ. Hence $M^\perp$ is essentially normal if and only if both $K(q_1)$ and $K(q_2)$ are finite dimensional, which is the case if and only if both q_1 and q_2 are finite Blaschke products. Moreover, in this case the commutators $[S_i^*, S_i], i = 1, 2$ are both of finite rank.

The essential normality of the Beurling type quotient module $H^2(\mathbb{D}^2) \ominus \theta H^2(\mathbb{D}^2)$ is an intriguing problem. The following theorem gives an elegant characterization.

Theorem 7.2 (Guo and K. Wang [42]) *Let $\theta \in H^2(\mathbb{D}^2)$ be an inner function. Then $H^2(\mathbb{D}^2) \ominus \theta H^2(\mathbb{D}^2)$ is essentially normal if and only if θ is a rational inner function of degree at most $(1, 1)$.*

Here, a rational function $p(z_1,z_2)/q(z_1,z_2)$ is said to be of degree (m,n) if p and q are coprime polynomials with maximal degree m in the variable z_1 and n in the variable z_2. For instance the rational function $(z_1^3z_2^2 - 4z_2^3)/(1 - z_1z_2^4)$ has degree $(3,4)$. Paper [12] observed a connection of this problem with Agler's decomposition ([3])

$$1 - \overline{\theta(\lambda_1,\lambda_2)}\theta(z_1,z_2) = (1 - \overline{\lambda_2}z_2)K_1(\lambda,z) + (1 - \overline{\lambda_1}z_1)K_2(\lambda,z), \tag{8.7}$$

where $K_i : \mathbb{D}^2 \times \mathbb{D}^2 \to \mathbb{C},\ i = 1,2$ are positive kernels, and obtained the following result.

Theorem 7.3 (Bickel and Liaw [12]) *Let θ be a two-variable inner function. Then on $H^2(\mathbb{D}^2) \ominus \theta H^2(\mathbb{D}^2)$ the commutator $[S_1^*, S_1]$ has rank n if and only if θ is a rational inner function of degree $(1,n)$ or $(0,n)$.*

Observe that Theorems 7.2 and 7.3 together indicate that when $H^2(\mathbb{D}^2) \ominus \theta H^2(\mathbb{D}^2)$ is essentially normal the two commutators $[S_i^*, S_i], i = 1,2$ are at most rank one. The kernel function K_i in Agler's decomposition (8.7) naturally gives rise to a S_i-invariant subspaces, $i = 1,2$. A detailed study about the two spaces is made in Bickel and Gorkin [11]. It is worth noting that in the polydisc case $n \geq 3$, Das, Gorai and Sarkar [23] observed that Beurling type quotient modules $H^2(\mathbb{D}^n) \ominus \theta H^2(\mathbb{D}^n)$ are never essentially normal.

For submodules generated by a homogeneous polynomial, the essential normality of N is studied in the papers [44, 45, 91, 92]. Every homogeneous polynomial $p(z_1,z_2)$ has the decomposition $p = p_1p_2$, where

$$p_1(z_1,z_2) = \prod_{|\alpha_i|=|\beta_i|} (\alpha_i z_1 - \beta_i z_2) \text{ and } p_2(z_1,z_2) = \prod_{|\alpha_i|\neq|\beta_i|} (\alpha_i z_1 - \beta_i z_2).$$

Clearly, the polynomial p_1 has the property that $Z(p_1) \cap \partial\mathbb{D}^2 \subset \mathbb{T}^2$. A polynomial with this property is sometimes called a polynomial with distinguished variety ([6]). The following complete characterizaion of essential normality for homogeneous quotient modules is obtained.

Theorem 7.4 (Guo and P. Wang [44]) *Let p be a two variable homogeneous polynomial with decomposition $p = p_1p_2$ as above. Then $N = [p]^\perp$ is essentially normal if and only if p_2 has one of the following forms:*

(1) p_2 is a nonzero constant;

(2) $p_2 = \alpha z_1 - \beta z_2$ with $|\alpha| \neq |\beta|$;

(3) $p_2 = c(z_1 - \alpha z_2)(z_2 - \beta z_1)$ with $|\alpha| < 1$ and $|\beta| < 1$.

In particular, Theorem 7.4 indicates that if p has distinguished variety then p_2 is a constant and hence $[p]^\perp$ is essentially normal. This connection between a homogeneous polynomial with distinguished variety and essential normality remains valid for the Hardy space over the polydisc $H^2(\mathbb{D}^n)$ ([91, 92]). These facts motivate the following question.

Problem 7. Let $p(z_1, z_2)$ be any polynomial with distinguished variety. Then is $[p]^\perp$ essentially normal?

Essential normality is one of the most important topics in multivariable operator theory. For more information related to the bidisc we refer readers to [24, 29, 43]. Other related studies can be found in [33, 41].

8.8 Two single companion operators

The two-variable nature of submodule M and the associated pairs (R_1, R_2) and (S_1, S_2) presents a challenge for our study. A fruitful idea is to find some single operators that are tightly related to the submodule and the pairs so that classical one-variable techniques can assist more substantially in their studies. Two such operators, namely fringe operator and core operator, have been defined ([46, 96]) and well-studied in recent years.

8.8.1 Fringe operator

Fringe operators are defined on the defect spaces $M_1 = M \ominus z_1 M$. In parallel, they can be defined on $M_2 = M \ominus z_2 M$ as well, but we shall not need them for the survey here.

Definition 8.1 ([96]) *Given a submodule M, the associated fringe operator F is the compression of the operator R_2 to the space M_1. More precisely,*

$$Ff = P_{M_1} z_2 f, \qquad f \in M_1.$$

One observes carefully that the definition of fringe operator relies on R_1 as well as R_2. Hence F is indeed a single operator but with a two variable nature. The following results summarize some elementary properties of fringe operators.

Proposition 8.2 *Let F be the fringe operator associated with a submodule M. Then*

(a) $range(F) = (z_1 M + z_2 M) \ominus z_1 M$,

(b) $kerF^* = M \ominus (z_1 M + z_2 M)$,

(c) $kerF = z_1(kerS_1 \cap kerS_2)$,

(d) $\sigma(F) = \overline{\mathbb{D}}$.

Corollary 8.3 *If M is a submodule in $H^2(\mathbb{D}^2)$ then the fringe operator F on M_1 is Fredholm if and only if the tuple (R_1, R_2) on M is Fredholm. Moreover, in this case we have $ind(F) = -ind(R_1, R_2)$.*

Proposition 8.4 *Let M be a submodule. Then on $M \ominus z_1 M$ we have*

(a) $I - F^*F = [R_2{}^*, R_1][R_1{}^*, R_2]$,

(b) $I - FF^* = [R_1{}^*, R_1][R_2{}^*, R_2][R_1{}^*, R_1]$.

One observes that if M is Hilbert-Schmidt then F is Fredholm, and hence it follows from Proposition 8.2 (a) that $z_1 M + z_2 M$ is closed. So far there is no known example of a submodule M for which $z_1 M + z_2 M$ is not closed. The following problem is mentioned in [96].

Problem 8. Is $z_1 M + z_2 M$ closed for evey submodule M?

An application of fringe operators is the proof of Theorem 4.6 (a). In the case when M is Hilbert-Schmidt, Proposition 8.4 implies that both $I - F^*F$ and $I - FF^*$ are trace class with

$$tr(I - FF^*) = \Sigma_0(M), \quad tr(I - F^*F) = \Sigma_1(M).$$

It then follows from a trace formula by Caldron (cf. [47] Lemma 7.1), Theorem 6.1, Proposition 8.2 and Corollary 8.3 above that

$$\Sigma_0(M) - \Sigma_1(M) = tr[F^*, F] = ind F^* = ind(R_1, R_2) = 1.$$

This proves Theorem 4.6 (a).

Although the fringe operator captures much information about a submodule, it is not a complete invariant, meaning that there exist submodules M and M' such that their associated fringe operators are unitarily equivalent but the two submodules are not ([98]). More in-depth study of fringe operators for some particular submodules were made in [50, 51, 52].

8.8.2 Core operator

It is well-known that $H^2(\mathbb{D}^2)$ has the reproducing kernel

$$K(z, \lambda) = (1 - \overline{\lambda_1} z_1)^{-1}(1 - \overline{\lambda_2} z_2)^{-1},$$

where $z = (z_1, z_2)$ and $\lambda = (\lambda_1, \lambda_2)$ are points in $\mathbb{D}^2$. Clearly, for all fixed $z \in \mathbb{T}^2$ we have

$$\lim_{r \to 1} K(rz, rz) = \infty.$$

This behavior is shared by the reproducing kernel $K^M(z, \lambda)$ of submodules M. Hence it makes good sense to expect that the quotient

$$G^M(z, \lambda) = \frac{K^M(z, \lambda)}{K(z, \lambda)}$$

shall behave relatively well on the distinguished boundary $\mathbb{T}^2 \times \mathbb{T}^2$. As a matter of fact, what is true is surprising. It is shown in [97] that $G^M(z, z)$ is subharmonic in z_1 and z_2. Moreover, we have

Theorem 8.5 (Guo and Yang [46]) *Let M be a submodule. Then*

$$\lim_{r\to 1^+} G^M(rz,rz) = 1$$

for almost every $z \in \mathbb{T}^2$.

This theorem bears the closest resemblance to Beurling's theorem for $H^2(\mathbb{D})$. In an effort to study the behavior of G^M, the core operator is defined as follows.

Definition 8.6 ([46]) *For a submodule M, we define its core operator as*

$$C^M(f)(z) = \int_{\mathbb{T}^2} G^M(z,\lambda) f(\lambda) dm(\lambda), \quad f \in M,$$

where m is the normalized Lebesgue measure on $\mathbb{T}^2$.

For convenience we shall write C^M as C whenever there is no risk of confusion. Core operators have played a key role in many recent studies, because it has very nice properties and is also closely linked with the commutators mentioned in Section 8.4. One verifies first that C is self-adjoint, it maps M into itself, and it is equal to 0 on $N = M^{\perp}$. Furthermore, it follows from Theorem 8.5 that when C is trace class, we have

$$\mathrm{Tr} C = \int_{\mathbb{T}^2} G^M(z,z) dm(z) = 1. \tag{8.8}$$

It is quite entertaining to work out the following three examples.

Example 8.7 *Let $M = \theta H^2(\mathbb{D}^2)$, where θ is an inner function. Then it can be verified that*

$$K^M(z,\lambda) = \theta(z)\overline{\theta(\lambda)} K(z,\lambda).$$

Hence

$$G^M(z,\lambda) = \theta(z)\overline{\theta(\lambda)}$$

and C is the rank-one projection $\theta \otimes \theta$.

As a matter of fact, this is the only case where C has rank 1 ([46]).

Example 8.8 *Let M be an INS-submodule. Then one can verify that*

$$G^M(z,\lambda) = q_1(z_1)\overline{q_1(\lambda_1)} + q_2(z_2)\overline{q_2(\lambda_2)} - q_1(z_1)q_2(z_2)\overline{q_1(\lambda_1)q_2(\lambda_2)}.$$

Hence

$$C = q_1 \otimes q_1 + q_2 \otimes q_2 - (q_1 q_2) \otimes (q_1 q_2)$$

and it has the following four eigenvalues:

$$0, 1, \pm\sqrt{(1-|q_1(0)|^2)(1-|q_2(0)|^2)}.$$

There is another type of submodule with rank 3 core operator.

Example 8.9 (Izuchi and Ohno [58]; K. J. Izuchi and K. H. Izuchi [53]) *Consider $L^2(\mathbb{T}^2)$. Let*

$$k_{rz_2}(z_1) = \frac{\sqrt{1-r^2}}{1-r\overline{z_2}z_1}, \quad 0 \le r < 1,$$

and set

$$L = H^2(\mathbb{D}^2) \oplus \bigoplus_{j=0}^{\infty} \mathbb{C} z_1^j \overline{z_2} k_{rz_2}(z_1).$$

Then one can verify that L is invariant under the multiplication by z_1 and z_2. Furthermore, it can be shown that there exists an inner function $\theta \in H^2(\mathbb{D}^2)$ such that $M = \theta L$ is a submodule in $H^2(\mathbb{D}^2)$. Moreover, the core operator

$$C = \theta \otimes \theta + \theta \overline{z_2} k_{rz_2} \otimes \theta \overline{z_2} k_{rz_2} - \theta k_{rz_2} \otimes \theta k_{rz_2}.$$

The following problem is open.

Problem 9. Characterize all submodules M for which $\operatorname{rank}(C) = 3$.

But is there a submodule M such that $\operatorname{rank}(C) = 2$? It was observed in [99] that on every submodule M we have

$$C = \Delta_{R^*} = I - R_1 R_1^* - R_2 R_2^* + R_1 R_2 R_1^* R_2^*. \tag{8.9}$$

It is now apparent that C is self-adjoint. It is also not hard to show that the operator norm $\|C\| = 1$. Moreover, (8.9) implies that if M_1 and M_2 are unitarily equivalent submodules then C^{M_1} and C^{M_2} are unitarily equivalent. An unexpected fact is as follows.

Lemma 8.10 *For every submodule M, the square C^2 is unitarily equivalent to the diagonal block matrix*

$$\begin{pmatrix} [R_1^*, R_1][R_2^*, R_2][R_1^*, R_1] & 0 \\ 0 & [R_1^*, R_2][R_2^*, R_1] \end{pmatrix}.$$

This indicates that a submodule M is Hilbert-Schmidt if and only if C is Hilbert-Schmidt, or equivalently, if and only if the function G^M has radial boundary value $\lim_{r\to 1} G^M(rz, r\lambda)$ in $L^2(\mathbb{T}^2 \times \mathbb{T}^2)$, where $z, \lambda \in \mathbb{T}^2$. Moreover, it follows from Lemma 8.10 that

$$\operatorname{Tr}(C^2) = \|G^M\|_2^2 = \Sigma_0(M) + \Sigma_1(M).$$

The decomposition of C^2 in Lemma 8.10 plays an important role in the classification of submodules in [102]. A more refined study of C's structure is contained in [104]. Lemma 8.10 enables us to prove the following

Theorem 8.11 ([99]) *Let M be a submodule such that its core operator C is of finite rank. Then*

$$\operatorname{rank}(C) = 2\operatorname{rank}([R_2^*, R_1]) + 1.$$

Now it becomes clear that rank(C), when finite, is always odd.

Example 8.12 *Let $K \subset \mathbb{D}^2$ be a finite collection of distinct points and let $|K|$ denote its cardinality. Consider*

$$M_K = \{f \in H^2(\mathbb{D}^2) \mid f(z) = 0, \forall z \in K\}.$$

It is shown in [99] that if K is a generic finite subset of $\mathbb{D}^2$ then $\operatorname{rank}([R_2^*, R_1]) = |K|$. *Hence* rank$(C)$ *can be any positive odd number.*

For a submodule M, it is not hard to check that the spaces $M \ominus (z_1M + z_2M)$ and $(z_1M \cap z_2M) \ominus z_1z_2M$ (if nontrivial) are the eigenspace of its core operator C corresponding to the eigenvalues 1 and -1, respectively. Since C is a contraction, the number 1 is its largest eigenvalue. It is shown in [102] that if $\lambda \in (-1, 1)$ is an eigenvalue of C then so is $-\lambda$ with the same multiplicity. Regarding the second largest eigenvalue of C, the following fact is discovered.

Theorem 8.13 (Azari Key, Lu and Yang [4]) *If M is a singly generated Hilbert-Schmidt submodule, then the second largest eigenvalue of C is equal to the operator norm $\|[R_1^*, R_2]\|$.*

In the case $M = [p]$, where p is a homogeneous polynomial, all eigenvalues of C can be computed through the following positive definite Toeplitz matrices.

$$A^n = \begin{pmatrix} ||p||^2 & \overline{\langle pw, pz\rangle} & \cdots & \overline{\langle pw^n, pz^n\rangle} \\ \langle pw, pz\rangle & ||p||^2 & \cdots & \overline{\langle pw^{n-1}, pz^{n-1}\rangle} \\ \langle pw^2, pz^2\rangle & \langle pw, pz\rangle & \cdots & \overline{\langle pw^{n-2}, pz^{n-2}\rangle} \\ \vdots & \vdots & & \vdots \\ \langle pw^n, pz^n\rangle & \langle pw^{n-1}, pz^{n-1}\rangle & \cdots & ||p||^2 \end{pmatrix}, \quad n \geq 1.$$

Note that A^n is a $(n+1) \times (n+1)$ matrix with rows and columns indexed by $\{0, 1, 2, \cdots, n\}$

Theorem 8.14 ([4]) *Given a homogeneous submodule $[p]$, its core operator C has eigenvalues*

$$0, 1, \pm\left(1 - \frac{(|D_n|^2 - |A_{0,n}^n|^2)^2}{D_{n-1}D_n^2D_{n+1}}\right)^{1/2}, \quad n \geq 1,$$

where $D_n = \det A^n$ and $A_{0,n}^n$ is the $(0,n)$-th minor of A^n.

It is a tempting problem whether one may obtain a simple estimate of $\operatorname{Tr}(C^2)$ in terms of p with the help of Theorem 8.14, since such an estimate will enable us to settle the next problem.

Problem 10. Does $\operatorname{Tr}(C^2)$ have an upper bound on $[p]$ as p varies in $\mathscr{R}$?

Somewhat surprisingly, in general there is no control of $\mathrm{Tr}(C^2)$ in terms of the rank of M. An example is given in [105] using two-inner-sequence-based submodules.

A subset $Z \in \mathbb{D}^2$ is called a zero set of $H^2(\mathbb{D}^2)$ if there is a nontrivial function $f \in H^2(\mathbb{D}^2)$ such that $Z = \{z \in \mathbb{D}^2 \mid f(z) = 0\}$. For such a zero set Z, we define

$$H_Z = \{h \in H^2(\mathbb{D}^2) | \, h(z) = 0 \; \forall z \in Z\}.$$

Clearly, H_Z is a nontrivial submodule in $H^2(\mathbb{D}^2)$. Submodules of this kind shall be called zero-based submodules. For example, the submodule $M = [z - w]$ is zero-based, but $[(z-w)^2]$ is not. One verifies that Rudin's submodules in Examples 2.2 and 2.3 are in fact both zero-based ([76]). When Z is "opposite" to distinguished variety we have the following result.

Theorem 8.15 (Qin and Yang [76]) *Let Z be a zero set of $H^2(\mathbb{D}^2)$ such that no point of $\mathbb{T}^2$ is a limit point of Z. Then H_Z is Hilbert Schmidt.*

We end this section with a conjecture in [76].

Conjecture 11. Every zero-based submodule is Hilbert-Schmidt.

8.9 Congruent submodules and their invariants

The unitary equivalence and similarity in Definition 2.5 appear to be the most natural equivalence relations for submodules in $H^2(\mathbb{D}^2)$. However, as suggested by Theorem 2.8 and Example 2.9, they are too rigid for the purpose of classification of submodules. To make this point more lucent let us consider the group $Aut(\mathbb{D}^2)$ of biholomorphic self-maps of $\mathbb{D}^2$. It is known ([78]) that $Aut(\mathbb{D}^2)$ is generated by the reflection $(z_1, z_2) \to (z_2, z_1)$ and the Möbius maps

$$(z_1, z_2) \to \left(\eta_1 \frac{z_1 - \lambda_1}{1 - \overline{\lambda_1} z_1}, \eta_2 \frac{z_2 - \lambda_2}{1 - \overline{\lambda_2} z_2} \right), \quad \lambda \in \mathbb{D}^2, \eta \in \mathbb{T}^2.$$

One observes that for every $x \in Aut(\mathbb{D}^2)$ the composition

$$L_x f(z) = f(x(z)), \quad f \in H^2(\mathbb{D}^2)$$

is a bounded invertible operator on $H^2(\mathbb{D}^2)$. Moreover, if M is a submodule then so is $L_x(M)$. We say that two submodules M and M' are $Aut(\mathbb{D}^2)$-equivalent if there is an $x \in Aut(\mathbb{D}^2)$ such that $M' = L_x(M)$. The submodules H_α and H_β in Example 2.9 are not unitarily equivalent or similar by the rigidity theorem, but they are $Aut(\mathbb{D}^2)$-equivalent because there exists an $x \in Aut(\mathbb{D}^2)$ such that $\beta = x(\alpha)$. On the other hand, if θ is an inner function then $\theta H^2(\mathbb{D}^2)$ is unitarily equivalent to $H^2(\mathbb{D}^2)$ but they are not $Aut(\mathbb{D}^2)$-equivalent. If we hold the belief that both unitary equivalence and $Aut(\mathbb{D}^2)$-equivalence are natural then, for the purpose of classification of submodules, we need an equivalence relation that is coarser than both of them. By Formula (8.9) unitarily equivalent submodules have unitarily equivalent core operators. The following fact about core operators thus motivates the definition of congruent submodules.

Proposition 9.1 *For every $x \in Aut(\mathbb{D}^2)$ and every submodule M, we have*

$$C^{L_x(M)} = L_x C^M L_x^*.$$

Definition 9.2 ([99]) *Two submodules M and M' are said to be congruent if C^M and $C^{M'}$ are congruent; i.e., there is a bounded invertible linear operator J from M to M' such that $C^{M'} = JC^M J^*$.*

Therefore, if two submodules are unitary equivalent or $Aut(\mathbb{D}^2)$-equivalent then they are congruent. After an analysis on the spectral picture of core operators, paper [102] gives the following classification of submodules.

Theorem 9.3 *Let M and M' be submodules with finite rank core operators. Then they are congruent if and only if C^M and $C^{M'}$ have the same rank.*

Theorem 9.3 and Example 8.12 imply that, up to congruence, every submodule with a finite rank core operator is of the type M_K for some finite subset $K \subset \mathbb{D}^2$. In an attempt to find invariants for congruent submodules with an infinite rank core operator, the notions of Lorentz group and little Lorentz group for submodules are defined and studied by Wu, Seto and the author [90].

Definition 9.4 *Let M be a submodule of $H^2(\mathbb{D}^2)$ and denote by $B^{-1}(M)$ the set of all invertible bounded linear operators on M. We call the set*

$$\mathscr{G}(M) = \{T \in B^{-1}(M) \mid T^* C^M T = C^M\}$$

the Lorentz group of M.

It is not hard to verify that $\mathscr{G}(M)$ is indeed a group.

Proposition 9.5 *If two submodules M and M' are congruent, then their Lorentz groups $\mathscr{G}(M)$ and $\mathscr{G}(M')$ are isomorphic.*

The converse of Proposition 9.5 is also true if the associated core operators are of finite rank. If we let $(H^\infty)^{-1}$ be the set of all invertible elements in the Banach algebra $H^\infty(\mathbb{D}^2)$, then for every $\varphi \in (H^\infty)^{-1}$ and every submodule M we have $R_\varphi \in B^{-1}(M)$, where R_φ is the restriction of the Toeplitz operator T_φ to M. Thus one can define a subset of $\mathscr{G}(M)$ as follows.

Definition 9.6 *Let M be a submodule, then the set*

$$\mathscr{G}_0(M) = \{\varphi \in (H^\infty)^{-1} \mid R_\varphi{}^* C R_\varphi = C\}$$

is called the little Lorentz group of M.

The set $\mathscr{G}_0(M)$ is indeed a non-trivial proper Abelian subgroup in $\mathscr{G}(M)$. Surprisingly, it turns out to be a good invariant with respect to unitary equivalence of submodules.

Proposition 9.7 *If two submodules M and M' are unitarily equivalent, then*

$$\mathscr{G}_0(M) = \mathscr{G}_0(M').$$

The converse of Proposition 9.7 is not true. A digression to subgroups of $(H^\infty)^{-1}$ is needed to show a counter-example. First of all, we have the following fact from [38].

Lemma 9.8 *Denote by $L^\infty_{\mathbb{R}}(\mathbb{T})$ the set of essentially bounded real-valued functions on $\mathbb{T}$. Then there is a surjective group homomorphism ρ from $(H^\infty(\mathbb{D}))^{-1}$ to $L^\infty_{\mathbb{R}}(\mathbb{T})$ with $\ker\rho = \mathbb{T}$.*

A two variable version of this lemma is shown in [90]. Since the set of nonzero complex numbers $\mathbb{C}_\times$ is a "trivial" subgroup in $(H^\infty)^{-1}$, we only look at subgroups in $(H^\infty)^{-1}/\mathbb{C}_\times$.

Example 9.9 *If J is an ideal in H^∞, then*

$$\mathscr{G}(J) := (1+J)\cap(H^\infty)^{-1}$$

is a group. To see it, we let $1+f$ and $1+g$ be in the set, where $f,\ g \in J$. Clearly,

$$(1+f)(1+g) \in \mathscr{G}(J).$$

Further,

$$(1+f)^{-1} = 1 - f(1+f)^{-1},$$

which is in $\mathscr{G}(J)$.

We first look at the one variable case. Consider the ideals $J^n = w^n H^\infty(\mathbb{D}),\ n \geq 0$. Although $\mathscr{G}(J^n)$ is a proper subgroup in $\mathscr{G}(J^{n-1})$ for each n, they are all isomorphic to each other ([90]).

Theorem 9.10 *$\mathscr{G}(J^0)$ is isomorphic to $\mathscr{G}(J^n)$ for each $n \geq 1$.*

On $\mathbb{D}^2$, similar subgroups can be defined. For non-negative integers $n_1,\ n_2$, we let

$$J^{n_1,n_2} = \{f \in H^\infty(\mathbb{D}^2) \mid \frac{\partial^i f}{\partial z^i}|_{(0,0)} = 0, \frac{\partial^j f}{\partial w^j}|_{(0,0)} = 0,\ 0 \leq i \leq n_1,\ 0 \leq j \leq n_2\}.$$

However, it is no longer clear whether the groups $\mathscr{G}(J^{n_1,n_2})$ are isomorphic.

Example 9.11 *Let*

$$M = z_1 H^2(\mathbb{D}^2) + z_2 H^2(\mathbb{D}^2)$$

and M' be as in Example 8.9. We have

$$\mathscr{G}_0(M) = \mathscr{G}_0(M') = \mathbb{T} \times \mathscr{G}(J^{1,1}).$$

Since it can be verified that M and M' are not unitarily equivalent, we see that the converse of Proposition 9.7 is not true.

The above exploration on the Lorentz group and little Lorentz group are preliminary at this stage. Whether the two groups can help to classify submodules with infinite-rank core operators remains to be seen. The following two problems may be worth looking into.

Problem 12. For an ideal $J \subset H^\infty$, is rank(J) an invariant for the group $\mathscr{G}(J)$?

Problem 13. For a submodule M, is $\mathscr{G}_0(M)$ maximal Abelian in $\mathscr{G}(M)$?

8.10 Concluding remarks

The progress of research on $H^2(\mathbb{D}^2)$ has been rapid in the past two decadess, and it is still actively ongoing to this day. This very sketchy survey is based on a lecture note for the 2018 summer school at Dalian University of Technology. Due to its time contraint, but more to the author's limited knowledge, some significant topics of the $H^2(\mathbb{D}^2)$ theory have not been included in this survey, most notably among which are commutant lifting, Hermitian bundles and interpolation. But interested readers may find information on these topics in [5, 10, 31] and some other references that are already included in this survey. In closing, the author would like to thank Y. Lu for the invitation to the summer school, Y. Yang and F. Azari Key for hospitality and their help in collecting some references and M. Seto for comments and suggestions. Finally, this article is dedicated to the memory of Keiji Izuchi who carefully checked its first draft at a most difficult time.

References

[1] P. R. Ahern and D. N. Clark, Invariant subspaces and analytic continuation in several variables, *J. Math. Mech.* **19** (1970), 963-969.

[2] O. Agrawal, D. N. Clark and R. G. Douglas, Invariant subspaces in the polydisk, *Pacific J. Math.* **121** (1986), 1-11.

[3] J. Agler, On the representation of certain holomorphic functions defined on a polydisc, *Oper. Theory Adv. Appl.* **48**, Birkhauser Verlag, Basel, 1990, 47-66.

[4] F. Azari Key, Y. Lu and R. Yang, Numerical invariants of homogeneous submodules in $H^2(\mathbb{D}^2)$, *New York J. of Math.* **23** (2017), 505-526.

[5] J. Agler and J. E. McCarthy, *Pick Interpolation and Hilbert Function Spaces*, Graduate Studies in Mathematics, Vol. **44**, American Mathematical Society, Providence, RI, 2002.

[6] J. Agler and J. E. McCarthy, Distinguished varieties, *Acta Math.* **194** (2005), no. 2, 133-153.

[7] A. Aleman, S. Richter and C. Sundberg, Beurlings theorem for the Bergman space, *Acta Math.* **177** (1996), 275-310.

[8] H. Bercovici, *Operator Theory and Arithmetic in H^∞*, Mathematical Surveys and Monographs, no. **26**, A.M.S. 1988, Providence, Rhode Island.

[9] A. Beurling, On two problems concerning linear transformations in Hilbert space, *Acta Math.* **81** (1948), 239-255.

[10] H. Bercovici, C. Foias, L. Kerchy and B. Sz.-Nagy, *Harmonic Analysis of Operators on Hilbert Space* (revised and enlarged edition), Universitext, Springer, 2010.

[11] K. Bickel and P. Gorkin, Compressions of the shift on the bidisk and their numerical ranges, *J. Oper. Theory* **79** (2018), no. 1, 225-265.

[12] K. Bickel and C. Liaw, Properties of Beurling-type submodules via Agler decompositions, *J. Funct. Anal.* **272** (2017), 83-111.

[13] M. S. Brodskii, *Triangular and Jordan Representations of Linear Operators*, Nauka, Moscow, 1969; Transl. Math. Monographs, Vol. 32, A.M.S., Providence, R. I., 1971.

[14] M. J. Cowen and R. G. Douglas, Complex geometry and operator theory, *Acta Math.* **141** (1978), 187-261.

[15] X. Chen and R. G. Douglas, Localization of Hilbert modules, *Mich. Math. J.* **39** (1992), 443-454.

[16] X. Chen and K. Guo, *Analytic Hilbert Modules*, Research Notes in Mathematics, vol. **433**, Chapman & Hall/CRC, Boca Raton, FL, 2003.

[17] A. Chattopadhyay, B K. Dias and J. Sarkar, Rank of a doubly commuting submodule is 2, *Proc. A.M.S.* **146** (2018), no. 3, 1181-1187.

[18] D. N. Clark, Restrictions of H^p functions in the polydisk, *Amer. J. Math.* **110** (1988), 1119-1152.

[19] R. Curto, P. Muhly and K. Yan, The C^*-algebra of a homogeneous ideal in two variables is type I, *Current Topics in Operator Algebras* (Nara, 1990). River Edge, NJ: World Sci. Publishing, 1991, 130-136.

[20] R. Curto, Applications of several complex variables to multiparameter spectral theory, *Surveys of Some Recent Results in Operator Theory*, 25-90, Pitman Res. Notes Math. Ser. **192**, Longman Sci. Tech., Harlow, 1988.

[21] C. C. Cowen and L. A. Rubel, A joint spectrum for shift invariant subspaces of H^2 of the polydisc, *Proc. R. Ir. Acad.* **80**A (1980), no.2, 233-243.

[22] M. Cotlar and C. Sadosky, A polydisk version of Beurlings characterization for invariant subspaces of finite multi-codimension, *Operator Theory for Complex and Hypercomplex Analysis* (Mexico City, 1994), 51-56, *Contemp. Math.* **212**, A.M.S, Providence, 1998.

[23] B. K. Das, S. Gorai and J. Sarkar, On quotient modules of $H^2(\mathbb{D}^n)$: essential normality and boundary representations, preprint.

[24] R. G. Douglas and G. Misra, Some calculations for Hilbert modules, *J. Orissa Math. Soc.* **12-15** (1993-96), 75-85.

[25] R. G. Douglas and G. Misra, Geometric invariants for resolutions of Hilbert modules, *Operator Theory: Advances and Applications*, Birkhäuser, 1993, 83-112.

[26] R. G. Douglas and G. Misra, Equivalence of quotient Hilbert module-II, *Trans. A.M.S.* **360** (2007), 2229-2264.

[27] R. G. Douglas, Canonical models, *Topics in Operator Theory*, Mathematics Surveys, no. **13**, A.M.S., Providence, Rhode Island, 1974.

[28] R. G. Douglas, *Banach Algebra Techniques in Operator Theory* (2nd ed.), Springer-Verlag, New York, 1998.

[29] R. G. Douglas, Essentially reductive Hilbert modules, *J. Operator Theory* **55** (2006), no. 1, 117-133.

[30] R. G. Douglas and C. Foais, Uniqueness of multivariable canocical methods, *Acta Sci. Math.* (Szeged) **57** (1993), no. 1, 79-81.

[31] R. G. Douglas and V. I. Paulsen, *Hilbert Modules over Function Algebras*, Pitman Research Notes in Mathematics Series **217**, Longman Scientific & Technical, Harlow, 1989.

[32] R. G. Douglas, V. I. Paulsen, C. Sah and K. Yan, Algebraic reduction and rigidity for Hilbert Modules, *Amer. J. Math.* **117** (1995), no. 1, 75-92.

[33] R. G. Douglas and K. Wang, A harmonic analysis approach to essential normality of principal submodules, *J. Funct. Anal.* **261** (2011), no. 11, 3155-3180.

[34] R. G. Douglas and K. Yan, On the rigidity of Hardy submodules, *Integr. Equ. Oper. Theory* **13** (1990), 350-363.

[35] R. G. Douglas and K. Yan, A multivariable Berger-Shaw theorem, *J. Operator Theory* **27** (1992), 205-217.

[36] R. G. Douglas and R. Yang, Operator theory in the Hardy space over the bidisc (I), *Integr. Equ. and Oper. Theory* **38** (2000), 207-221.

[37] M. Englis, Density of algebras generated by Toeplitz operators on Bergman spaces, *Ark. Math.* **30** (1992), no. 1, 227-240.

[38] J. B. Garnett, *Bounded Analytic Functions*, Pure and Applied Mathematics, vol. **96**, Academic Press Inc. [Harcourt Brace Jovanovich Publishers], New York, 1981.

[39] P. Ghatage and V. Mandrekar, On Beurling type invariant subspaces of $L^2(\mathbb{T}^2)$ and their equivalence, *J. Oper. Theory* **20** (1988), no. 1, 83-89.

[40] K. Guo, S. Sun, D. Zheng and C. Zhong, Multiplication operators on the Bergman space via the Hardy space over the bidisc, *J. Reine Andew. Math.* **628** (2009), 129-168.

[41] K. Guo and K. Wang, Essentially normal Hilbert modules and K-homology, *Math. Ann.* **340** (2008), no. 4, 907-934.

[42] K. Guo and K. Wang, Beurling-type quotient modules over the bidisc and boundary representations, *J. Funct. Anal.* **257** (2009), no. 10, 3218-3238.

[43] K. Guo, K. Wang and G. Zhang, Trace formulas and p-essentially normal properties of quotient modules on the bidisc, *J. Oper. Theory* **67** (2012), no. 2, 511-535.

[44] K. Guo and P. Wang, Essential normal Hilbert modules and K-homology (III), Homogeneous quotient modules of Hardy modules on the bidisc, *Sci. China Ser. A* **50** (2007), no. 3, 387-411.

[45] K. Guo and P. Wang, Essential normal Hilbert modules and K-homology (IV), Quasi-homogeneous quotient modules of Hardy modules on the polydisks, *Sci. China Ser. A* **55** (2012), no. 8, 1613-1626.

[46] K. Guo and R. Yang, The core function of submodules over the bidisc, *Indiana Univ. Math. J.* **53** (2004), no. 1, 205-222.

[47] L. Hörmander, The Weyl calculus of pseudo-differential operators, *Acta. Math.* **32** (1979), 359-443.

[48] K. J. Izuchi, K. H. Izuchi and Y. Izuchi, Blaschke products and the rank of backward shift invariant subspaces over the bidisc, *J. Funct. Anal.* **261** (2011), no. 6, 1457-1468.

[49] K. J. Izuchi, K. H. Izuchi and Y. Izuchi, Splitting invariant subspaces in the Hardy space over the bidisk, *J. Aust. Math. Soc.* **102** (2017), 205-223.

[50] K. J. Izuchi, K. H. Izuchi and Y. Izuchi, Ranks of invariant subspaces of the Hardy space over the bidisk, *J. Reine Angew. Math.* **659** (2011), 101-139.

[51] K. J. Izuchi, K. H. Izuchi and Y. Izuchi, Fredholm indices of some fringe operators over the bidisc, *Acta Sci. Math.* (Szeged) **83** (2017), 441-455.

[52] K. J. Izuchi, K. H. Izuchi and Y. Izuchi, Zero based invariant subspaces and fringe operators over the bidisc, *J. Korean Math. Soc.* **53** (2016), no. 4, 847-868.

[53] K. J. Izuchi and K. H. Izuchi, Rank-one commutators on invariant subspaces of the Hardy space on the bidisc, *J. Math. Anal. Appl.* **316** (2006), 1-8.

[54] K. J. Izuchi and K. H. Izuch, Ranks of cross commutators on backward shift invariant subspaces over the bidisc, *Rocky Mountain J. Math.* **40** (2010), no. 3, 929-942.

[55] K. J. Izuchi and K. H. Izuchi, Rank-one cross commutators on backward shift invariant subspaces on the bidisk, *Acta Math. Sinica* (English ser.) **25** (2009), 693-714.

[56] K. J. Izuchi, T. Nakazi and M. Seto, Backward shift invariant subspaces in the bidisc II, *J. Oper. Theory* **51** (2004), 361-376.

[57] K. J. Izuchi, T. Nakazi and M. Seto, Backward shift invariant subspaces in the bidisc III, *Acta Sci. Math.* (Szeged) **70** (2004), 727-749.

[58] K.Izuchi and S. Ohno, Self adjoint commutators and invariant subspaces on the torus, *J. Oper. Theory* **31** (1994), 189-204.

[59] K. H. Izuchi, Cyclicity of reproducing kernels in weighted Hardy spaces over the bidisc, *J. Funct. Anal.* **272** (2017), 546-558.

[60] C. A. Jacewicz, On a principal invariant subspace of the Hardy space on the torus, *Proc. Amer. Math. Soc.* **31** (1972), 127-129.

[61] B. B. Koca, Two types of invariant subspaces in the polydisc, *Results Math.* **71** (2017), 1297-1305.

[62] B. B. Koca and N. Sadik, Invariant subspaces generated by a single function in the polydisk, *Math. Notes* **102** (2017), no. 2, 193-197.

[63] P. Lax, Translation invariant spaces, *Acta Math.* **101** (1959),163-178.

[64] M. S. Livsic, *Operators, Osillations, Waves, and Open Systems*, Nauka, Moscow, 1966.

[65] S. Luo, K. Izuchi and R. Yang, Hilbert-Schmidtness of some finitely generated submodules in $H^2(\mathbb{D}^2)$, *J. Math. Anal. Appl.* **465** (2018), 531-546.

[66] Y. Lu and Y. Yang, Invariant subspaces of a Hardy space on the polydisk *Adv. Math.* (China) **41** (2012), no. 3, 313-319.

[67] Y. Lu, R. Yang and Y. Yang, An index formula for the two variable Jordan block, *Proc. Amer. Math. Soc.* **139** (2011), no. 2, 511-520.

[68] V. Mandrekar, On the validity of Beurling theorems in polydisk, *Proc. Amer. Math. Soc.* **103** (1988), 145-148.

[69] T. Nakazi, Szegö's theorem on a bidisc, *Trans. Amer. Math. Soc.* **328** (1991), 421-432.

[70] T. Nakazi, Homogeneous polynomials and invariant subspaces in the polydiscs, *Arch. Math.* **58** (1992), 56-63.

[71] T. Nakazi, Invariant subspaces in the bidisc and commutators, *J. Austral. Math. Soc.* **56** (1994), 232-242.

[72] T. Nakazi, On an invariant subspace whose common zero set is the zeros of some function, *Nihokai Math. J.* **11** (2000), 1-9.

[73] T. Nakazi and M. Seto, Double commuting compressed shifts and generalized interpolation in the Hardy space over the bidisk, *Intgr. Equ. Oper. Theory* **56** (2006), no. 4, 543-558.

[74] N.K. Nikol'skii, *Treatise on the Shift Operator*, A Series of Comprehensive Studies in Mathematics **273**, Springer-Verlag, 1986.

[75] Y. Qin and R. Yang, A characterization of submodules via Beurling-Lax-Halmos theorem, *Proc. Amer. Math. Soc.* **142** (2014), no. 10, 3505-3510.

[76] Y. Qin and R. Yang, A note on Rudin's pathological submodules in $H^2(D^2)$, *Intgr. Equ. Oper. Theory* **88** (2017), no. 3, 363-372.

[77] M. Rosenblum and J. Rovnyak, *Hardy Classes and Operator Theory*, Dover Publications, Inc., New York 1997.

[78] W. Rudin, *Function Theory in Polydiscs*, W. A. Benjamin Inc., New York, 1969.

[79] W. Rudin, Invariant subspaces of H^2 on a torus, *J. Funct. Anal.* **61** (1985), 378-384.

[80] J. Sarkar, Jordan blocks of $H^2(\mathbb{D}^n)$, *J. Oper. Theory* **72** (2014), no. 2, 371-385.

[81] J. Sarkar, Submodules of the Hardy module over polydisc, *Israel J. Math.* **205** (2015), no. 1, 317-336.

[82] M. Seto, Infinite sequences of inner functions and submodules in $H^2(\mathbb{D}^2)$, *J. Oper. Theory* **61** (2009), 75-86.

[83] M. Seto and R. Yang, Inner sequence based invariant subspace in $H^2(\mathbb{D}^2)$, *Proc. Amer. Math. Soc.* **135** (2007), no. 8, 2519-2526.

[84] S. Sun and D. Zheng, Beurling type theorem on the Bergman space via the Hardy space over the bidisc, *Sci. China Ser. A* **52** (2009), no. 11, 2517-2529.

[85] S. Sun, D. Zheng and C. Zhong, Classification of reducing subspaces of a class of multiplication operators on the Bergman space via the Hardy space over the bidisc, *Canad. J. Math* **62** (2010), no. 2, 415-438.

[86] J. Taylor, The analytic functional calculus for several commuting operators, *Acta Math.* **125** (1970), 1-38.

[87] J. L. Taylor, A joint spectrum for several commutative operators, *J. Funct. Anal.* **6** (1970), 172-191.

[88] J. L. Taylor, A general framework for multi-operator functional calculus, *Adv. Math.* **9** (1972), 137-182.

[89] P. Wang, The essential normality of N_η-type quotient module of Hardy module on the polydisc, *Proc. Amer. Math. Soc.* **142** (2014), no. 1, 151-156.

[90] Y. Wu, M. Seto and R. Yang, Krein space representation and Lorentz groups of analytic Hilbert modules, *Sci. China Math.* **61** (2018), no. 4, 745-768.

[91] P. Wang and C. Zhao, Essential normality of quasi-homogeneous quotient modules over the polydisc, to appear in *Sci. China Ser. A.*

[92] P. Wang and C. Zhao, Essential normality of homogenous quotient modules over the polydisc: distinguished variety case, *Integr. Equ. Oper. Theory* **90** (2018), no. 1, Art. 13, 24pp.

[93] P. Wang and C. Zhao, Essentially normal homogeneous quotient modules on the polydisc, *Adv. Math.* **339** (2018), 404-425.

[94] R. Yang, The Berger-Shaw theorem in the Hardy module over the bidisc, *J. Oper. Theory* **42** (1999), no. 2, 379-404.

[95] R. Yang, Operator theory in the Hardy module over the bidisc (II), *Integr. Equ. Oper. Theory* **42** (2002), 99-124.

[96] R. Yang, Operator theory in the Hardy space over the bidisc (III), *J. Funct. Anal.* **186** (2001), 521-545.

[97] R. Yang, Beurling's phenomenon in two variables, *Integr. Equ. Oper. Theory* **48** (2004), no. 3, 411-423.

[98] R. Yang, Hilbert-Schmidt submodules and issue of unitary equivalence, *J. Oper. Theory* **53** (2005), no. 1, 169-184.

[99] R. Yang, The core operator and congruent submodules, *J. Funct. Anal.* **228** (2005), no. 2, 469-489.

[100] R. Yang, On two variable Jordan blocks, *Acta Sci. Math.* (Szeged) **69** (2003), 739-754.

[101] R. Yang, On two variable Jordan blocks (II), *Integr. Equ. Oper. Theory* **56** (2006), 431-449.

[102] R. Yang, A note on classification of submodules in $H^2(\mathbb{D})$, *Proc. Amer. Math. Soc.* **137** (2009), no. 8, 2655-2659.

[103] Y. Yang, Two-inner-sequence-based invariant subspace in $H^2(\mathbb{D}^2)$, *Integr. Equ. Oper. Theory* **77** (2013), 279-290.

[104] Y. Yang, Commutator on Hardy module over bidisc, *Houston J. Math.* **41** (2015), no. 2, 611-620.

[105] Y. Yang and R. Yang, A note on multi-variable Berger-Shaw theorem, *Houston J. Math.* **41** (2015), no. 1, 273-278.

Chapter 9

Weighted Composition Operators on Some Analytic Function Spaces

Ruhan Zhao
Department of Mathematics, SUNY Brockport, Brockport, NY 14420, USA
rzhao@brockport.edu

CONTENTS

9.1 Introduction

The study of composition operators started from the late 1960's. Since it is a natural mathematical object, and because of its bounteous connections with other mathematical areas, more and more mathematicians from various areas are getting interested in the theory of composition operators. More recently, weighted composition operators appeared on the stage. As a combination of composition operators and pointwise multiplication operators, weighted composition operators arise naturally. For example, surjective isometries on Hardy spaces H^p and Bergman spaces A^p, $1 \leq p < \infty$, $p \neq 2$, are given by weighted composition operators. See [13] and [19].

Let us recall the definition of weighted composition operators here. Let $D = \{z : |z| < 1\}$ be the unit disk in the complex plane $\mathbb{C}$ and let $H(D) := \{f : f \text{ is analytic on } D\}$. Let u be an analytic function on the unit disk D and φ be an analytic self-map of D. The weighted composition operator uC_φ is defined as follows: for an analytic function f on D,

$$(uC_\varphi)f(z) = u(z)f(\varphi(z)).$$

Obviously, when $u = 1$, the operator uC_φ is the composition operator C_φ; when φ is the identity map, then the operator uC_φ is the pointwise multiplication operator defined by $M_u f = uf$. The extra weight function u for uC_φ presents a big challenge for the study of these operators, especially when the study involves derivatives of the functions. However, there has been a great development on the study of weighted composition operators in recent years. Boundedness and compactness of weighted composition operators on several classical function spaces have been completely characterized, and estimates of the essential norms for these spaces have been also obtained.

In this survey we summarize some recent results on weighted composition operators on some classical function spaces, including Hardy spaces, weighted Bergman spaces, weighted Banach spaces of analytic functions, and Bloch type spaces. We will focus on those spaces on the unit disk D.

9.2 Preliminaries

Let us introduce some spaces of analytic functions that are involved as well as some other basic concepts that are needed.

9.2.1 Hardy spaces

Let $0 < p < \infty$ and let f be analytic on the unit disk D. We say f is in the Hardy space H^p if

$$\|f\|_{H^p} = \sup_{0<r<1} \left(\frac{1}{2\pi} \int_0^{2\pi} |f(re^{i\theta})|^p \, d\theta \right)^{1/p} < \infty.$$

It is known that, for any $0 < p < \infty$, if $f \in H^p$ then $\lim_{r \to 1} f(re^{i\theta}) = f(e^{i\theta})$ almost everywhere on the unit circle ∂D. The limit function is usually also denoted by f. Then it is known that

$$\|f\|_{H^p} = \left(\frac{1}{2\pi} \int_0^{2\pi} |f(e^{i\theta})|^p \, d\theta \right)^{1/p} = \left(\int_{\partial D} |f(z)|^p \, d\sigma(z) \right)^{1/p},$$

where $d\sigma(z)$ is the normalized Lebesgue measure on ∂D.

9.2.2 Weighted Bergman spaces

Let $0 < p < \infty$ and $-1 < \alpha < \infty$, let $L^{p,\alpha}$ denote the weighted Lebesgue spaces which contain measurable functions f on D such that

$$\|f\|_{p,\alpha}^p = \int_D |f(z)|^p \, dA_\alpha(z) < \infty,$$

where

$$dA_\alpha(z) = (\alpha+1)(1-|z|^2)^\alpha \, dA(z) = \frac{\alpha+1}{\pi}(1-|z|^2)^\alpha \, dxdy/$$

is the normalized weighted area measure on D. The weighted Bergman space is defined as

$$A_\alpha^p = L^{p,\alpha} \cap H(D),$$

with the same norm as above. If $\alpha = 0$, we simply write $L^{p,\alpha}$ and A_α^p as L^p and A^p respectively and $\|f\|_p$ for the norm of f in these spaces.

We refer to [41] for basic properties of Hardy and weighted Bergman spaces.

9.2.3 Weighted Banach spaces of analytic functions

Let v be a strictly positive, continuous and bounded function on D. We will call such a function v as a *weight function* or simply a *weight*. The weight v is called *radial* if $v(z) = v(|z|)$ for all $z \in D$. We are interested in radial weights which are non-increasing with respect to $|z|$ and such that $\lim_{|z|\to 1} v(z) = 0$.

We define the general weighted Banach space of analytic functions as follows:

$$H_v^\infty = \{f \in H(D) : \sup_{z\in D} v(z)|f(z)| < \infty\}.$$

Clearly, H_v^∞ is a Banach space under the norm

$$\|f\|_v = \sup_{z\in D} v(z)|f(z)|.$$

It is well known that the norm convergence in H_v^∞ implies uniform convergence on compact subsets of D.

For a given weight v its *associated weight* $\tilde{v}$ is defined as follows:

$$\tilde{v}(z) = \left(\sup_{z\in D} |f(z)| : f \in H_v^\infty, \|f\|_v \le 1 \right)^{-1}.$$

Here are some basic properties of $\tilde{v}$ which is Lemma 2.1 in [16].

Lemma 2.1 *Let v be a radial, non-increasing weight function such that* $\lim_{r\to 1} v(r) = 0$. *Then* *(i)* $\tilde{v}$ *is radial, continuous and* $\tilde{v} \geq v > 0$,

(ii) $\|\cdot\|_v = \|\cdot\|_{\tilde{v}}$,

(iii) $\tilde{v}(r)$ *is non-increasing for* $r \in (0,1)$,

(iv) $\lim_{r\to 1} \tilde{v}(r) = 0$,

(v) $\varphi(t) := -\log \tilde{v}(e^t)$ *is convex for* $t \in (-\infty, 0)$.

9.2.4 Bloch type spaces

Let v be a weight. The Bloch-type space B_v is defined as follows:

$$B_v = \{f \in H(D) : \|f\|_{B_v} = \sup_{z\in D} v(z)|f'(z)| < \infty\},$$

with the norm given by

$$\||f\||_{B_v} = |f(0)| + \sup_{z\in D} v(z)|f'(z)|.$$

In particular, when

$$v(z) = v_\alpha(z) := (1-|z|^2)^\alpha$$

for $0 < \alpha < \infty$, we get the Bloch-type space, or the α-Bloch space B^α, which is the space that consists of all analytic functions f on D such that

$$\|f\|_{B^\alpha} = \sup_{z\in D} |f'(z)|(1-|z|^2)^\alpha < \infty$$

with the norm

$$\||f\||_{B^\alpha} = |f(0)| + \|f\|_{B^\alpha}.$$

When $\alpha = 1$, the α-Bloch space is the classical Bloch space, and is denoted by B. We also use the Bloch-type space defined by $\tilde{B}_v = \{f \in B_v : f(0) = 0\}$ with the norm $\|\cdot\|_{B^v}$. It is known that if $\alpha > 1$, B^α is the same as $H^\infty_{\alpha-1}$ with equivalent norms. For more information on Bloch type spaces, we refer to [42].

9.2.5 Bounded and compact operators

Recall that for Banach spaces X and Y, a linear operator $T : X \to Y$ is *bounded*, if TB_X is bounded in Y, where B_X is the unit ball in X; The operator T is *compact*, if the closure of TB_X is a compact set in Y. It is well-known that an linear operator T is compact from X to Y if and only if for any sequence $\{f_n\}$ that is weakly convergent to 0 in X we have $\lim_{n\to\infty} \|Tf_n\|_Y = 0$.

A point evaluation on a Banach space X of analytic functions is defined as $K_z(f) = f(z)$ for $f \in X$. To say a point evaluation K_z is a bounded (or continuous) functional, it means there is a constant C (which may depend on z) such that $|f(z)| \leq C\|f\|_X$.

Tjani proved the following useful result on compact operators in [34].

Lemma 2.2 *Let X, Y be two Banach spaces of analytic functions on D. Suppose*

(i) the point evaluation functionals on X are continuous;

(ii) the closed unit ball of X is a compact subset of X in the topology of uniform convergence on compact sets;

(iii) $T : X \to Y$ *is continuous when X and Y are given the topology of uniform convergence on compact sets.*

Then T is a compact operator if and only if given a bounded sequence (f_n) *in X such that* $f_n \to 0$ *uniformly on compact sets, the sequence* (Tf_n) *converges to 0 in the norm of Y.*

It is easy to check that the weighted Bergman space A^p_α $(1 \leq p < \infty)$, the Hardy space H^p $(1 \leq p < \infty)$, the spaces H^∞_α and B^α all satisfy (i) and (ii) and $uC_\varphi : X \to Y$ satisfy condition (iii) when X and Y are two of the above function spaces. Thus we have the following corollary.

Corollary 2.3 *Let X and Y be two of the following spaces:* A^p_α $(1 \leq p < \infty)$, H^p $(1 \leq p < \infty)$, H^∞_α *and* B^α *Then* $uC_\varphi : X \to Y$ *is a compact operator if and only if for any bounded sequence* (f_n) *in X with* $f_n \to 0$ *uniformly on compact sets of D as* $n \to \infty$, *we have* $\|uC_\varphi f_n\|_Y \to 0$ *as* $n \to \infty$.

9.2.6 Essential norms

Suppose that T is a bounded operator between Banach spaces X and Y. The essential norm $\|T\|_e$ of T is defined as the distance from T to the space of compact operators from X to Y. Hence T is compact if and only if $\|T\|_e = 0$.

9.3 Carleson measures

One of our main tools is the Carleson measure on the function spaces. The Carleson measure was introduced by Carleson [5] for studying the problem of interpolation by bounded analytic functions and for solving the famous corona problem. Later on it was found that the Carleson measure is very useful in many problems in the function space theory and the operator theory. Variations of the Carleson measure have been also introduced and studied. Let μ be a positive Borel measure on D. Let X be a Banach space of analytic functions on D. Let $q > 0$. We say that μ is an (X,q)-Carleson measure, if there is a constant $C > 0$ such that, for any $f \in X$,

$$\int_D |f(z)|^q \, d\mu(z) \leq C\|f\|^q_X.$$

Let I be an arc in the unit circle ∂D. We use $|I|$ to denote the normalized length of I such that $|\partial D| = 1$. Let $S(I)$ be the Carleson square defined by

$$S(I) = \{z \in D : 1 - |I| \le |z| < 1, z/|z| \in I\}.$$

For $a \in D$ let $\sigma_a(z) = (a-z)/(1-\bar{a}z)$ be the Möbius transformation that interchanges 0 and a. Then clearly $\sigma_a'(z) = -(1-|a|^2)/(1-\bar{a}z)^2$. We state some characterizations of the (A_α^p, q)-Carleson measures here. The following result is well-known.

Theorem 3.1 *Let μ be a positive Borel measure on D. Let $0 < p \le q < \infty$ and $-1 < \alpha < \infty$. Then the following statements are equivalent:*

(i) There is a constant $C_1 > 0$ such that, for any $f \in A_\alpha^p$,

$$\int_D |f(z)|^q \, d\mu(z) \le C_1 \|f\|_{p,\alpha}^q.$$

(ii) There is a constant $C_2 > 0$ such that, for any arc $I \in \partial D$,

$$\mu(S(I)) \le C_2 |I|^{(2+\alpha)q/p}.$$

(iii) There is a constant $C_3 > 0$ such that, for every $a \in D$,

$$\int_D |\sigma_a'(z)|^{(2+\alpha)q/p} \, d\mu(z) \le C_3.$$

The result was proved by several authors. The equivalence of (i) and (ii) can be found in [15] and [21], and a proof of the equivalence of (ii) and (iii) can be found in [4]. Notice that the constants C_1, C_2 and C_3 in this theorem can be made to be comparable, which means there are positive constants M_1 and M_2, independent of μ, such that

$$\frac{1}{M_1} C_1 \le C_2 \le M_1 C_1, \qquad \frac{1}{M_1} C_1 \le C_3 \le M_1 C_1.$$

To check this fact, one may refer to the proof of Theorem 6.2.2 in [41, p.109-110] and [4]. We define

$$\|\mu\| = \sup_{I \subset \partial D} \frac{\mu(S(I))}{|I|^{(2+\alpha)q/p}}.$$

Then all of the above constants $C_1 - C_3$ can be made to be comparable to $\|\mu\|$.

For the case $p > q$, we need the following concepts. For a positive measure μ on D, $-1 < \alpha < \infty$, and a fixed number $r \in (0,1)$, define

$$\widehat{\mu_{r,\alpha}}(z) = \frac{\mu(D(z,r))}{|D(z,r)|^{1+\alpha/2}}, \qquad B_\alpha(\mu)(z) = \int_D |\sigma_z'(w)|^{2+\alpha} \, d\mu(w),$$

where $D(z,r) = \{w \in D \,|\, |\sigma_z(w)| < r\}$ is the pseudohyperbolic disk with center z and radius r and $|D(z,r)|$ denotes the Lebesgue area measure of $D(z,r)$. $B_\alpha(\mu)$ is called the *weighted Berezin transform* of the measure μ. When $\mu(w) = h(w)\,dA_\alpha(w)$ for some measurable function h on D, we write

$$B_\alpha(h)(z) = \int_D |\sigma_z'(w)|^{2+\alpha} h(w)\,dA_\alpha(w),$$

and call it the *weighted Berezin transform* of h.

We have the following result.

Theorem 3.2 *Let μ be a positive measure on D. Let $0 < q < p < \infty$ and $-1 < \alpha < \infty$. Then the following statements are equivalent.*

(i) *μ is a (A^p_α, q)-Carleson measure.*

(ii) *For any fixed $r \in (0,1)$, $\widehat{\mu_{r,\alpha}} \in L^{p/(p-q),\alpha}$.*

(iii) *$B_\alpha(\mu) \in L^{p/(p-q),\alpha}$.*

The equivalence of (i) and (ii) is given by Luecking ([22] and [24]) for the case $\alpha = 0$. For $-1 < \alpha < \infty$, the result can be similarly proved as in [24]. The equivalence of (ii) and (iii) is given in [10].

For the case of Hardy spaces, the following result is due to Carleson [5] (for $p = q$) and Duren [12] (for $p \le q$).

Theorem 3.3 *Let $0 < p \le q < \infty$. Let μ be a positive measure on D. Then μ is an (H^p, q)-Carleson measure if and only if there is a constant $K > 0$ such that for any arc $I \subset \partial D$,*

$$\mu(S(I)) \le K|I|^{q/p}. \tag{9.1}$$

For the case $0 < q < p < \infty$, the following result is due to I. V. Videnskii [35] and D. Luecking [23]. Let $\zeta \in \partial D$. The Stolz region $\Gamma(\zeta)$ generated by ζ is the interior of the convex hull of $\{\zeta\} \cup (\alpha\overline{D})$. Here $0 < \alpha < 1$ is arbitrary, but fixed.

Theorem 3.4 *Let $0 < q < p < \infty$. Let μ be a positive measure on D. Let*

$$H(\zeta) = \int_{\Gamma(\zeta)} \frac{d\mu(z)}{1 - |z|^2}.$$

Then μ is a (H^p, q)-Carleson measure if and only if $H(\zeta) \in L^{p/(p-q)}(\partial D)$.

9.4 Weighted composition operators between weighted Bergman spaces

In this section we consider weighted composition operators on weighted Bergman spaces. The main sources of this section are [9] and [10].

Our first result concerns bounded weighted composition operators mapping A^p_α into A^q_β for $p \le q$. For the unweighted case, this means mapping a larger Bergman space into a smaller one. Our results will be expressed in terms of the integral operator

$$I_{\varphi,\alpha,\beta}(u)(a) = \int_D \left(\frac{1-|a|^2}{|1-\bar{a}\varphi(w)|^2} \right)^{(2+\alpha)q/p} |u(w)|^q \, dA_\beta(w).$$

Theorem 4.1 *Let u be an analytic function on D and φ be an analytic self-map of D. Let $0 < p \le q < \infty$, and $\alpha, \beta > -1$. Then the weighted composition operator uC_φ is bounded from A^p_α into A^q_β if and only if*

$$\sup_{a\in D} I_{\varphi,\alpha,\beta}(u)(a) < \infty. \tag{9.2}$$

Proof By definition, uC_φ is bounded from A^p_α into A^q_β if and only if for any $f \in A^p_\alpha$,

$$\|(uC_\varphi)f\|^q_{q,\beta} \le C\|f\|^q_{p,\alpha},$$

that is

$$\int_D |u(z)|^q |f(\varphi(z))|^q \, dA_\beta(z) \le C\|f\|^q_{p,\alpha}. \tag{9.3}$$

Letting $w = \varphi(z)$ we get

$$\int_D |f(w)|^q \, d\mu_u(w) \le C\|f\|^q_{p,\alpha},$$

where $\mu_u = \nu_u \circ \varphi^{-1}$ and $d\nu_u(z) = |u(z)|^q \, dA_\beta(z)$. But (9.3) means $d\mu_u$ is an (A^p_α, q)-Carleson measure. By Theorem 3.1, this is equivalent to

$$\sup_{a\in D} \int_D |\sigma_a'(w)|^{(2+\alpha)q/p} \, d\mu_u(w) < \infty.$$

Changing the variable w back to z we get (9.2). The proof is complete.

We have the following estimates for the essential norm of uC_φ.

Theorem 4.2 *Let u be an analytic function on D and φ be an analytic self-map of D. Let $1 < p \le q < \infty$, and $\alpha, \beta > -1$. Let uC_φ be bounded from A^p_α into A^q_β. Then there is an absolute constant $C \ge 1$ such that*

$$\limsup_{|a|\to 1} I_{\varphi,\alpha,\beta}(u)(a) \le \|uC_\varphi\|^q_e \le C \limsup_{|a|\to 1} I_{\varphi,\alpha,\beta}(u)(a).$$

For proving this result, we need the following two lemmas.

Lemma 4.3 *Let $0 < r < 1$. Let μ be a positive Borel measure on D. Let*

$$N_r^* = \sup_{|a|\ge r} \int_D |\sigma_a'(z)|^{(2+\alpha)q/p} \, d\mu(z).$$

If μ is an (A^p_α, q)-Carleson measure for $0 < p \le q < \infty$, then so is $\mu_r = \mu|_{D\setminus D_r}$. Moreover, $\|\mu_r\| \le MN_r^*$, where M is an absolute constant.

Proof Let $0 < r < 1$ and let $D_r = \{z \in D : |z| < r\}$. Denote

$$N_r = \sup_{|I| \leq 1-r} \frac{\mu(S(I))}{|I|^{(2+\alpha)q/p}}.$$

Take any arc $I \subset \partial D$. Suppose $|I| = \gamma(1-r)$ for some constant $\gamma > 0$.

If $0 < \gamma \leq 1$ then obviously $S(I) \subset D \setminus D_r$, and so

$$\mu_r(S(I)) = \mu(S(I)) \leq N_r |I|^{(2+\alpha)q/p}.$$

Now consider the case $\gamma > 1$. Let $[\gamma]$ be the greatest integer less than or equal to γ. Then $[\gamma] + 1 > \gamma$ and $([\gamma]+1)/\gamma \leq 2$. Let $n = [\gamma] + 1$. Then obviously $n \leq 2\gamma$. In this case, it is possible to cover I by n smaller arcs $I_1, I_2, ..., I_n$ such that $|I_k| = 1 - r$, $k = 1, 2, ..., n$. Thus

$$\begin{aligned} \mu_r(S(I)) &= \mu(S(I) \cap (D \setminus D_r)) \leq \sum_{k=1}^{n} \mu(S(I_k)) \leq \sum_{k=1}^{n} N_r |I_k|^{(2+\alpha)q/p} \\ &\leq N_r \left(\sum_{k=1}^{n} |I_k| \right)^{(2+\alpha)q/p} = N_r (n(1-r))^{(2+\alpha)q/p} \\ &\leq N_r [2\gamma(1-r)]^{(2+\alpha)q/p} = 2^{(2+\alpha)q/p} N_r |I|^{(2+\alpha)q/p}. \end{aligned}$$

Together with the estimate above this implies that $\|\mu_r\| \leq 2^{(2+\alpha)q/p} N_r$. Now let

$$N_r^* = \sup_{|a| \geq r} \int_D |\sigma_a'(z)|^{(2+\alpha)q/p} \, d\mu(z);$$

then we just need to prove that $N_r \leq M N_r^*$ for an absolute constant $M > 0$. Take any arc $I \subset \partial D$ with $|I| \leq 1 - r$. Let $a = (1 - |I|)e^{i\theta}$, where $e^{i\theta}$ is the center of I. Then $|a| = 1 - |I| \geq r$. By a geometric consideration, it can be proved that for any $z \in S(I)$, $|\sigma_a'(z)| \geq 4/(25|I|)$. Thus,

$$\begin{aligned} \frac{\mu(S(I))}{|I|^{(2+\alpha)q/p}} &\leq \left(\frac{25}{4} \right)^{(2+\alpha)q/p} \int_{S(I)} |\sigma_a'(z)|^{(2+\alpha)q/p} \, d\mu(z) \\ &\leq \left(\frac{25}{4} \right)^{(2+\alpha)q/p} N_r^*. \end{aligned}$$

Taking supremum over all arcs I with $|I| \leq 1 - r$ we get

$$\|\mu_r\| \leq 2^{(2+\alpha)q/p} N_r \leq \left(\frac{25}{2} \right)^{(2+\alpha)q/p} N_r^*.$$

The result is proved with $M = \left(\frac{25}{2}\right)^{(2+\alpha)q/p}$.

For $f(z) = \sum_{k=0}^{\infty} a_k z^k$ analytic on D, let $K_n f(z) = \sum_{k=0}^{n} a_k z^k$ and $R_n = I - K_n$, where $If = f$ is the identity map. Hence $R_n f(z) = \sum_{k=n+1}^{\infty} a_k z^k$. Then we have

Lemma 4.4 *If uC_φ is bounded from A^p_α into A^q_β for $0 < p \le q < \infty$, then*

$$\|uC_\varphi\|_e \le \liminf_{n\to\infty} \|uC_\varphi R_n\|.$$

Proof Since $(R_n + K_n)f = f$ and K_n is compact, we have for each n,

$$\|uC_\varphi\|_e \le \|uC_\varphi R_n + uC_\varphi K_n\|_e \le \|uC_\varphi R_n\|_e \le \|uC_\varphi R_n\|.$$

Therefore $\|uC_\varphi\|_e \le \liminf_{n\to\infty} \|uC_\varphi R_n\|$.

Proof of Theorem 4.2 First we prove the upper estimate. By Lemma 4.4,

$$\|uC_\varphi\|_e \le \liminf_{n\to\infty} \|uC_\varphi R_n\| \le \liminf_{n\to\infty} \sup_{\|f\|_{p,\alpha}\le 1} \|(uC_\varphi R_n)f\|_{q,\beta}.$$

However, for any fixed $0 < r < 1$,

$$\begin{aligned} \|(uC_\varphi R_n)f\|^q_{q,\beta} &= \int_D |u(z)|^q |(R_n f)(\varphi(z))|^q \, dA_\beta(z) \\ &= \int_D |R_n f(w)|^q \, d\mu_u(w) \\ &= \int_{D\setminus D_r} |R_n f(w)|^q \, d\mu_u(w) + \int_{D_r} |R_n f(w)|^q \, d\mu_u(w) \\ &= I_1 + I_2, \end{aligned} \tag{9.4}$$

where μ_u is the pull-back measure induced by φ defined before. Since uC_φ is bounded from A^p_α into A^q_β, μ_u is an (A^p_α, q)-Carleson measure. Let

$$\langle f, g\rangle_\alpha = \int_D f(z)\overline{g(z)}(1-|z|^2)^\alpha \, dA(z)$$

be the integral pairing that identifies $(A^p_\alpha)^*$ and $A^{p'}_\alpha$, where p' is the conjugate index of p satisfying $\frac{1}{p} + \frac{1}{p'} = 1$. Then it is easy to see that $\langle R_n f, g\rangle_\alpha = \langle f, R_n g\rangle_\alpha$ for all n, $f \in A^p_\alpha$ and $g \in A^{p'}_\alpha$. Thus

$$|R_n f(w)| = \left| \int_D f(z)\overline{R_n K_w(z)}(1-|z|^2)^\alpha \, dA(z) \right| \le \|f\|_{p,\alpha} \|R_n K_w\|_{p',\alpha}.$$

For a fixed $0 < r < 1$ and $|w| \le r$, $z \in D$, we use the Taylor expansion

$$K_w(z) = \sum_{k=0}^{\infty} (k+1)\bar{w}^k z^k$$

to obtain the estimate

$$|R_n K_w(z)| \le \sum_{k=n+1}^{\infty} r^k (k+1).$$

Hence for any $\varepsilon > 0$, if n large enough then

$$\int_D |R_n K_w(z)|^{p'}(1-|z|^2)^\alpha \, dA(z) < \varepsilon^{p'},$$

which implies that, for n large enough,

$$|R_n f(w)| \le \varepsilon \|f\|_{p,\alpha}.$$

Thus

$$I_2 \le \varepsilon^q \|f\|_{p,\alpha}^q \mu_u(D_r) \le \varepsilon^q \|f\|_{p,\alpha}^q \|u\|_{q,\beta}^q.$$

Hence, for a fixed r,

$$\sup_{\|f\|_{p,\alpha}\le 1} I_2 \to 0, \qquad n \to \infty.$$

On the other hand, if we denote by $\mu_{u,r} = \mu_u|_{D\setminus D_r}$, then, by Theorem 3.1 and Lemma 4.3,

$$I_1 = \int_{D\setminus D_r} |R_n f(w)|^q \, d\mu_{u,r}(w) \le K\|\mu_{u,r}\| \|R_n f\|_{p,\alpha}^q \le KCMN_r^* \|f\|_{p,\alpha}^q,$$

where K, C and M are constants independent of u and r, and N_r^* is defined as in Lemma 4.3. Taking supremum in (9.4) over analytic functions f in the unit ball of A_α^p, and letting $n \to \infty$, we get

$$\liminf_{n\to\infty} \sup_{\|f\|_{p,\alpha}\le 1} \|(uC_\varphi R_n)f\|_{p,\alpha}^q \le \liminf_{n\to\infty} KCMN_r^* = KCMN_r^*.$$

Thus $\|uC_\varphi\|_e^q \le KCMN_r^*$. Letting $r \to 1$ we get

$$\begin{aligned}
\|uC_\varphi\|_e^q &\le KCM \lim_{r\to 1} N_r^* = KCM \limsup_{|a|\to 1} \int_D |\sigma_a'(w)|^{(2+\alpha)q/p} \, d\mu_u(w) \\
&= KCM \limsup_{|a|\to 1} \int_D |\sigma_a'(\varphi(z))|^{(2+\alpha)q/p} |u(z)|^q \, dA_\beta(z) \\
&= KCM \limsup_{|a|\to 1} I_{\varphi,\alpha,\beta}(u)(a),
\end{aligned}$$

which gives us the desired upper bound.

Now let us prove the lower estimate. Consider the normalized kernel function

$$k_a(z) = -\sigma_a'(z) = \frac{1-|a|^2}{(1-\bar{a}z)^2}.$$

Let $f_a = k_a^{(2+\alpha)/p}$. Then $\|f_a\|_{p,\alpha} = 1$, and $f_a \to 0$ uniformly on compact subsets of D as $|a| \to 1$. It is easy to see that this implies that $f_a \to 0$ weakly in A_α^p as $|a| \to 1$. Fix a compact operator $\mathscr{K}$ from A_α^p into A_β^q. Then $\|\mathscr{K} f_a\|_{q,\beta} \to 0$ as $|a| \to 1$. Therefore,

$$\begin{aligned}
\|uC_\varphi - \mathscr{K}\| &\ge \limsup_{|a|\to 1} \|(uC_\varphi - \mathscr{K})f_a\|_{q,\beta} \\
&\ge \limsup_{|a|\to 1} \left(\|(uC_\varphi)f_a\|_{q,\beta} - \|\mathscr{K} f_a\|_{q,\beta}\right) \\
&= \limsup_{|a|\to 1} \|(uC_\varphi)f_a\|_{q,\beta}.
\end{aligned}$$

Thus

$$\|uC_\varphi\|_e^q \geq \limsup_{|a|\to 1} \|(uC_\varphi)f_a\|_{q,\beta}^q = \limsup_{|a|\to 1} I_{\varphi,\alpha,\beta}(u)(a).$$

The proof is complete.

The following corollary is now immediate.

Corollary 4.5 *Let u be an analytic function on D and φ be an analytic self-map of D. Let $1 < p \leq q < \infty$, and $\alpha, \beta > -1$. Let uC_φ be bounded from A^p_α into A^q_β. Then the weighted composition operator uC_φ is compact from A^p_α into A^q_β if and only if*

$$\limsup_{|a|\to 1} I_{\varphi,\alpha,\beta}(u)(a) = 0.$$

Let

$$\sigma_z(w) = \frac{z-w}{1-\bar{z}w}$$

be a Möbius transformation on D. Let φ be an analytic self-map of the unit disk. Let $-1 < \alpha, \beta < \infty$. The weighted φ-*Berezin transform* of a measurable function h is defined as follows.

$$\begin{aligned} B_{\varphi,\alpha,\beta}(h)(z) &= \int_D |\sigma_z'(\varphi(w))|^{2+\alpha} h(w)\, dA_\beta(w) \\ &= \int_D \frac{(1-|z|^2)^{2+\alpha} h(w)}{|1-\bar{z}\varphi(w)|^{4+2\alpha}}\, dA_\beta(w). \end{aligned}$$

If $\varphi(z) = z$, $B_{\varphi,\alpha,\alpha}$ is just the usual weighted Berezin transform B_α.

For the case $q < p$, we have the following characterization of the boundedness of uC_φ.

Theorem 4.6 *Let φ be an analytic self map of the unit disk D and u be an analytic function on D. Let $1 \leq q < p < \infty$, and let $-1 < \alpha, \beta < \infty$. Then the following statements are equivalent.*

(i) uC_φ *is bounded from* A^p_α *to* A^q_β;

(ii) uC_φ *is compact from* A^p_α *to* A^q_β;

(iii) $B_{\varphi,\alpha,\beta}(|u|^q) \in L^{p/(p-q),\alpha}$.

Proof Let $d\nu_u(z) = |u(z)|^q\, dA_\beta(z)$ and $\mu_u = \nu_u \circ \varphi^{-1}$ be the pull-back measure of ν_u. Then uC_φ is bounded from A^p_α to A^q_β if and only if for any $f \in A^p_\alpha$,

$$\int_D |u(z)|^q |f(\varphi(z))|^q\, dA_\beta(z) \leq C\|f\|_{p,\alpha}^q,$$

or

$$\int_D |f(w)|^q\, d\mu_u(w) \leq C\|f\|_{p,\alpha}^q.$$

Thus μ_u is an (A^p_α, q)-Carleson measure. By Theorem 3.2, this is equivalent to that $B_\alpha(\mu_u) \in L^{p/(p-q),\alpha}$. Thus (i) and (iii) are equivalent since $B_\alpha(\mu_u) = B_{\varphi,\alpha,\beta}(|u|^q)$.

The equivalence of (i) and (ii) follows from a general result of Banach space theory. It is known that, for $1 \le q < p < \infty$, every bounded operator from ℓ^p to ℓ^q is compact (see, for example [20, p. 31, Theorem I.2.7]). Since the Bergman space A^p_α is isomorphic to ℓ^p (see, [37, p. 89, Theorem 11]), we get (ii) from (i) directly from the above result. On the other hand, it is obvious that (ii) implies (i).

9.5 Weighted composition operators between Hardy spaces

In this section we consider weighted composition operators between Hardy spaces. All the results in this section are from [10].

We note here that the essential norm of composition operators on the Hardy space H^2 was first obtained by Shapiro in [32]. However, the method in [32] does not seem to work for weighted composition operators due to the fact that the formula in [32] is obtained using an alternative expression of the H^2 norm in terms of derivatives of the functions in H^2.

The boundedness and compactness of weighted composition operators between Hardy spaces were characterized in [7] and [8], using a Carleson measure condition. Here we present analogous results as in the case of weighted Bergman spaces using the following integral operator

$$I_{\varphi,-1}(u)(a) = \int_{\partial D} \left(\frac{1-|a|^2}{|1-\bar{a}\varphi(w)|^2} \right)^{\frac{q}{p}} |u(w)|^q \, d\sigma(w),$$

where $a \in D$, ∂D is the unit circle and $d\sigma$ is the normalized arc length measure on ∂D.

Theorem 5.1 *Let u be an analytic function on D and φ be an analytic self-map of D. Let $0 < p \le q < \infty$. Then the weighted composition operator uC_φ is bounded from H^p into H^q if and only if*

$$\sup_{a \in D} I_{\varphi,-1}(u)(a) < \infty.$$

Using Theorem 3.3, the proof of Theorem 5.1 is similar to that of Theorem 4.1, We omit the details here.

We also have the following estimates for the essential norm of uC_φ.

Theorem 5.2 *Let u be an analytic function on D and φ be an analytic self-map of D. Let $1 < p \le q < \infty$. Let uC_φ be bounded from H^p into H^q. Then there is an absolute* *constant $C \geq 1$ such that*

$$\limsup_{|a|\to 1} I_{\varphi,-1}(u)(a) \leq \|uC_\varphi\|_e^q \leq C \limsup_{|a|\to 1} I_{\varphi,-1}(u)(a).$$

In particular, uC_φ is compact from H^p into H^q if and only if

$$\limsup_{|a|\to 1} I_{\varphi,-1}(u)(a) = 0.$$

The proof this theorem is also similar to that of Theorem 4.2, using a modified version of Lemma 4.3 (with $\alpha = -1$) and Lemma 4.4. We omit he details here.

Now let us consider the case $p > q$.

Theorem 5.3 *Let φ be an analytic self map of the unit disk D and u be an analytic function on D. Let $1 \leq q < p < \infty$. Then uC_φ is bounded from H^p to H^q if and only if*

$$\int_0^{2\pi} \left(\int_{\Gamma(\theta)} \frac{d\mu_u(w)}{1-|w|^2} \right)^{\frac{p}{p-q}} d\theta < \infty.$$

where $\mu_u = \nu_u \circ \varphi^{-1}$ and $d\nu_u(z) = |u(z)|^q d\sigma(z)$ with $d\sigma(z)$ the normalized measure of ∂D, and $\Gamma(\theta)$ is the Stolz angle at θ, which is defined for real θ as the convex hull of the set $\{e^{i\theta}\} \cup \{z : |z| < \sqrt{1/2}\}$.

Proof The operator uC_φ is bounded from H^p to H^q means, for any $f \in H^p$,

$$\int_{\partial D} |u(z) f(\varphi(z))|^q d\sigma(z) \leq C\|f\|_{H^p}^q.$$

Changing variable $w = \varphi(z)$ we get

$$\int_D |f(w)|^q d\mu_u(z) \leq C\|f\|_{H^p}^q,$$

which means $d\mu_u$ is an (H^p, q)-Carleson measure. By Theorem 3.4, this is equivalent to

$$\int_0^{2\pi} \left(\int_{\Gamma(\theta)} \frac{d\mu_u(z)}{1-|z|^2} \right)^{\frac{p}{p-q}} d\theta < \infty.$$

The result is proved.

Remark. From this result we can easily see that if u has no zeros in D, then uC_φ is bounded from H^p to H^q if and only if $u^q C_\varphi$ is bounded from $H^{p/q}$ to H^1.

The following result characterizes compactness in this case. For the case of composition operators (when $u = 1$), the result was first proved in [14].

Theorem 5.4 *Let φ be an analytic self map of the unit disk D and u be an analytic function on D that is not identically zero. Let $1 \leq q < p < \infty$. Let uC_φ be bounded from H^p into H^q. Then uC_φ is compact from H^p into H^q if and only if $|\varphi(z)| < 1$ a.e. on ∂D.*

Proof We first prove necessity. Let $1 \leq q < p < \infty$. It is well-known that the sequence $\{z^n\}$ is an H^p-weakly null sequence. Thus the compactness of uC_φ from H^p to H^q implies that $\|uC_\varphi z^n\|_{H^q} \to 0$ as $n \to \infty$, i.e.,

$$\lim_{n\to\infty} \int_0^{2\pi} |u(e^{i\theta})|^q |\varphi(e^{i\theta})|^{nq}\, d\theta = 0.$$

Because uC_φ is bounded from H^p to H^q, it is clear that $u = uC_\varphi 1 \in H^q$. Hence the convergence condition above means that $\{\xi \in \partial D : |\varphi(\xi)| = 1\}$ has measure 0.

Next we prove sufficiency. Assume that uC_φ is compact from H^p to H^q. We first prove the result for $q = 1$. Suppose $|\varphi(z)| < 1$ a.e. on ∂D. Let $\{f_n\} \subset H^p$ be an arbitrary weakly null sequence. This implies that $\{f_n\}$ converges to 0 uniformly on compact subsets of D. Since uC_φ is bounded from H^p to H^1, it takes a weakly null sequence in H^p into a weakly null sequence in H^1. Hence $u(f_n \circ \varphi) \to 0$ weakly in H^1. Since $|\varphi(z)| < 1$ a.e. on ∂D, it follows that $u(f_n \circ \varphi) \to 0$ a.e. on ∂D. This means that $u(f_n \circ \varphi) \to 0$ in measure. By the Dunford-Pettis Theorem (see [11]), we have $\|uC_\varphi f_n\|_{H^1} \to 0$. Hence uC_φ is completely continuous and, by reflexivity of H^p, uC_φ is compact.

Next, we prove the sufficiency for any q satisfying $1 \leq q < p < \infty$. Let us first assume that u is an outer function. Suppose $\{f_n\}$ is a sequence in the unit ball of H^p. For each n, we write $f_n = B_n F_n$, where B_n is inner, F_n is outer. Clearly, both sequences $\{B_n\}$ and $\{F_n\}$ are contained in the unit ball of H^p. The local boundedness of these sequences shows that they are normal families; we can use Montel's Theorem to extract subsequence $\{B_{n_j}\}$ and $\{F_{n_j}\}$ that converge uniformly on compact subsets of D. Put $G_j = F_{n_j}^q$. Then G_j is in the unit ball of $H^{p/q}$. Now recall that uC_φ is bounded from H^p to H^q. From the remark after Theorem 5.3, this is equivalent to saying that $u^q C_\varphi$ is bounded from $H^{p/q}$ to H^1. Since $|\varphi(z)| < 1$ a.e. on ∂D by assumption, it follows from the above discussion that $u^q C_\varphi$ is compact from $H^{p/q}$ to H^1. Therefore there is a subsequence $\{G_{j_k}\}$ of $\{G_j\}$ such that the sequence $\{u^q(G_{j_k} \circ \varphi)\}$ converges in the norm of H^1. Also, the fact that $|\varphi(z)| < 1$ a.e. on ∂D implies $\{u^q(G_{j_k} \circ \varphi)\}$ converges almost everywhere on ∂D. Vitali's Convergence Theorem implies

$$\lim_{\sigma(E)\to 0} \sup_k \int_E |u|^q |G_{j_k} \circ \varphi|\, d\sigma = 0,$$

where σ denotes the normalized Lebesgue measure on ∂D. This implies

$$\lim_{\sigma(E)\to 0} \sup_k \int_E |u|^q |f_{n_{j_k}} \circ \varphi|^q\, d\sigma \leq \lim_{\sigma(E)\to 0} \sup_k \int_E |u|^q |G_{j_k} \circ \varphi|\, d\sigma = 0.$$

Again, since $|\varphi(z)| < 1$ a.e. on ∂D, $\{u^q(f_{n_{j_k}} \circ \varphi)\}$ converges almost everywhere on ∂D. Using Vitali's Theorem again, we conclude that $u(f_{n_{j_k}} \circ \varphi)$ converges in H^q. Hence uC_φ is compact from H^p to H^q.

In general, if u is not outer, we can factor $u = B_u F_u$ where B_u is inner and F_u is outer. It is clear that uC_φ is compact from H^p to H^q if and only if $F_u C_\varphi$ is compact from H^p to H^q. By the proof above, this is equivalent to saying that $|\varphi| < 1$ a.e. on ∂D.

9.6 Weighted composition operators between weighted spaces of analytic functions

The following result was given in [27] and [6]. We refer to [27] for its proof.

Theorem 6.1 *Let v and w be radial, non-increasing weights satisfying* $\lim_{r\to 1} v(r) = 0$. *Let u be analytic on D and φ be an analytic self-map of D. Then*

(i) uC_φ *is bounded from* H_v^∞ *to* H_w^∞ *if and only if*

$$\sup_{z\in D} \frac{w(z)|u(z)|}{\tilde{v}(\varphi(z))} < \infty.$$

(ii)

$$\|uC_\varphi\|_e = \lim_{r\to 1} \sup_{|\varphi(z)|>r} \frac{w(z)|u(z)|}{\tilde{v}(\varphi(z))}.$$

In [16], the authors gave an alternative characterization for bounded weighted composition operators between weighted Banach spaces of analytic functions and estimated the essential norms in terms of φ^n as follows. Here we use $\mathbb{N}_0$ to denote $\mathbb{N}\cup\{0\}$, where $\mathbb{N}$ is the set of all positive integers.

Theorem 6.2 *Let v and w be radial, non-increasing weights satisfying* $\lim_{r\to 1} v(r) = 0$. *Let u be analytic on D and φ be an analytic self-map of D. Then*

(i) uC_φ *is bounded from* H_v^∞ *to* H_w^∞ *if and only if*

$$\sup_{n\in\mathbb{N}_0} \frac{\|u\varphi^n\|_w}{\|\xi^n\|_v} < \infty. \tag{9.5}$$

(ii)

$$\|uC_\varphi\|_e = \limsup_{n\to\infty} \frac{\|u\varphi^n\|_w}{\|\xi^n\|_v}. \tag{9.6}$$

To prove this result, we need the following instrumental lemma, which was originally given in [18].

Lemma 6.3 *Let v be a radial, non-increasing weight satisfying* $\lim_{r\to 1} v(r) = 0$. *Let*

$$\bar{v}(z) := \left(\sup_{n\in\mathbb{N}_0} \frac{|z|^n}{\|\xi^n\|_v}\right)^{-1},$$

where we understand that $0^0 = 1$. *Suppose that* $\psi(s) := -\log v(e^s)$ *is convex for* $s \in (-\infty, 0)$. *Then v is equivalent to both* $\tilde{v}$ *and* $\bar{v}$ *in* $\mathbb{D}$.

Proof Let $b(t) := \sup_{0<r<1} v(r)r^t$ for $t \in [0,\infty)$. Then $b(t)$ is non-increasing and $b(t) > 0$ for all $t \geq 0$, and $b(n) = \|\xi^n\|_v$ for $n \in \mathbb{N}_0$. Obviously,

$$\frac{1}{b(t)} \leq \frac{1}{v(r)r^t}$$

for all $t \in [0,\infty)$. Thus,

$$\sup_{t\in[0,\infty)} \frac{r^t}{b(t)} \leq \frac{1}{v(r)}$$

for all $r \in (0,1)$. Take an arbitrary $r_0 \in (0,1)$ and put $s_0 = \log r_0$, so $-\infty < s_0 < 0$. Since ψ is convex and non-decreasing in $(-\infty,0)$ we can find a constant $c = c(s_0)$ such that

$$c(s-s_0) + \psi(s_0) \leq \psi(s)$$

for all $s \in (-\infty,0)$. Thus

$$b(c) = \exp\left(\sup_{s<0}(\log v(e^s) + cs)\right) \leq e^{cs_0}e^{-\psi(s_0)} = r_0^c v(r_0)$$

and we conclude that

$$\sup_{t\in[0,\infty)} \frac{r_0^t}{b(t)} \geq \frac{r_0^c}{b(c)} \geq \frac{1}{v(r_0)}.$$

Thus we actually have

$$\sup_{t\in[0,\infty)} \frac{r^t}{b(t)} = \frac{1}{v(r)}. \tag{9.7}$$

We now show that we can find $C > 0$ such that for all $r \in (0,1)$, $v(r) \leq \tilde{v}(r) \leq \bar{v}(r) \leq Cv(r)$. For any $z \in D$, we clearly have $v(z) \leq \tilde{v}(z)$ and

$$\frac{1}{\bar{v}(z)} = \sup_{n\in\mathbb{N}_0} \frac{|z|^n}{\|\xi^n\|_v} \leq \sup\{|f(z)|, |f| \leq 1/v, f \in H(D)\} = \frac{1}{\tilde{v}(z)}.$$

For $r \in (\frac{1}{2},1)$, by (9.7) we have

$$\begin{aligned}\frac{1}{v(r)} &= \sup_{t\in(0,\infty)} \frac{r^t}{b(t)} \leq \sup_{n\in\mathbb{N}_0} \frac{r^n}{b(n+1)} \\ &\leq \frac{1}{r}\sup_{n\in\mathbb{N}_0} \frac{r^{n+1}}{b(n+1)} \leq \frac{1}{r}\frac{1}{\bar{v}(r)} \leq \frac{2}{\bar{v}(r)}.\end{aligned}$$

For $r \in [0,\frac{1}{2}]$, since v is non-increasing, we get

$$v(2^{-1}) \leq v(r) \leq \tilde{v}(r) \leq \bar{v}(r) \leq b(0) = v(0).$$

Thus $\bar{v}(r) \leq \frac{v(0)}{v(2^{-1})}v(r)$ for all $r \in [0,\frac{1}{2}]$. The proof is complete.

Proof of Theorem 6.2 (i) By Lemma 6.3 we find a constant $C > 0$ such that

$$\sup_{z\in D}\frac{w(z)|u(z)|}{\tilde{v}(\varphi(z))}\leq C\sup_{z\in D}\sup_{n\in\mathbb{N}_0}\frac{w(z)|u(z)||\varphi(z)|^n}{\|\xi^n\|_v}\leq C\sup_{z\in D}\sup_{n\in\mathbb{N}_0}\frac{\|u\varphi^n\|_w}{\|\xi^n\|_v}$$

which together with part (i) of Theorem 6.1 proves the sufficiency in (i). For the necessity of the condition, let $f_n(z)=z^n/\|\xi^n\|_v\in H_v^0$ for each $n\in\mathbb{N}_0$. Then

$$\|uC_\varphi\|_{H_v^\infty\to H_w^\infty}\geq\sup_{n\in\mathbb{N}_0}\|uC_\varphi(f_n)\|_w=\sup_{n\in\mathbb{N}_0}\frac{\|u\varphi^n\|_w}{\|\xi^n\|_v}.$$

This proves (i).

(ii) We first get the lower bound. Let $f_n(z)=z^n/\|\xi^n\|_v\in H_v^0$. Then (f_n) converges to zero in every compact subsets of D, from which we know that (f_n) converges to zero weakly in H_v^∞. Let $T:H_v^\infty\to H_w^\infty$ be a compact operator. Then we get

$$\begin{aligned}\|uC_\varphi-T\|_{H_v^\infty\to H_w^\infty}&\geq\limsup_{n\to\infty}\|uC_\varphi(f_n)\|_w\\&=\limsup_{n\to\infty}\frac{w(z)|u(z)||\varphi(z)|^n}{\|\xi^n\|_v}\\&=\limsup_{n\to\infty}\frac{\|u\varphi^n\|_w}{\|\xi^n\|_v},\end{aligned}$$

so

$$\|uC_\varphi\|_e\geq\limsup_{n\to\infty}\frac{\||u\varphi^n\|_w}{\|\xi^n\|_v}.$$

To get the upper bound for the essential norm, fix $n\in\mathbb{N}_0$. From the proof of Lemma 6.3 it follows that

$$\frac{1}{\tilde{v}(s)}\leq\frac{1}{s\tilde{v}(s)}$$

for every $s\in(0,1)$. Applying this to the supremum in Theorem 6.1 (ii) for some r gives

$$\begin{aligned}\sup_{|\varphi(z)|>r}\frac{w(z)|u(z)|}{\tilde{v}(\varphi(z))}&\leq\sup_{|\varphi(z)|>r}\frac{1}{|\varphi(z)|}\sup_{n\in\mathbb{N}_0}\frac{w(z)|u(z)||\varphi(z)|^n}{\|\xi^n\|_v}\\&\leq\frac{1}{r}\sup_{n>N}\frac{\|u\varphi^n\|_w}{\|\xi^n\|_v}+\frac{1}{r}\sup_{|\varphi(z)|>r}\sup_{0\leq n\leq N}\frac{w(z)|u(z)|}{\|\xi^n\|_v}.\end{aligned}$$

By boundedness of $uC_\varphi:H_v^\infty\to H_w^\infty$, there is a constant $C>0$ such that $w(z)|u(z)|\leq C\tilde{v}(\varphi(z))$ for each $z\in D$. Therefore, for some n_0 between 0 and N, the second term on the right of the above equation is less than

$$\frac{C}{r}\sup_{|\varphi(z)|>r}\frac{\tilde{v}(\varphi(z))}{\|\xi^{n_0}\|_v}\to 0$$

as $r\to 1$ by Lemma 2.1 (iv). Thus

$$\|uC_\varphi\|_e=\lim_{r\to 1}\sup_{|\varphi(z)|>r}\frac{w(z)|u(z)|}{\tilde{v}(\varphi(z))}\leq\sup_{n>N}\frac{\|u\varphi^n\|_w}{\|\xi^n\|_v}$$

for all $N \in N_0$. Letting $N \to \infty$ we get the required upper bound. The proof of the theorem is complete.

9.7 Weighted composition operators between Bloch type spaces

The situation becomes more complicated for weighted composition operators between Bloch-type spaces, due to a derivative in the definition of this type of spaces.

Boundedness and compactness of weighted composition operators on the Bloch space B were first characterized in [29]. The results were generalized in [28] to the case of weighted composition operators between possibly different Bloch type spaces. We present the following result in [28] here, and refer to [28] for its proof.

Theorem 7.1 *Let $\varphi : D \to D$ be analytic, let u be analytic on D, and let α and β be positive real numbers.*

(i) If $0 < \alpha < 1$, then uC_φ maps B^α boundedly into B^β if and only if $u \in B^\beta$ and

$$\sup_{z\in D} |u(z)| \frac{(1-|z|^2)^\beta}{(1-|\varphi(z)|^2)^\alpha} |\varphi'(z)| < \infty.$$

(ii) If $\alpha = 1$, then uC_φ maps B boundedly into B^β if and only if

$$\sup_{z\in D} |u'(z)|(1-|z|^2)^\beta \log \frac{e}{1-|\varphi(z)|^2} < \infty$$

and

$$\sup_{z\in D} |u(z)| \frac{(1-|z|^2)^\beta}{1-|\varphi(z)|^2} |\varphi'(z)| < \infty.$$

(iii) If $\alpha > 1$, then uC_φ maps B^α boundedly into B^β if and only if

$$\sup_{z\in D} |u'(z)| \frac{(1-|z|^2)^\beta}{(1-|\varphi(z)|^2)^{\alpha-1}} < \infty$$

and

$$\sup_{z\in D} |u(z)| \frac{(1-|z|^2)^\beta}{(1-|\varphi(z)|^2)^\alpha} |\varphi'(z)| < \infty.$$

In order to simplify the notation in the statement of the next result, we write

$$A = \lim_{s\to 1} \sup_{|\varphi(z)|\geq s} \frac{|u(z)||\varphi'(z)|(1-|z|^2)^\beta}{(1-|\varphi(z)|^2)} \tag{9.8}$$

and

$$B = \lim_{s\to 1} \sup_{|\varphi(z)|>s} |u'(z)|(1-|z|^2)^\beta \log \frac{e}{(1-|\varphi(z)|^2)}. \tag{9.9}$$

The following result for the essential norms of weighted composition operators between Bloch-type spaces was given in [25]. We refer to [25] for its proof.

Theorem 7.2 *Let $\varphi : D \to D$ be analytic, let u be analytic on D, and let α and β be positive real numbers. Suppose the weighted composition operator uC_φ is bounded from B^α to B^β. Then*

(i) If $0 < \alpha < 1$, then

$$\|uC_\varphi\|_e = \lim_{s\to 1} \sup_{|\varphi(z)|>s} |u(z)||\varphi'(z)| \frac{(1-|z|^2)^\beta}{(1-|\varphi(z)|^2)^\alpha}.$$

(ii) If $\alpha = 1$, then

$$\max\{A, \frac{1}{6}B\} \le \|uC_\varphi\|_e \le A+B.$$

In [25], the result for the case $\alpha > 1$ is not given. However, by modifying the proof of (ii) of the above theorem (or Theorem 4 in [25]), one can get the following result, which was also appeared in [18]. A proof for the lower bound can be found in the proof of Theorem 8 in [26] (see the proofs of (21) and (22) there). We omit the details of the proofs here.

Theorem 7.3 *Let $\varphi : D \to D$ be analytic, let u be analytic on D, and let α and β be positive real numbers. Suppose the weighted composition operator uC_φ is bounded from B^α to B^β. If $\alpha > 1$ then*

$$\max\{M(\alpha)C, N(\alpha)D\} \le \|uC_\varphi\|_e \le C+D,$$

where

$$C = \lim_{s\to 1} \sup_{|\varphi(z)|>s} \frac{|u(z)||\varphi'(z)|(1-|z|^2)^\beta}{(1-|\varphi(z)|^2)^\alpha},$$

$$D = \lim_{s\to 1} \sup_{|\varphi(z)|>s} \frac{|u'(z)|(1-|z|^2)^\beta}{(1-|\varphi(z)|^2)^{\alpha-1}},$$

and

$$M(\alpha) = \frac{1}{2^{1+\alpha}(3\alpha+2)}, \quad N(\alpha) = \frac{1}{2^{1+\alpha}3\alpha(\alpha+1)}.$$

Similar to the case of weighted spaces of analytic functions, one may also use φ^n to describe boundedness, compactness and essential norms for weighted composition operators between Bloch-type spaces. We need some preparations first. We need the following two integral operators. Let u be an analytic function on D. For every $f \in H(D)$, define

$$I_u f(z) = \int_0^z f'(\zeta)u(\zeta)\,d\zeta, \qquad J_u f(z) = \int_0^z f(\zeta)u'(\zeta)\,d\zeta.$$

The operators J_u, sometimes referred as Cesàro type operators or Riemann-Stieltjes integral operators, were first used by Ch. Pommerenke in [31] to characterize BMOA functions. They were first systematically studied by A. Aleman and A. G. Siskakis in [2]. They proved that J_u is bounded on the Hardy space H^p, $1 \leq p < \infty$, if and only if $g \in BMOA$. Thereafter there have been many papers on these operators. See, [1], [3], [17], [30], [33], and [38] for a few examples. The operators I_u, as companions of J_u, have been also studied, see, for example, [39].

For $0 < \alpha < \infty$, let $v_\alpha(z) = (1-|z|^2)^\alpha$ and

$$v_{\log} = (\log(\frac{e}{1-|z|^2}))^{-1}.$$

Then v_α and $v_{\log}$ are weights that satisfy the conditions of Lemma 2.1. For a weight v, consider the bounded operator $S_v : \tilde{B}_v \to H_v^\infty$ defined by $S_v(h) = h'$ and $S_v^{-1} : H_v^\infty \to \tilde{B}_v$ defined by

$$S_v^{-1}(h)(z) = \int_0^z h(\xi)\,d\xi.$$

Then $S_v \circ S_v^{-1} = \mathrm{id}_{H_v^\infty}$, $S_v^{-1} \circ S_v = \mathrm{id}_{\tilde{B}_v}$, and S_v and S_v^{-1} are isometric onto maps. We need the following lemmas.

Lemma 7.4 *Let v be a weight. Then*

(i) $n\|z^{n-1}\|_v = \|z^n\|_{B_v}$ *for all* $n \in \mathbb{N}$.

(ii) $\|z^n\|_{v_\alpha} = (\frac{2\alpha}{n+2\alpha})^\alpha(\frac{n}{n+2\alpha})^{n/2}$ *for all* $n \in \mathbb{N}$.

(iii) $\lim_{n\to\infty}(n+1)^\alpha\|z^n\|_{v_\alpha} = \lim_{n\to\infty} n^{\alpha-1}\|z^n\|_{B^\alpha} = (\frac{2\alpha}{e})^\alpha$.

(iv) $\lim_{n\to\infty}(\log n)\|z^n\|_{v_{\log}} = \lim_{n\to\infty}\frac{\log(n-1)}{n}\|z^n\|_{B_{v_{\log}}} = 1$.

The result can be proved by straightforward computations. We refer to [18] for its proof.

Lemma 7.5 *Let* $n \in \mathbb{N}$ *be fixed, and let v be a weight. Suppose that* $u\varphi'\varphi^{n-1} \in H_v^\infty$ *and* $u'\varphi^n \in H_v^\infty$. *Then*

$$\|u\varphi'\varphi^{n-1}\|_v = \frac{1}{n}\|I_u(\varphi^n)\|_{B_v}, \qquad \|u'\varphi^n\|_v = \|J_u(\varphi^n)\|_{B_v}.$$

Proof It is straightforward that

$$\|u\varphi'\varphi^{n-1}\|_v = \left\|S_v^{-1}\left(u\left(\frac{1}{n}\varphi^n\right)'\right)\right\|_{B_v} = \frac{1}{n}\|I_u(\varphi^n)\|_{B_v}$$

and also

$$\|u'\varphi^n\|_v = \|S_v^{-1}(u'\varphi^n)\|_{B_v} = \|J_u(\varphi^n)\|_{B_v}.$$

The following result was first given in [26] for the cases $0 < \alpha < 1$ and $\alpha > 1$. The case $\alpha = 1$ was left open in [26], but has been solved in [18]. When $\alpha = \beta = 1$ and $u = 1$, the result of (ii) (which is for the composition operator $C_\varphi : B \to B$) was first given in [36]. Note that in this case $I_u(\varphi^n) = \varphi^n - \varphi^n(0)$ and $J_u(\varphi^n) = 0$.

Theorem 7.6 *Let φ be an analytic self map of D, let u be analytic on D, and let α and β be positive real numbers.*

(i) If $0 < \alpha < 1$, then uC_φ maps B^α boundedly into B^β if and only if $u \in B^\beta$ and

$$\sup_{n\geq 1} n^{\alpha-1}\|I_u(\varphi^n)\|_{B^\beta} < \infty. \tag{9.10}$$

(ii) If $\alpha = 1$, then uC_φ maps B^α boundedly into B^β if and only if $u \in B^\beta$,

$$\sup_{n\geq 1} \|I_u(\varphi^n)\|_{B^\beta} < \infty \tag{9.11}$$

and

$$\sup_{n\geq 1}(\log n)\|J_u(\varphi^n)\|_{B^\beta} < \infty. \tag{9.12}$$

(iii) If $\alpha > 1$, then uC_φ maps B^α boundedly into B^β if and only if

$$\sup_{n\geq 1} n^{\alpha-1}\|I_u(\varphi^n)\|_{B^\beta} < \infty \tag{9.13}$$

and

$$\sup_{n\geq 1} n^{\alpha-1}\|J_u(\varphi^n)\|_{B^\beta} < \infty. \tag{9.14}$$

Proof The proof here is adapted from [18].

(i) By of Theorem 7.1 (i), we know that, for $0 < \alpha < 1$, uC_φ maps B^α boundedly into B^β if and only if $\varphi \in B^\beta$ and

$$\sup_{z\in D} \frac{|u(z)||\varphi'(z)|(1-|z|^2)^\beta}{(1-|\varphi(z)|^2)^\alpha} < \infty.$$

By Theorem 6.1 (i), this last condition means exactly that $u\varphi' C_\varphi$ maps $H^\infty_{v_\alpha}$ boundedly into $H^\infty_{v_\beta}$, which by Theorem 6.2, Lemma 7.4 and Lemma 7.5, is equivalent to

$$\sup_{n\geq 1} \frac{\|u\varphi'\varphi^{n-1}\|_{v_\beta}}{\|z^{n-1}\|_{v_\alpha}} = \sup_{n\geq 1} n^{\alpha-1}\|I_u(\varphi^n)\|_{B^\beta} < \infty.$$

(ii) By Theorem 7.1 (ii), we know that, for $\alpha = 1$, uC_φ maps B boundedly into B^β if and only if

$$\sup_{z\in\mathbb{D}} |u'(z)|(1-|z|^2)^\beta \log\frac{e}{1-|\varphi(z)|^2} < \infty$$

and

$$\sup_{z\in\mathbb{D}} |u(z)|\frac{(1-|z|^2)^\beta}{1-|\varphi(z)|^2}|\varphi'(z)| < \infty.$$

By Theorem 6.1, the first condition means that $u'C_\varphi$ maps $H^\infty_{v_{\log}}$ boundedly into $H^\infty_{v_\beta}$, and the second condition means that $u\varphi' C_\varphi$ maps $H^\infty_{v_1}$ boundedly into $H^\infty_{v_\beta}$, which by Theorem 6.2, Lemma 7.4 and Lemma 7.5, are equivalent to

$$\sup_{n\geq 1} \frac{\|J_u(\varphi^n)\|_{B^\beta}}{\|z^n\|_{v_{\log}}} \approx \max\left(\|u\|_{B^\beta}, \sup_{n\geq 1}(\log n)\|J_u(\varphi^n)\|_{B^\beta} \right) < \infty,$$

and

$$\sup_{n\geq 1} \|I_u(\varphi^n)\|_{B^\beta} < \infty.$$

(iii) By Theorem 7.1 (iii), we know that, for $\alpha > 1$, uC_φ maps B^α boundedly into B^β if and only if

$$\sup_{z\in\mathbb{D}} |u'(z)| \frac{(1-|z|^2)^\beta}{(1-|\varphi(z)|^2)^{\alpha-1}} < \infty$$

and

$$\sup_{z\in\mathbb{D}} |u(z)| \frac{(1-|z|^2)^\beta}{(1-|\varphi(z)|^2)^{\alpha}} |\varphi'(z)| < \infty.$$

By Theorem 6.1, the first condition means that $u' C_\varphi$ maps $H^\infty_{v_{\alpha-1}}$ boundedly into $H^\infty_{v_\beta}$, and the second condition means that $u\varphi' C_\varphi$ maps $H^\infty_{v_\alpha}$ boundedly into $H^\infty_{v_\beta}$, which again by Theorem 6.2, Lemma 7.4 and Lemma 7.5, are equivalent to

$$\sup_{n\geq 1} n^{\alpha-1} \|I_u(\varphi^n)\|_{B^\beta} < \infty$$

and

$$\sup_{n\geq 1} n^{\alpha-1} \|J_u(\varphi^n)\|_{B^\beta} < \infty.$$

The proof is complete.

Finally, let us look at the estimates of essential norms of uC_φ from B^α to B^β in terms of φ^n. For simplifying the notations, we denote

$$A^* = \frac{e}{2} \limsup_{n\to\infty} \|I_u(\varphi^n)\|_{B^\beta},$$

$$B^* = \limsup_{n\to\infty} (\log n)\|J_u(\varphi^n)\|_{B^\beta},$$

$$C^* = \left(\frac{e}{2\alpha}\right)^\alpha \limsup_{n\to\infty} n^{\alpha-1} \|I_u(\varphi^n)\|_{B^\beta}$$

and

$$D^* = \left(\frac{e}{2(\alpha-1)}\right)^{\alpha-1} \limsup_{n\to\infty} n^{\alpha-1} \|J_u(\varphi^n)\|_{B^\beta}.$$

The following result was also first given in [26] for the cases $0 < \alpha < 1$ and $\alpha > 1$. The case $\alpha = 1$ was given in [18]. For the case $\alpha = \beta = 1$ and $u = 1$, which is the case of the composition operator $C_\varphi : B \to B$, the result of (ii) was first given in [40].

Theorem 7.7 *Let φ be an analytic self map of D, let u be analytic on D, and let α and β be positive real numbers. Suppose uC_φ maps B^α boundedly into B^β. Then for uC_φ as an operator from B^α to B^β, we have the following results.*

(i) If $0 < \alpha < 1$ then

$$\|uC_\varphi\|_e = \left(\frac{e}{2\alpha}\right)^\alpha \limsup_{n\to\infty} n^{\alpha-1}\|I_u(\varphi^n)\|_{B^\beta}. \tag{9.15}$$

(ii) If $\alpha = 1$ then

$$\max\left(A^*, \frac{1}{6}B^*\right) \le \|uC_\varphi\|_e \le A^* + B^*. \tag{9.16}$$

(iii) If $\alpha > 1$, then

$$\max\left(\frac{1}{2^{1+\alpha}(3\alpha+2)}C^*, \frac{1}{2^{1+\alpha}3\alpha(\alpha+1)}D^*\right) \le \|uC_\varphi\|_e \le C^* + D^*. \tag{9.17}$$

Proof (i) Applying Theorem 7.2 (i), Theorem 6.1 (ii), Theorem 6.2 (ii), Lemma 7.5 and Lemma 7.4 consecutively we obtain

$$\begin{aligned}\|uC_\varphi\|_e &= \lim_{s\to 1}\sup_{|\varphi(z)|>s} \frac{|u(z)||\varphi'(z)|(1-|z|^2)^\beta}{(1-|\varphi(z)|^2)^\alpha}\\ &= \|u\varphi' C_\varphi\|_{e,H^\infty_{v_\alpha}\to H^\infty_{v_\beta}} = \limsup_{n\to\infty}\frac{\|u\varphi'\varphi^{n-1}\|_{v_\beta}}{\|z^{n-1}\|_{v_\alpha}}\\ &= \limsup_{n\to\infty}\frac{\|I_u(\varphi^n)\|_{B^\beta}}{n\|z^{n-1}\|_{v_\alpha}} = \left(\frac{e}{2\alpha}\right)^\alpha \limsup_{n\to\infty} n^{\alpha-1}\|I_u(\varphi^n)\|_{B^\beta}.\end{aligned}$$

(ii) From Theorem 7.2 (ii) we know that

$$\max\{A, \frac{1}{6}B\} \le \|uC_\varphi\|_e \le A + B.$$

As in the proof of (i) above, we have

$$A = \lim_{s\to 1}\sup_{|\varphi(z)|>s}\frac{|u(z)||\varphi'(z)|(1-|z|^2)^\beta}{(1-|\varphi(z)|^2)} = \frac{e}{2}\limsup_{n\to\infty}\|I_u(\varphi^n)\|_{B^\beta} = A^*,$$

and by Theorem 6.1 (ii), Theorem 6.2 (ii), Lemma 7.5 and Lemma 7.4,

$$\begin{aligned}B &= \lim_{s\to 1}\sup_{|\varphi(z)|>s}|u'(z)|(1-|z|^2)^\beta \log\frac{e}{(1-|\varphi(z)|^2)}\\ &= \|u'C_\varphi\|_{e,H^\infty_{v_{\log}}\to H^\infty_{v_\beta}} = \limsup_{n\to\infty}\frac{\|u'\varphi^n\|_{v_\beta}}{\|z^n\|_{v_{\log}}}\\ &= \limsup_{n\to\infty}\frac{\|J_u(\varphi^n)\|_{B^\beta}}{\|z^n\|_{v_{\log}}} = \limsup_{n\to\infty}(\log n)\|J_u(\varphi^n)\|_{B^\beta} = B^*.\end{aligned}$$

Thus (ii) is true.

(iii) From Theorem 7.3 we know that

$$\max\{M(\alpha)C, N(\alpha)D\} \leq \|uC_\varphi\|_e \leq C+D.$$

As in the previous proof, applying Theorem 6.1 (ii), Theorem 6.2 (ii), Lemma 7.5 and Lemma 7.4 consecutively we obtain

$$\begin{aligned} C &= \lim_{s\to 1} \sup_{|\varphi(z)|>s} \frac{|u(z)||\varphi'(z)|(1-|z|^2)^\beta}{(1-|\varphi(z)|^2)^\alpha} = \|u\varphi' C_\varphi\|_{e,H^\infty_{v_\alpha}\to H^\infty_{v_\beta}} \\ &= \limsup_{n\to\infty} \frac{\|u\varphi'\varphi^{n-1}\|_{v_\beta}}{\|z^{n-1}\|_{v_\alpha}} = \limsup_{n\to\infty} \frac{\|I_u(\varphi^n)\|_{B^\beta}}{n\|z^{n-1}\|_{v_\alpha}} \\ &= \left(\frac{e}{2\alpha}\right)^\alpha \limsup_{n\to\infty} n^{\alpha-1}\|I_u(\varphi^n)\|_{B^\beta} \\ &= C^*, \end{aligned}$$

and

$$\begin{aligned} D &= \lim_{s\to 1} \sup_{|\varphi(z)|>s} \frac{|u'(z)|(1-|z|^2)^\beta}{(1-|\varphi(z)|^2)^{\alpha-1}} = \|u' C_\varphi\|_{e,H^\infty_{v_{\alpha-1}}\to H^\infty_{v_\beta}} \\ &= \limsup_{n\to\infty} \frac{\|u'\varphi^n\|_{v_\beta}}{\|z^n\|_{v_{\alpha-1}}} = \limsup_{n\to\infty} \frac{\|J_u(\varphi^n)\|_{B^\beta}}{\|z^n\|_{v_{\alpha-1}}} \\ &= \left(\frac{e}{2(\alpha-1)}\right)^{\alpha-1} \limsup_{n\to\infty} n^{\alpha-1}\|J_u(\varphi^n)\|_{B^\beta} \\ &= D^*. \end{aligned}$$

Thus (iii) is true, and the proof is complete.

Acknowledgment

The author's research is supported by the National Natural Science Foundation of China (Grant Number 11720101003).

References

[1] A. Aleman and J. A. Cima, An integral operator on H^p and Hardy's inequality, *J. Anal. Math.* **85** (2001), 157-176.

[2] A. Aleman and A. G. Siskakis, An integral operator on H^p, *Complex Variables* **28** (1995), 149-158.

[3] A. Aleman and A. G. Siskakis, Integration operators on Bergman spaces, *Indiana Univ. Math. J.* **46** (1997), 337–356.

[4] R. Aulaskari, D. Stegenga and J. Xiao, Some subclasses of $BMOA$ and their characterization in terms of Carleson measures, *Rocky Mountain J. Math.* **26** (1996), 485-506.

[5] L. Carleson, Interpolation by bounded analytic functions and the corona problem, *Ann. of Math.* **76** (1962) 547-559.

[6] M. Contreras and A. Hernandez-Diaz, Weighted composition operators in weighted Banach spaces of analytic functions, *J. Austral. Math. Soc.* **69** (2000), 41-60.

[7] M. D. Contreras and A. G. Hernandez-Diaz, Weighted composition operators on Hardy spaces, *J. Math. Anal. Appl.* **263** (2001), 224-233.

[8] M. D. Contreras and A. G. Hernandez-Diaz, Weighted composition operators between different Hardy spaces, *Integral Equ. Oper. Theory* **46** (2003), 165-188.

[9] Ž. Čučković and R. Zhao, Weighted composition operators on the Bergman space, *J. London Math. Soc.* **70** (2004), 499-511.

[10] Ž. Čučković and R. Zhao, Weighted composition operators between different weighted Bergman spaces and different Hardy spaces, *Illinois J. Math.* **51** (2007), 479-498.

[11] N. Dunford and J. Schwartz, *Linear Operators*, Part 1. Wiley, New York, 1957.

[12] P. Duren, Extension of a theorem of Carleson, *Bull. Amer. Math. Soc.* **75** (1969), 143-146.

[13] F. Forelli, The isometries of H^p, *Can. J. Math.* **16** (1964), 721-728.

[14] T. E. Goebeler, Jr., Composition operators acting between Hardy spaces *Integral Equa. Oper. Theory* **41** (2001), 389-395.

[15] W. W. Hastings, A Carleson measure theorem for Bergman spaces *Proc. Amer. Math. Soc.* **52** (1975), 237-241.

[16] O. Hyvärinen, M. Kemppainen, M. Lindström, A. Rautio and E. Saukko, The essential norm of weighted composition operators on weighted Banach spaces of analytic functions, *Integral Equ. Oper. Theory* **72** (2012), 151-157.

[17] Z. Hu, Extended Cesàro operators on mixed norm spaces, *Proc. Amer. Math. Soc.* **131** (2003), 2171–2179.

[18] O. Hyvärinen and M. Lindström, Estimates of essential norms of weighted composition operators between Bloch-type spaces, *J. Math. Analy. Appl.* **393** (2012), 38-44.

[19] C. J. Kolaski, Isometries of the weighted Bergman spaces, *Can. J. Math.* **34** (1982), 910-915.

[20] J. Lindenstrauss and L. Tzafriri, *Classical Banach Spaces*, Lecture Notes in Math. **338**, Springer-Verlag, Berlin, 1973.

[21] D. H. Luecking, Forward and reverse Carleson inequalities for functions in Bergman spaces and their derivatives, *Amer. J. Math.* **107** (1985), 85-111.

[22] D. H. Luecking, Trace ideal criteria for Toeplitz operators, *J. Funct. Anal.* **73** (1987), 345-368.

[23] D. H. Luecking, Embedding derivatives of Hardy spaces into Lebesgue spaces, *Proc. London Math. Soc. (3)* **63** (1991) 595-619.

[24] D. H. Luecking, Embedding theorems for spaces of analytic functions via Khinchine's inequality, *Michigan Math. J.* **40** (1993), 333-358.

[25] B. MacCluer and R. Zhao, Essential norms of weighted composition operators between Bloch-type spaces, *Rocky Mountain J. Math.* **33** (2003), 1437-1458.

[26] J. S. Manhas and R. Zhao, New estimates of essential norms of weighted composition operators between Bloch type spaces, *J. Math. Anal. Appl.* **389** (2012), 32-47.

[27] A. Montes-Rodriguez, Weighted composition operators on weighted Banach spaces of analytic functions, *J. London Math. Soc.* **61** (2000), 872-884.

[28] S. Ohno, K. Stroethoff and R. Zhao, Weighted composition operators between Bloch-type space, *Rocky Mountain J. Math.* **33** (2003), 191-215.

[29] S. Ohno and R. Zhao, Weighted composition operators on the Bloch space, *Bull. Austral. Math. Soc.* **63** (2001), 177-185.

[30] J. Pau and J. Á. Peláez, Embedding theorems and integration operators on Bergman spaces with rapidly decreasing weights, *J. Funct. Anal.* **259** (2010), 2727-2756.

[31] Ch. Pommerenke, Schlichte Funktionen und analytische Funktionen von beschränkten mittlerer Oszillation, *Comm. Math. Helv.* **52** (1977), 591-602.

[32] J. Shapiro, The essential norms of a composition operator, *Ann. Math.* **125** (1987), 375-404.

[33] A. G. Siskakis and R. Zhao, A Volterra type operator on spaces of analytic functions, *Function Spaces*, Contemp. Math. **232**, Amer. Math. Soc., Providence, RI, 1999, 299-311.

[34] M. Tjani, Compact composition operators on Besov spaces, *Trans. Amer. Math. Soc.* **355** (2003), 4683-4698.

[35] I. V. Videnskii, On an analogue of Carleson measures, *Dokl. Akad. Nauk SSSR* **298** (1988), 1042-1047.

[36] H. Wulan, D. Zheng and K. Zhu, Composition operators on BMOA and the Bloch space, *Proc. Amer. Math. Soc.* **137** (2009), 3861-3868.

[37] P. Wojtaszczyk, *Banach Spaces for Analysts*, Cambridge Studies in Advanced Mathematics, 25, Cambridge University Press, 1991.

[38] J. Xiao, Riemann-Stieltjes operators on weighted Bloch and Bergman spaces of the unit ball, *J. London Math. Soc.* **70** (2004), 199-214.

[39] R. Yoneda, Pointwise multipliers from $BMOA^{\alpha}$ to $BMOA^{\beta}$, *Complex Variables* **19** (2004), 1045-1061.

[40] R. Zhao, Essential norms of composition operators between Bloch type spaces, *Proc. Amer. Math. Soc.* **138** (2010), 2537-2546.

[41] K. Zhu, *Operator Theory in Function Spaces*, Marcel Dekker, New York, 1990.

[42] K, Zhu, Bloch type spaces of analytic functions, *Rocky Mountain J. Math.* **23** (1993), 1143-1177.

Chapter 10

Toeplitz Operators on the Bergman Space and the Berezin Transform

Xianfeng Zhao
College of Mathematics and Statistics, Chongqing University, 401331, China
xianfengzhao@cqu.edu.cn

Dechao Zheng
Center of Mathematics, Chongqing University, Chongqing 401331, China, and Department of Mathematics, Vanderbilt University, Nashville, TN 37240, USA
dechao.zheng@vanderbilt.edu

CONTENTS

10.1 Introduction

Let dA denote the Lebesgue measure on the open unit disk $\mathbb{D}$ in the complex plane $\mathbb{C}$, normalized so that the measure of the disk $\mathbb{D}$ is 1. The complex space $L^2(\mathbb{D}, dA)$ is a Hilbert space with the inner product

$$\langle f, g \rangle = \int_{\mathbb{D}} f(z)\overline{g(z)}dA(z).$$

The Bergman space $L_a^2(\mathbb{D})$ is the set of those functions in $L^2(\mathbb{D}, dA)$ that are analytic on $\mathbb{D}$. Thus the Bergman space is a closed subspace of $L^2(\mathbb{D}, dA)$ and so there is an orthogonal projection P from $L^2(\mathbb{D}, dA)$ onto $L_a^2(\mathbb{D})$. For $\varphi \in L^\infty(\mathbb{D}, dA)$, the Toeplitz operator T_φ with symbol φ is defined by $T_\varphi f = P(\varphi f)$.

A central problem in the theory of Toeplitz operators is to establish relationships between the fundamental properties of those operators and analytic and geometric properties of their symbols [1], [2], [3], [7], [15] and [28].

Using the reproducing kernel

$$K_z(w) = \frac{1}{(1-\overline{z}w)^2} \quad (z, w \in \mathbb{D})$$

for the Bergman space $L_a^2(\mathbb{D})$, we express the Toeplitz operator to be an integral operator:

$$\begin{aligned} T_\varphi f(z) &= \int_{\mathbb{D}} \varphi(w) f(w) \overline{K_z(w)} dA(w) \\ &= \int_{\mathbb{D}} \frac{\varphi(w) f(w)}{(1-\overline{w}z)^2} dA(w). \end{aligned}$$

For a bounded operator S on $L_a^2(\mathbb{D})$, the *Berezin transform* of S is the function $\tilde{S}(z)$ on $\mathbb{D}$ defined by $\tilde{S}(z) = \langle Sk_z, k_z \rangle$, where k_z is the normalized reproducing kernel $\frac{K_z}{\|K_z\|}$. The Berezin transform $\tilde{\varphi}$ of a function $\varphi \in L^\infty(\mathbb{D})$ is defined to be the Berezin transform of the Toeplitz operator T_φ. The Berezin transform has played an important role in studying the Toeplitz operators on the Bergman space [28]. A nice survey of previously known results connecting the Berezin transform with Toeplitz operators can be found in [16].

In this paper, we will survey some recent results for the Toeplitz operators on the Bergman space of the unit disk via the Berezin transform.

In Section 10.3, we will use the Berezin transforms to study the positivity of Toeplitz operators with bounded symbols on the Bergman space. We will show that the positivity of a Toeplitz operator on the Bergman space is not completely determined by the positivity of the Berezin transform of its symbol. Moreover, we construct a quadratic polynomial of $|z|$ such that the minimal value of the Berezin transform of this polynomial is positive, but the corresponding Toeplitz operator may not be positive. The main results on the positivity of Toeplitz operators in Section 10.3 are contained in [24].

The invertibility problem for a Toeplitz operator on the Bergman space has been investigated by many people [8], [10], [11], [12] and [20]. For a nonnegative symbol φ, Luecking [12] obtained a necessary and sufficient condition for T_φ to be invertible on $L_a^2(\mathbb{D})$. Based on Luecking's results, Faour [8] gave a necessary condition for T_φ to be invertible if φ is a continuous function on the closed unit disk and satisfies that $|\varphi(z_1)| \geqslant |\varphi(z_2)|$ whenever $|z_1| \leqslant |z_2|$. In general, Karaev [11] obtained some sufficient conditions on the invertibility of a linear bounded operator via the Berezin transform and atomic decomposition. Using Karaev's results, Gürdal and Söhret [10] established a sufficient condition on the invertibility of Toeplitz operators. Most recently, Čučković and Vasaturo [4] obtained a necessary and sufficient condition for the invertibility of Toeplitz operators whose symbols are averaging functions of certain Carleson measures. Moreover, Guo and the authors studied the invertibility and the spectrum of Toeplitz operators with the symbol $\overline{z}+p$, where p is an analytic polynomial on the unit disk; see [9], [25] and [27].

In Section 10.4, we will use the Berezin transform and the n-th Berezin transform to study the invertibility of Toeplitz operators with bounded nonnegative symbols, harmonic symbols and continuous symbols on the Bergman space. For bounded nonnegative symbols, by means of Luecking's characterization on the reverse Carleson measure for the Bergman space [12] we will show that the Toeplitz operator is invertible on the Bergman space if and only if its symbol is invertible in $L^\infty(\mathbb{D})$ (Theorem 4.3). For Toeplitz operators with bounded harmonic symbols, we will establish a sufficient condition of the Chang-Tolokonnikov type such that these operators are invertible on the Bergman space (Theorem 4.5). Finally, we also consider the invertibility of Toeplitz operators with continuous symbols on the Bergman space in Theorem 4.10. Part of the results on the invertibility of Toeplitz operators on the Bergman space are contained in [26].

10.2 Basic properties of Toeplitz operators and the Berezin transform

In this section, we give a brief review on basic properties of Toeplitz operators and the Berezin transforms. Let us begin with some basic properties of the Bergman Toeplitz operators.

Proposition 2.1 *Suppose that α and β are complex constants, and φ and ψ are bounded functions on the open unit disk $\mathbb{D}$. Then we have*

(a) $T_\varphi^* = T_{\overline{\varphi}}$;

(b) $T_{\alpha\varphi+\beta\psi} = \alpha T_\varphi + \beta T_\psi$;

(c) $T_\varphi \geqslant 0$ *if* $\varphi \geqslant 0$ *a.e.*;

(d) $\|T_\varphi\| \leqslant \|\varphi\|_\infty$.

If in addition, $\bar{\varphi}$ or ψ is analytic, then we have

(e) $T_\varphi T_\psi = T_{\varphi\psi}$.

Proof For (a), we have by the definition of the Toeplitz operator that

$$\langle f, T_\varphi^* g\rangle = \langle T_\varphi f, g\rangle = \langle P(\varphi f), g\rangle = \langle \varphi f, g\rangle = \langle f, \overline{\varphi} g\rangle = \langle f, T_{\overline{\varphi}} g\rangle$$

for all f and g in $L_a^2(\mathbb{D})$. This gives $T_\varphi^* = T_{\overline{\varphi}}$.

For each function $f \in L_a^2(\mathbb{D})$ and $\alpha, \beta \in \mathbb{C}$,

$$\begin{aligned} T_{\alpha\varphi+\beta\psi} f &= P[(\alpha\varphi + \beta\psi)f] = P(\alpha\varphi f + \beta\psi f) \\ &= \alpha P(\varphi f) + \beta P(\psi f) = (\alpha T_\varphi + \beta T_\psi) f. \end{aligned}$$

This gives (b).

If $\varphi \geqslant 0$ a.e., then

$$\langle T_\varphi f, f\rangle = \int_{\mathbb{D}} \varphi(z)|f(z)|^2 dA(z) \geqslant 0$$

for all $f \in L_a^2(\mathbb{D})$. Thus T_φ is positive on the Bergman space. This gives (c).

Part (d) follows from the fact that the norm of the Bergman projection P is 1.

To get (e), first we assume that $\bar{\varphi}$ is analytic. Then for each f and g in $L_a^2(\mathbb{D})$,

$$\begin{aligned}\langle T_\varphi T_\psi f, g\rangle &= \langle T_\psi f, T_\varphi^* g\rangle = \langle T_\psi f, T_{\bar{\varphi}}^* g\rangle \\ &= \langle \psi f, \overline{\varphi} g\rangle = \langle \varphi\psi f, g\rangle = \langle T_{\varphi\psi} f, g\rangle.\end{aligned}$$

The third equality follows from that $\overline{\varphi} g$ is in the Bergman space $L^2(\mathbb{D})$. This gives that $T_\varphi T_\psi = T_{\varphi\psi}$. Similarly we can obtain the above equality in the case that ψ is analytic on the unit disk. This completes the proof.

The next result is about the Fredholm properties of Toeplitz operators with continuous symbols on the Bergman space; one can consult [17] or [28] for more information.

Proposition 2.2 *Suppose that $\varphi \in C(\overline{\mathbb{D}})$. Then the essential spectrum of the Toeplitz operator T_φ is given by $\sigma_e(T_\varphi) = \varphi(\partial\mathbb{D})$. Moreover, if T_φ is a Fredholm operator, then the Fredholm index of T_φ is given by*

$$\text{index}(T_\varphi) = \dim \text{Ker}(T_\varphi) - \dim \text{Ker}(T_\varphi^*) = -\text{wind}\big(\varphi(\partial\mathbb{D}), 0\big),$$

where $\text{wind}\big(\varphi(\partial\mathbb{D}), 0\big)$ *is the winding number of the closed oriented curve $\varphi(\partial\mathbb{D})$ with respect to the origin, which is defined by*

$$\text{wind}\big(\varphi(\partial\mathbb{D}), 0\big) = \frac{1}{2\pi i}\int_{\varphi(\partial\mathbb{D})} \frac{dz}{z}.$$

The Berezin transform is a useful tool to study operator theory on any reproducing kernel Hilbert space. In this paper we will focus on the Bergman space $L_a^2(\mathbb{D})$. From the definition of the Berezin transform, for a function φ in $L^1(\mathbb{D})$, its Berezin transform $\tilde{\varphi}$ is given by

$$\tilde{\varphi}(z) = \widetilde{T_\varphi}(z) = \langle T_\varphi k_z, k_z\rangle = \langle \varphi k_z, k_z\rangle = \int_{\mathbb{D}} \varphi(w)|k_z(w)|^2 dA(w). \tag{10.1}$$

By a change of variables, we obtain

$$\widetilde{\varphi}(z) = \int_{\mathbb{D}} \varphi(\varphi_z(w)) dA(w) \quad \big(z \in \mathbb{D}\big), \tag{10.2}$$

where

$$\varphi_z(w) = \frac{z - w}{1 - \bar{z}w} \quad \big(z, w \in \mathbb{D}\big)$$

is the Möbius transform from the unit disk $\mathbb{D}$ to $\mathbb{D}$.

The Berezin transform of a bounded function φ (or a Toeplitz operator T_φ) has the following important properties, which will be used many times in the rest of this paper. We mention here that (f) of the following proposition was proved by Nazarov [14] and (g) is the main result in [2].

Proposition 2.3 *Let φ and ψ be two functions in $L^\infty(\mathbb{D})$. Then we have:*

(a) $\widetilde{\varphi+\psi} = \widetilde{\varphi} + \widetilde{\psi}$ *and* $\widetilde{\overline{\varphi}} = \overline{\widetilde{\varphi}}$;

(b) $\|\widetilde{\varphi}\|_\infty \leqslant \|\varphi\|_\infty$;

(c) $\widetilde{\varphi} \geqslant 0$ *on* $\mathbb{D}$ *if* $T_\varphi \geqslant 0$;

(d) If $\varphi \in C(\overline{\mathbb{D}})$, *then* $\widetilde{\varphi}$ *is also in* $C(\overline{\mathbb{D}})$ *and* $(\varphi - \widetilde{\varphi})|_{\partial\mathbb{D}} = 0$;

(e) $\widetilde{\varphi} = 0$ *if and only if* $\varphi = 0$ *a.e. on* $\mathbb{D}$. *Consequently,* $T_\varphi = 0$ *if and only if* $\varphi = 0$ *a.e., and* T_φ *is self-adjoint if and only if* φ *is real-valued;*

(f) There is no constant $M > 0$ *such that* $\|T_\varphi\| \leqslant M\|\widetilde{\varphi}\|_\infty$ *for all* $\varphi \in L^\infty(\mathbb{D})$;

(g) T_φ *is compact on the Bergman space if and only if* $\widetilde{\varphi}(z) \to 0$ *as* $|z| \to 1^-$.

Proof The properties (a)-(d) can be proved directly by the definition of the Berezin transform. Indeed,

$$\widetilde{\varphi+\psi}(z) = \langle(\varphi+\psi)k_z, k_z\rangle = \langle\varphi k_z, k_z\rangle + \langle\widetilde{\psi} k_z, k_z\rangle = \widetilde{\varphi}(z) + \widetilde{\psi}(z)$$

for $z \in \mathbb{D}$. Moreover,

$$\widetilde{\overline{\varphi}}(z) = \langle\overline{\varphi}k_z, k_z\rangle = \langle k_z, \varphi k_z\rangle = \overline{\langle\varphi k_z, k_z\rangle} = \overline{\widetilde{\varphi}}(z)$$

for all $z \in \mathbb{D}$. This gives (a).

The conclusion in (b) follows from that $\|k_z\| = 1$ and

$$|\widetilde{\varphi}(z)| = |\langle\varphi k_z, k_z\rangle| \leqslant \|\varphi k_z\| \cdot \|k_z\| \leqslant \|\varphi\|_\infty$$

for all $z \in \mathbb{D}$.

T_φ is positive implies that $\widetilde{\varphi}(z) = \langle T_\varphi k_z, k_z\rangle \geqslant 0$ for $z \in \mathbb{D}$. This gives (c).

To show (d), for any fixed $z_0 \in \partial\mathbb{D}$, we have

$$\lim_{z\to z_0} \varphi_z(w) = z_0$$

for all $w \in \mathbb{D}$. Since φ is continuous on the closed unit disk, by the Lebesgue dominated convergence theorem, we obtain

$$\widetilde{\varphi}(z_0) = \lim_{z\to z_0} \int_{\mathbb{D}} \varphi(\varphi_z(w))dA(w) = \varphi(z_0).$$

This gives that $\widetilde{\varphi} = \varphi$ on the boundary $\partial\mathbb{D}$. Thus we obtain (d).

Next, we are going to show (e); i.e., the Berezin transform is one-to-one. The proof that the Berezin transform of a general operator is injective can be found in many references such as [16] and [28], which requires a classical result in several complex variables. However, for (e) the case of the Berezin transform on the Bergman space, one can prove the injective directly via the properties of the reproducing kernel.

Note that the normalized reproducing kernel is expressed as the following power series:

$$k_z(w) = (1-|z|^2)^2 \sum_{n=0}^{\infty} (n+1)(\bar{z}w)^n.$$

By the definition of the Berezin transform we have

$$\begin{aligned}
\widetilde{\varphi}(z) &= \langle \varphi k_z, k_z \rangle \\
&= (1-|z|^2)^2 \left\langle \varphi(w) \sum_{n=0}^{\infty} (n+1)(\bar{z}w)^n, \sum_{m=0}^{\infty} (m+1)(\bar{z}w)^m \right\rangle \\
&= \left(1 + z^2\bar{z}^2 - 2z\bar{z}\right) \sum_{n=0}^{\infty} \sum_{m=0}^{\infty} (n+1)(m+1)\bar{z}^n z^m \left\langle \varphi(w)w^n, w^m \right\rangle.
\end{aligned}$$

Taking partial derivatives with respect to $\bar{z}$ and z and then evaluating at 0 give

$$\frac{\partial^{m+n}\widetilde{\varphi}}{\partial z^m \partial \bar{z}^n}(0) = \left\langle \varphi(w)w^n, w^m \right\rangle$$

for integers $n, m \geqslant 0$. If $\widetilde{\varphi} = \widetilde{T_\varphi} = 0$, then we have

$$0 = \left\langle \varphi(w)w^n, w^m \right\rangle = \left\langle T_\varphi w^n, w^m \right\rangle$$

for all nonnegative integers n and m. This gives that $T_\varphi = 0$ since $\left\{w^n : n \geqslant 0\right\}$ is an orthogonal basis of $L_a^2(\mathbb{D})$. Moreover, since the linear span of $\{w^n \bar{w}^m : n, m \geq 0\}$ is dense in $L^2(\mathbb{D})$ and

$$\left\langle T_\varphi w^n, w^m \right\rangle = \int_{\mathbb{D}} \varphi(w) w^n \bar{w}^m dA(w),$$

we conclude that $\varphi = 0$ a.e. on the unit disk.

If $\varphi = 0$, then we of course have $T_\varphi = 0$. Conversely, $T_\varphi = 0$ implies that $\widetilde{\varphi} = 0$, which gives $\varphi = 0$ since the Berezin transform is one-to-one. Moreover, note that T_φ is self-adjoint if and only if $T_{\overline{\varphi}-\varphi} = 0$, which gives that $\varphi = \overline{\varphi}$, as desired.

Since the proofs of (f) and (g) are too long, we do not include them here. For more details, see [14] for (f) and [2] for (g). This completes the proof.

To study the invertibility of a Bergman Toeplitz operator, we need to introduce the n-th Berezin transform of a Toeplitz operator T_φ with φ a bounded function on the unit disk. For more details about the n-th Berezin transfrom, see [18] and [19].

Definition 2.4 *Let $\varphi \in L^\infty(\mathbb{D})$ and n be a nonnegative integer. The n-th Berezin transform of T_φ (or φ) is defined by*

$$\begin{aligned} B_n(T_\varphi)(z) &= B_n(\varphi)(z) \\ &= (n+1)\int_{\mathbb{D}} \varphi(w)(1-|w|^2)^n \frac{(1-|z|^2)^{n+2}}{|1-\bar{z}w|^{2n+4}} dA(w) \\ &= (n+1)\int_{\mathbb{D}} \varphi(\varphi_z(w))(1-|w|^2)^n dA(w). \end{aligned}$$

Note that the 0-th Berezin transform is the usual Berezin transform. Since $(n+1)(1-|w|^2)^n dA(w)$ is a probability measure that tends to concentrate its mass at 0 when $n \to \infty$, $B_n(\varphi)$ is an average of φ satisfying $\|B_n(\varphi)\|_\infty \leqslant \|\varphi\|_\infty$ for all $\varphi \in L^\infty(\mathbb{D})$ and all $n \geqslant 0$.

The useful properties of the n-th Berezin transform are listed in the following proposition. The proof of these properties can be found in [18], [19] and [28], so we omit the details here.

Proposition 2.5 *Let φ be in $L^\infty(\mathbb{D})$ and n be a nonnegative integer. The n-th Berezin transform has the following properties:*

(a) $B_n B_k(T_\varphi) = B_k B_n(T_\varphi)$ *for all* $n, k \geqslant 0$;

(b) *For each* $k \geqslant 0$, $B_n B_k(T_\varphi)$ *converges uniformly to* $B_k(T_\varphi)$ *in* $\mathbb{D}$ *as* $k \to \infty$;

(c) $\lim_{n\to\infty} \|T_{B_n\varphi} - T_\varphi\| = 0$;

(d) $B_n(\varphi) = \varphi$ *if* φ *is harmonic on* $\mathbb{D}$;

(e) $\lim_{n\to\infty} \int_{\mathbb{D}} |B_n(\varphi)(z) - \varphi(z)| dA(z) = 0$;

(f) *If* $\varphi \in C(\overline{\mathbb{D}})$, *then* $\lim_{n\to\infty} \|B_n(\varphi) - \varphi\|_\infty = 0$.

10.3 Positivity of Toeplitz operators via the Berezin transform

In this section we will study the positivity of Bergman Toeplitz operators with bounded symbols via the Berezin transforms. Proposition 2.3 (c) tells us that the Berezin transform $\widetilde{\varphi}$ is nonnegative when $T_\varphi \geqslant 0$. This leads to the following natural question:

Question 3.1 *Is T_φ positive on the Bergman space if its Berezin transform $\widetilde{\varphi}$ is nonnegative on the disk $\mathbb{D}$?*

Theorem 3.2 tells us that the answer to the above question for Bergman Toeplitz operators with harmonic symbols is positive. However, surprisingly we will show that the positivity of a Toeplitz operator with a quadratic polynomial of $|z|$ on the Bergman space is not completely determined by the positivity of the Berezin transform of its symbol.

The answer to an analogous question for Toeplitz operators on the Hardy space H^2 is also affirmative by means of a well-known spectral theorem for the self-adjoint Hardy Toeplitz operators and the harmonic extension. We can use the harmonic extension to prove the following theorem on the Bergman space.

Theorem 3.2 *Let $\varphi \in L^\infty(\mathbb{D})$ be a harmonic function. T_φ is positive on the Bergman space if and only if $\varphi(z) \geqslant 0$ for all $z \in \mathbb{D}$.*

Proof We need only to show the necessity since the sufficiency part follows immediately from (c) of Proposition 2.3. Assume that T_φ is positive. Then $\langle T_\varphi k_z, k_z\rangle \geqslant 0$ for $z \in \mathbb{D}$. Thus this implies $\widetilde{\varphi}(z) \geqslant 0$ for $z \in \mathbb{D}$. Since for a bounded harmonic function φ on the unit disk, the mean value property of harmonic functions and (10.2) give $\varphi(z) = \widetilde{\varphi}(z)$, we obtain $\varphi(z) \geqslant 0$. This completes the proof.

For a general function φ in $L^\infty(\mathbb{D})$, the following theorem shows that the above theorem fails.

Theorem 3.3 *The positivity of Toeplitz operators on the Bergman space is not completely determined by the positivity of the Berezin transform of their symbols.*

Proof Suppose that the positivity of Toeplitz operators on the Bergman space is completely determined by the positivity of the Berezin transform of their symbols. This implies that for any real-valued functions φ in L^∞, if $\widetilde{\varphi}(z) \geqslant 0$ on $\mathbb{D}$, then T_φ is positive. We will show that this implies $\|T_\varphi\| \leqslant 2\|\widetilde{\varphi}\|_\infty$ for all φ in $L^\infty(\mathbb{D})$. This contradicts the Nazarov theorem ((f) of Proposition 2.3) that there is no constant $M > 0$ such that $\|T_\varphi\| \leqslant M\|\widetilde{\varphi}\|_\infty$ for all $\varphi \in L^\infty(\mathbb{D})$.

To do this, first we consider that φ is a real-valued function in $L^\infty(\mathbb{D})$. Since both $\|\widetilde{\varphi}\|_\infty + \varphi(z)$ and $\|\widetilde{\varphi}\|_\infty - \varphi(z)$ are nonnegative a.e. on the unit disk, we have

$$\widetilde{\left(\|\widetilde{\varphi}\|_\infty \mp \varphi\right)}(z) \geqslant 0$$

for $z \in \mathbb{D}$ and $T_{\|\widetilde{\varphi}\|_\infty \mp \varphi} \geqslant 0$. Thus we have $\|\widetilde{\varphi}\|_\infty \geqslant \pm T_\varphi$. So we have that for any f in $L_a^2(\mathbb{D})$,

$$\|\widetilde{\varphi}\|_\infty \|f\|^2 \geqslant \pm\langle T_\varphi f, f\rangle,$$

which gives

$$\|\widetilde{\varphi}\|_\infty \|f\|^2 \geqslant |\langle T_\varphi f, f\rangle|.$$

Since for a self-adjoint operator T_φ,

$$\|T_\varphi\| = \sup_{\|f\| \leqslant 1,\ f \in L_a^2} |\langle T_\varphi f, f\rangle|,$$

we obtain $\|T_\varphi\| \leqslant \|\widetilde{\varphi}\|_\infty$.

Next for a general function φ in $L^\infty(\mathbb{D})$, we also have

$$\|\widetilde{\mathrm{Re}(\varphi)}\|_\infty \leqslant \|\widetilde{\varphi}\|_\infty, \qquad \|\widetilde{\mathrm{Im}(\varphi)}\|_\infty \leqslant \|\widetilde{\varphi}\|_\infty.$$

Thus we get

$$\begin{aligned}\|T_\varphi\| = \|T_{\mathrm{Re}(\varphi)} + iT_{\mathrm{Im}(\varphi)}\| &\leqslant \|T_{\mathrm{Re}(\varphi)}\| + \|T_{\mathrm{Im}(\varphi)}\| \\ &\leqslant \|\widetilde{\mathrm{Re}(\varphi)}\|_\infty + \|\widetilde{\mathrm{Im}(\varphi)}\|_\infty \leqslant 2\|\widetilde{\varphi}\|_\infty.\end{aligned}$$

This completes the proof of Theorem 3.3.

In the rest of this section, we consider Toeplitz operators with some radial symbols. Indeed, we will construct a quadratic polynomial φ of $|z|$ on $\mathbb{D}$ such that the Berezin transform $\widetilde{\varphi}$ is positive but the corresponding Toeplitz operator is not positive.

Let $e_n(z) = \sqrt{n+1}z^n$. Then $\{e_n\}_{n=0}^\infty$ is an orthonormal basis of the Bergman space $L_a^2(\mathbb{D})$. For the special function $\varphi(z) = |z|^2 + a|z| + b$ we can find the relationship between the positivity of the Toeplitz operator T_φ and the Berezin transform $\widetilde{\varphi}$ by its matrix representation. The matrix representation of T_φ with respect to this basis is a diagonal matrix.

Lemma 3.4 *Let $\varphi(z) = |z|^2 + a|z| + b$ $(a, b \in \mathbb{R})$. Then the matrix representation of the Toeplitz operator T_φ under the basis $\{e_n\}$ is given by*

$$\mathrm{diag}\left(\left\{\frac{2n+2}{2n+4} + a\frac{2n+2}{2n+3} + b\right\}_{n=0}^\infty\right).$$

Proof For each $k \in \mathbb{N}$, we have

$$\begin{aligned}T_{|z|^k}e_n(z) &= \langle T_{|z|^k}e_n, K_z\rangle = \langle |w|^k e_n, K_z\rangle \\ &= \sqrt{n+1}\int_{\mathbb{D}} |w|^k \frac{w^n}{(1-\bar{w}z)^2}dA(w) \\ &= \sqrt{n+1}\int_{\mathbb{D}} |w|^k w^n \sum_{j=0}^\infty (j+1)\bar{w}^j z^j dA(w) \\ &= \sqrt{n+1}\frac{1}{\pi}\int_0^{2\pi}\int_0^1 r^k \cdot r^n \cdot e^{in\theta}\sum_{j=0}^\infty (j+1)r^j e^{-j(i\theta)}z^j \, rdrd\theta \\ &= \frac{2\sqrt{n+1}(n+1)}{2n+k+2}z^n = \frac{2n+2}{2n+k+2}e_n(z).\end{aligned}$$

Thus the matrix representation of $T_{|z|^k}$ is a diagonal matrix under the basis $\{e_n\}_{n=0}^\infty$. And so is the matrix representation of

$$T_\varphi = T_{|z|^2+a|z|+b} = T_{|z|^2} + aT_{|z|} + bI$$

since it is a linear combination of $T_{|z|^k}$. In fact we have

$$T_\varphi e_n(z) = \left[\frac{2n+2}{2n+4} + a\frac{2n+2}{2n+3} + b\right] e_n(z).$$

This gives the matrix representation of the Toeplitz operator $T_{|z|^2+a|z|+b}$.

The above matrix representation of T_φ immediately gives the following criterion on the positivity of T_φ for $\varphi = |z|^2 + a|z| + b$.

Lemma 3.5 *Let $\varphi(z) = |z|^2 + a|z| + b$ $(a, b \in \mathbb{R})$. Then T_φ is positive on the Bergman space $L_a^2(\mathbb{D})$ if and only if*

$$1 + a + b \geqslant \frac{a}{2n+3} + \frac{2}{2n+4}$$

for all $n \geqslant 0$.

Little is known concerning the positivity of the Toeplitz operator with symbol φ in $C(\overline{\mathbb{D}})$, but when ψ is a polynomial of $|z|$ with degree 1, we will show that the positivity of the Toeplitz operator T_ψ is completely determined by the positivity of the Berezin transform but is not determined by its symbol. In other words, even if $\varphi(z)$ is positive on the unit disk, the Toeplitz operator T_φ may not be positive.

Theorem 3.6 *Let $\psi(z) = |z| - a$, where $a \in \mathbb{R}$. Then the following are equivalent:*

(i) $T_\psi \geqslant 0$;

(ii) $a \leqslant \frac{2}{3}$;

(iii) $\widetilde{\psi}(z) \geqslant 0$ *for all* $z \in \mathbb{D}$.

Proof First we show that (i) $\Leftrightarrow$ (ii). The proof of Lemma 3.4 gives that the matrix of T_ψ is a diagonal operator with diagonal entries

$$\left\{\frac{2n+2}{2n+3} - a\right\}_{n=0}^{\infty}.$$

Thus we get

$$T_\psi \geqslant 0 \Longleftrightarrow a \leqslant \frac{2n+2}{2n+3} \quad (\forall n \geqslant 0).$$

Since the sequence

$$\left\{\frac{2n+2}{2n+3}\right\}_{n=0}^{\infty}$$

is increasing, we see that

$$a \leqslant \frac{2n+2}{2n+3}$$

for all $n \geqslant 0$ if and only if $a \leqslant \frac{2}{3}$, which gives (i) $\Rightarrow$ (ii).

The implication (i) $\Rightarrow$ (iii) was proved in (c) of Proposition 2.3. To complete the proof we need only to verify that (iii) $\Rightarrow$ (ii). A simple computation gives

$$\widetilde{\psi}(z) = 2(1-|z|^2)^2\left[\sum_{n=0}^{\infty}\frac{(n+1)^2}{2n+3}|z|^{2n}\right] - a.$$

Letting $z = 0$ in the above formula, we have

$$\widetilde{\psi}(0) = \frac{2}{3} - a.$$

If $\widetilde{\psi}(z) \geqslant 0$ on the unit disk, we get $a \leqslant \frac{2}{3}$, which proves the theorem.

For φ a quadratic polynomial of $|z|$, we have the following result on the positivity of T_φ.

Theorem 3.7 *Let $\varphi(z) = |z|^2 + a|z| + b$ $(a, b \in \mathbb{R})$ and suppose that*

$$a \in \mathbb{R}\setminus(-2, -\frac{5}{4}).$$

Then T_φ is a positive Toeplitz operator on $L_a^2(\mathbb{D})$ if and only if $\widetilde{\varphi}(z)$ is a nonnegative function on $\mathbb{D}$.

Proof Let $\varphi(z) = |z|^2 + a|z| + b$ with $a, b \in \mathbb{R}$ and assume that $\widetilde{\varphi}(z) \geqslant 0$ for all $z \in \mathbb{D}$. Elementary calculations give us

$$\begin{aligned}\widetilde{\varphi}(z) &= 2(1-|z|^2)^2\left[\sum_{n=0}^{\infty}\frac{(n+1)^2}{2n+4}|z|^{2n} + a\sum_{n=0}^{\infty}\frac{(n+1)^2}{2n+3}|z|^{2n}\right] + b\\ &= \left[2 - \frac{1}{|z|^2} - \frac{(1-|z|^2)^2}{|z|^4}\log(1-|z|^2)\right]\\ &\quad + \frac{a}{2}\left[3 - \frac{1}{|z|^2} + \frac{(1-|z|^2)^2}{2|z|^3}\log\frac{1+|z|}{1-|z|}\right] + b.\end{aligned}$$

Letting $|z| = 0$ in the above power series, we obtain

$$\widetilde{\varphi}(0) = 2\left(\frac{1}{4} + \frac{a}{3}\right) + b = \frac{2a}{3} + b + \frac{1}{2}.$$

On the other hand, since $\varphi(z) = |z|^2 + a|z| + b$ is continuous on the closure of the unit disk, we obtain that $\widetilde{\varphi}$ is also continuous on the closed disk $\overline{\mathbb{D}}$ and $\widetilde{\varphi} = \varphi$ on $\partial\mathbb{D}$ (see (d) of Proposition 2.3). Thus we get $\widetilde{\varphi}(1) = \varphi(1) = 1 + a + b$. So we have

$$\widetilde{\varphi}(0) = b + \frac{2a}{3} + \frac{1}{2}$$

and $\widetilde{\varphi}(1) = b + a + 1$. Therefore,

$$b + \frac{2a}{3} + \frac{1}{2} \geqslant 0$$

and $b + a + 1 \geqslant 0$.

By Lemma 3.5, we have

$$T_\varphi \geqslant 0 \Longleftrightarrow 1 + a + b \geqslant \frac{a}{2n+3} + \frac{2}{2n+4} \quad (n \geqslant 0).$$

We need only to show

$$1 + a + b \geqslant \frac{a}{2n+3} + \frac{2}{2n+4}$$

for all $n \geqslant 0$.

To do this, we consider the following three cases.

Case I. Suppose $-\infty < a \leqslant -2$. In this case, we have

$$a \leqslant -2 + \frac{2}{2n+4}$$

for all $n \geqslant 0$. Thus

$$\frac{a}{2n+3} + \frac{2}{2n+4} \leqslant 0.$$

for all $n \geqslant 0$. Since $1 + a + b \geqslant 0$, we obtain

$$1 + a + b \geqslant \frac{a}{2n+3} + \frac{2}{2n+4}$$

for all $n \geqslant 0$.

Case II. Suppose $-\frac{9}{8} \leqslant a < +\infty$. In this case, we have

$$a \geqslant -\frac{3}{2} + \frac{3}{4(n+2)}$$

for all $n \geqslant 0$. This implies

$$\frac{a}{2n+3} + \frac{2}{2n+4} - \frac{a}{3} - \frac{1}{2} \leqslant 0 \quad (n \geqslant 0).$$

Thus we have

$$b + \frac{2a}{3} + \frac{1}{2} \geqslant 0 \geqslant \frac{a}{2n+3} + \frac{2}{2n+4} - \frac{a}{3} - \frac{1}{2}$$

for all $n \geqslant 0$, to get

$$1 + a + b \geqslant \frac{a}{2n+3} + \frac{2}{2n+4}$$

for all $n \geqslant 0$, as desired.

Case III. Suppose $-\frac{5}{4} \leqslant a \leqslant -\frac{9}{8}$. First, we observe that

$$3 \leqslant \frac{\sqrt{\frac{-a}{2}}}{1-\sqrt{\frac{-a}{2}}} \leqslant 5.$$

Next we want to find the maximal term of the sequence

$$\left\{\frac{a}{2n+3}+\frac{2}{2n+4}\right\}_{n=0}^{\infty}.$$

To do this, let F be the function

$$F(x) = \frac{a}{x}+\frac{2}{x+1},$$

where $x = 2n+3 \geqslant 3$. A simple calculation gives that $F(x)$ is increasing if

$$x < \frac{\sqrt{\frac{-a}{2}}}{1-\sqrt{\frac{-a}{2}}}$$

and $F(x)$ is decreasing if

$$x \geqslant \frac{\sqrt{\frac{-a}{2}}}{1-\sqrt{\frac{-a}{2}}}.$$

This implies that the maximal term of the above sequence is

$$\begin{aligned}\max\left\{\frac{a}{2n+3}+\frac{2}{2n+4} : n \geqslant 0\right\} &= F(3) \quad \left(\text{since } a \geqslant -\frac{5}{4} \text{ and } F(3) \geqslant F(5)\right)\\ &= \frac{a}{3}+\frac{1}{2}.\end{aligned}$$

Since $\widetilde{\varphi}$ is nonnegative, the expression of $\widetilde{\varphi}$ gives

$$b+\frac{2a}{3}+\frac{1}{2} \geqslant 0.$$

Thus we obtain

$$1+a+b \geqslant \frac{a}{3}+\frac{1}{2} = \max\left\{\frac{a}{2n+3}+\frac{2}{2n+4} : n \geqslant 0\right\},$$

to complete the proof.

On the other hand, we have the following counter example of the above theorem even for the quadratic polynomial of $|z|$, which implies that the answer to Question 3.1 is negative.

Theorem 3.8 *There exist $a \in (-2, -\frac{5}{4})$, $b \in \mathbb{R}$ and a positive constant δ such that the Berezin transform of $\varphi(z) = |z|^2 + a|z| + b$ satisfies $\widetilde{\varphi}(z) \geqslant \delta > 0$ for all z in $\mathbb{D}$, but the Toeplitz operator T_φ is not positive on the Bergman space $L_a^2(\mathbb{D})$.*

Proof The main idea of this proof is to estimate the minimal value of $\widetilde{\varphi}(z)$ on the unit disk. To do so, let $x = |z|^2$. Then x is in $(0,1)$ if z is in the unit disk. Simple calculations give

$$\begin{aligned}\sum_{n=0}^{\infty} \frac{(n+1)^2}{2n+4} x^n &= \frac{1}{4} + \sum_{n=1}^{\infty} \frac{n^2+2n+1}{2n+4} x^n \\ &= \frac{1}{4} + \frac{1}{2} \sum_{n=1}^{\infty} \left(n + \frac{1}{n+2}\right) x^n \\ &= \frac{1}{4} + \frac{1}{2} \sum_{n=1}^{\infty} n x^n + \frac{1}{2} \sum_{n=1}^{\infty} \frac{1}{n+2} x^n \\ &= \frac{1}{4} + \frac{x}{2(1-x)^2} + \frac{1}{2} \sum_{n=1}^{\infty} \frac{1}{n+2} x^n\end{aligned}$$

and

$$\begin{aligned}\sum_{n=0}^{\infty} \frac{(n+1)^2}{2n+3} x^n &= \frac{1}{3} + \sum_{n=1}^{\infty} \frac{n^2+2n+1}{2n+3} x^n \\ &= \frac{1}{3} + \frac{1}{2} \sum_{n=1}^{\infty} \frac{(n^2+\frac{3}{2}n) + (\frac{1}{2}n+1)}{n+\frac{3}{2}} x^n \\ &= \frac{1}{3} + \frac{1}{2} \sum_{n=1}^{\infty} n x^n + \frac{1}{4} \sum_{n=1}^{\infty} x^n + \sum_{n=1}^{\infty} \frac{1}{8n+12} x^n \\ &= \frac{1}{3} + \frac{x}{2(1-x)^2} + \frac{x}{4(1-x)} + \sum_{n=1}^{\infty} \frac{1}{8n+12} x^n.\end{aligned}$$

Combining the above two series with the proof of Theorem 3.7 gives

$$\begin{aligned}\widetilde{\varphi}(z) &= 2(1-x)^2 \left[\sum_{n=0}^{\infty} \frac{(n+1)^2}{2n+4} x^n + a \sum_{n=0}^{\infty} \frac{(n+1)^2}{2n+3} x^n\right] + b \\ &= \left(\frac{1}{2} + \frac{a}{6}\right) x^2 + \frac{a}{6} x + \left(\frac{1}{2} + \frac{2}{3} a + b\right) + \frac{(1-x)^2}{2} \sum_{n=1}^{\infty} \frac{(4+a)n + (2a+6)}{(n+2)(2n+3)} x^n \\ &\geqslant \left(\frac{1}{2} + \frac{a}{6}\right) x^2 + \frac{a}{6} x + \left(\frac{1}{2} + \frac{2}{3} a + b\right) + \frac{(1-x)^2}{2} \sum_{n=1}^{2} \frac{(4+a)n + (2a+6)}{(n+2)(2n+3)} x^n \\ &= \frac{1}{420} \left[(30a+105)x^4 - (18a+70)x^3 + (16a+35)x^2 + (112a+140)x\right]\end{aligned}$$

$$
\begin{aligned}
&+\left(\frac{1}{2}+\frac{2}{3}a+b\right)\\
&=\frac{30a+105}{420}\left[x^4-\frac{18a+70}{30a+105}x^3+\frac{16a+35}{30a+105}x^2+\frac{112a+140}{30a+105}x\right]\\
&+\left(\frac{2a}{3}+b+\frac{1}{2}\right),
\end{aligned}
$$

where the inequality above follows from that $a>-2$.

Letting

$$r=\frac{1}{10+\frac{35}{a}},$$

we have that r is in $(-\frac{2}{15},-\frac{1}{18})$ (since $-2<a<-\frac{5}{4}$) and

$$
\begin{aligned}
G(x)&:=x^4-\frac{18a+70}{30a+105}x^3+\frac{16a+35}{30a+105}x^2+\frac{112a+140}{30a+105}x\\
&=x^4+\left(\frac{2r}{3}-\frac{2}{3}\right)x^3+\left(\frac{1}{3}+2r\right)x^2+\left(\frac{4}{3}+24r\right)x.
\end{aligned}
$$

Thus

$$
\begin{aligned}
\widetilde{\varphi}(z)&\geqslant\frac{30a+105}{420}G(x)+\left(\frac{2a}{3}+b+\frac{1}{2}\right)\\
&\geqslant\frac{30a+105}{420}\inf_{x\in[0,1]}G(x)+\left(\frac{2a}{3}+b+\frac{1}{2}\right).
\end{aligned}
$$

To finish the proof, we need to choose $a\in(-\frac{14}{9},-\frac{5}{4})$ and a real constant b such that

$$\inf_{z\in\mathbb{D}}\widetilde{\varphi}(z)>0$$

and T_φ is not positive. As we show above, we need the following inequality:

$$\delta:=\frac{30a+105}{420}\inf_{x\in[0,1]}G(x)+\left(\frac{2a}{3}+b+\frac{1}{2}\right)>0.$$

This gives

$$-\frac{2}{3}a-\frac{1}{2}-\frac{30a+105}{420}\inf_{x\in[0,1]}G(x)<b.$$

By Lemma 3.5, we have that T_φ is positive if and only if

$$1+a+b\geqslant\max\left\{\frac{a}{2n+3}+\frac{2}{2n+4}:n\geqslant 0\right\}.$$

If T_φ is not positive, the above inequality gives

$$1+a+b<\max\left\{\frac{a}{2n+3}+\frac{2}{2n+4}:n\geqslant 0\right\}.$$

Equivalently, we need to find a real constant b and $-\frac{14}{9} < a < -\frac{5}{4}$ such that

$$-\frac{2}{3}a - \frac{1}{2} - \frac{30a+105}{420} \inf_{x\in[0,1]} G(x) < b < \max\left\{\frac{a}{2n+3} + \frac{2}{2n+4} : n \geqslant 0\right\} - a - 1.$$

Indeed, we will show that

$$\max\left\{\frac{a}{2n+3} + \frac{2}{2n+4} : n \geqslant 0\right\} + \frac{30a+105}{420} \inf_{x\in[0,1]} G(x) > \frac{a}{3} + \frac{1}{2}$$

if we take

$$a = -\frac{35}{26} \in \left(-\frac{14}{9}, -\frac{5}{4}\right),$$

that is,

$$r = \frac{1}{10 + \frac{35}{a}} = -\frac{1}{16}.$$

Let $K(x) = \frac{1}{4}G'(x)$ $(x \in (0,1))$ and recall that

$$F(x) = \frac{a}{x} + \frac{2}{x+1} \quad (x = 2n+3 \geqslant 3).$$

Then we have

$$3 \leqslant \frac{\sqrt{\frac{-a}{2}}}{1 - \sqrt{\frac{-a}{2}}} \leqslant 5$$

and $F(5) \geqslant F(3)$. This gives us

$$\max\left\{\frac{a}{2n+3} + \frac{2}{2n+4} : n \geqslant 0\right\} = F(5) = \frac{a}{5} + \frac{1}{3}.$$

Thus we need to show that

$$\frac{30a+105}{420} \inf_{x\in[0,1]} G(x) > \left(\frac{a}{3} + \frac{1}{2}\right) - \left(\frac{a}{5} + \frac{1}{3}\right) = \frac{a}{15} + \frac{1}{6},$$

which is equivalent to

$$\inf_{x\in[0,1]} G(x) > -\frac{1}{12}.$$

A simple computation gives

$$K(0) = \frac{1}{3} + 6 \times \left(-\frac{1}{16}\right) = -\frac{1}{24} < 0$$

and

$$K\left(\frac{1}{2}\right) = \frac{53r}{8} + \frac{5}{12} = \frac{1}{384} > 0.$$

Thus there exists a point $x_0 \in (0, \frac{1}{2})$ such that $K(x_0) = 0$.

One can show easily that x_0 is the unique point where $G(x)$ reaches its minimal value when $r = -\frac{1}{16}$. Moreover, x_0 satisfies the following equation:

$$x_0^3 + \frac{r-1}{2}x_0^2 + \left(r + \frac{1}{6}\right)x_0 + \left(6r + \frac{1}{3}\right) = 0,$$

and hence

$$x_0^4 + \frac{r-1}{2}x_0^3 + \left(r + \frac{1}{6}\right)x_0^2 + \left(6r + \frac{1}{3}\right)x_0 = 0.$$

So we have

$$\inf_{x\in[0,1]} G(x) = \frac{r-1}{6}x_0^3 + \frac{1+6r}{6}x_0^2 + (1+18r)x_0.$$

Let

$$L(t) = \frac{r-1}{6}t^3 + \frac{1+6r}{6}t^2 + (1+18r)t \quad \big(t \in (0,1)\big).$$

Taking derivative of $L(t)$ gives

$$L'(t) = \frac{r-1}{2}t^2 + \frac{1+6r}{3}t + (1+18r) < 0$$

for all $t \in (0,1)$. This yields

$$\begin{aligned}\inf_{x\in[0,1]} G(x) = L(x_0) &\geqslant L\left(\frac{1}{2}\right) \quad \left(\text{since } 0 < x_0 < \frac{1}{2}\right)\\ &= \frac{445}{48}r + \frac{25}{48} = -\frac{15}{256} \quad \left(\text{using } r = -\frac{1}{16}\right)\\ &> -\frac{1}{12},\end{aligned}$$

as required. This completes the proof of the theorem.

10.4 Invertibility of Toeplitz operators via the Berezin transform

In this section, we will study the invertibility of Toeplitz operators with nonnegative symbols and bounded harmonic symbols via the Berezin transforms. For Toeplitz operators with nonnegative symbols, we show that the Toeplitz operator is invertible on $L_a^2(\mathbb{D})$ if and only if its Berezin transform is invertible in $L^\infty(\mathbb{D})$. Furthermore, we also consider the invertibility of Toeplitz operators with bounded harmonic symbols and study the Douglas question for these operators. More precisely, we will establish a Chang-Tolokonnikov type theorem for Toeplitz operators with bounded harmonic symbols.

Let us begin with the bounded nonnegative symbol. In order to study the invertibility of Toeplitz operators with nonnegative symbols, we review the characterization for the invertibility of Toeplitz operators with bounded nonnegative symbols (see [12]).

Lemma 4.1 *(Luecking) Let φ be a bounded nonnegative measurable function on $\mathbb{D}$. Then the following conditions are equivalent:*

(1) The Toeplitz operator T_φ is invertible on $L_a^2(\mathbb{D})$;

(2) There exists a constant $\eta > 0$ such that

$$\int_{\mathbb{D}} |\varphi(z)f(z)|^2 dA(z) \geqslant \eta \int_{\mathbb{D}} |f(z)|^2 dA(z)$$

for all $f \in L_a^2(\mathbb{D})$;

(3) There exist $r > 0$, $\delta > 0$ and $0 < \varepsilon < 1$ such that

$$m(G \cap D(a,\varepsilon)) > \delta m(D(a,\varepsilon))$$

for all $a \in \mathbb{D}$, where $G = \{z \in \mathbb{D} : \varphi(z) > r\}$ and $D(a,r)$ is a pseudo-hyperbolic disk:

$$D(a,r) = \{w \in \mathbb{D} : |\varphi_a(w)| < r\}.$$

Here m denotes the area measure on the complex plane.

Lemma 4.1 gives the following sufficient condition for the Toeplitz operator with bounded symbol to be invertible on the Bergman space.

Theorem 4.2 *Let φ be a function in $L^\infty(\mathbb{D})$. If T_φ is invertible, then $\widetilde{|\varphi|}$ is invertible in $L^\infty(\mathbb{D})$. However, this condition is not sufficient.*

Proof Let φ be in $L^\infty(\mathbb{D})$. Suppose that T_φ is invertible. Then T_φ is bounded below on $L_a^2(\mathbb{D})$. Thus there exists a constant $\varepsilon > 0$ such that

$$\varepsilon \|f\|_2 \leqslant \|T_\varphi f\|_2 \leqslant \|\varphi f\|_2 = \||\varphi| f\|_2$$

for all f in $L_a^2(\mathbb{D})$. So (2) in Lemma 4.1 holds for $T_{|\varphi|}$. By Lemma 4.1, we have that the positive Toeplitz operator $T_{|\varphi|}$ is invertible to get that there is a constant $\delta > 0$ such that

$$\widetilde{|\varphi|}(z) = \langle T_{|\varphi|} k_z, k_z \rangle \geqslant \delta \langle k_z, k_z \rangle = \delta$$

for all $z \in \mathbb{D}$. This gives that $\widetilde{|\varphi|}$ is invertible.

For the second part of the theorem, we need to construct a function $\varphi \in L^\infty(\mathbb{D})$ such that $\widetilde{|\varphi|}$ is invertible in $L^\infty(\mathbb{D})$ but the Toeplitz operator T_φ is not invertible on the Bergman space. Let $\varphi(z) = z^2$. From the proof of Theorem 3.7, we obtain

$$\begin{aligned}
\widetilde{|\varphi|}(z) &= \int_{\mathbb{D}} |\varphi(w)| \cdot |k_z(w)|^2 dA(w) \\
&= \int_{\mathbb{D}} |wk_z(w)|^2 dA(w) \\
&= 2(1-|z|^2)^2 \sum_{n=0}^{\infty} \frac{(n+1)^2}{2n+4} |z|^{2n} \\
&= 2(1-|z|^2)^2 \left[\frac{1}{4} + \frac{|z|^2}{2(1-|z|^2)^2} + \frac{1}{2} \sum_{n=1}^{\infty} \frac{|z|^{2n}}{n+2}\right] \\
&= \frac{1}{2} + \frac{1}{2}|z|^4 + (1-|z|^2)^2 \sum_{n=1}^{\infty} \frac{|z|^{2n}}{n+2} \\
&\geqslant \frac{1}{2}
\end{aligned}$$

for all $z \in \mathbb{D}$. This gives that $\widetilde{|\varphi|}$ is bounded below on $\mathbb{D}$. However, T_φ is not invertible on the Bergman space since 1 is an eigenvector of the adjoint $T_{z^2}^*$ of T_φ corresponding to the eigenvalue 0, i.e., $T_{z^2}^* 1 = 0$. This completes the proof.

Based on the above two results, we obtain a necessary and sufficient condition for Toeplitz operators with bounded nonnegative symbols to be invertible on the Bergman space in terms of the Berezin transform.

Theorem 4.3 *Let φ be a nonnegative function in $L^\infty(\mathbb{D})$. Then T_φ is invertible on the Bergman space if and only if $\widetilde{\varphi}$ is invertible in $L^\infty(\mathbb{D})$.*

Proof If T_φ is invertible on the Bergman space and φ is nonnegative, Theorem 4.2 gives that $\widetilde{\varphi} = \widetilde{|\varphi|}$ is invertible in $L^\infty(\mathbb{D})$.

Conversely, suppose $\widetilde{\varphi}$ is invertible in $L^\infty(\mathbb{D})$. Then there exists a constant $\delta > 0$ such that $\widetilde{\varphi}(z) \geqslant \delta$ for all $z \in \mathbb{D}$. Lemma 4.1 implies that we need only to verify Condition (3). To do so, we choose $r = \frac{\delta}{4}$ and define the measurable set

$$G = \{z \in \mathbb{D} : \varphi(z) > r\}$$

as Condition (3) in Lemma 4.1.

For each $a \in \mathbb{D}$ and $\varepsilon \in (0,1)$, we observe that

$$\begin{aligned}
\frac{4\|\varphi\|_\infty}{(1-|a|)^2} m(G \cap D(a,\varepsilon)) &\geqslant \int_{G \cap D(a,\varepsilon)} \frac{4\varphi(z)}{(1-|a|)^2} dA(z) \\
&\geqslant \int_{G \cap D(a,\varepsilon)} \varphi(z) |k_a(z)|^2 dA(z)
\end{aligned}$$

$$
\begin{aligned}
&= \int_G \varphi(z)|k_a(z)|^2 dA(z) - \int_{G\setminus D(a,\varepsilon)} \varphi(z)|k_a(z)|^2 dA(z) \\
&\geqslant \int_G \varphi(z)|k_a(z)|^2 dA(z) - \int_{\mathbb{D}\setminus D(a,\varepsilon)} \varphi(z)|k_a(z)|^2 dA(z) \\
&\geqslant \int_G \varphi(z)|k_a(z)|^2 dA(z) - \|\varphi\|_\infty \int_{\mathbb{D}\setminus D(a,\varepsilon)} |k_a(z)|^2 dA(z),
\end{aligned}
$$

where the second inequality comes from

$$
|k_a(z)|^2 \leqslant \frac{4}{(1-|a|)^2}
$$

for each $a \in \mathbb{D}$. Since

$$
\begin{aligned}
\widetilde{\varphi}(a) &= \int_{\mathbb{D}} \varphi(z)|k_a(z)|^2 dA(z) \\
&= \int_G \varphi(z)|k_a(z)|^2 dA(z) + \int_{\mathbb{D}\setminus G} \varphi(z)|k_a(z)|^2 dA(z) \\
&\geqslant \delta \quad (a \in \mathbb{D}),
\end{aligned}
$$

we obtain

$$
\begin{aligned}
\frac{4\|\varphi\|_\infty}{(1-|a|)^2} m(G\cap D(a,\varepsilon)) &\geqslant \delta - \int_{\mathbb{D}\setminus G} \varphi(z)|k_a(z)|^2 dA(z) - \|\varphi\|_\infty \int_{\mathbb{D}\setminus D(a,\varepsilon)} |k_a(z)|^2 dA(z) \\
&\geqslant \delta - r\int_{\mathbb{D}\setminus G} |k_a(z)|^2 dA(z) - \|\varphi\|_\infty \int_{\mathbb{D}\setminus D(a,\varepsilon)} |k_a(z)|^2 dA(z) \\
&\geqslant \delta - r\int_{\mathbb{D}} |k_a(z)|^2 dA(z) - \|\varphi\|_\infty \int_{\mathbb{D}\setminus D(a,\varepsilon)} |k_a(z)|^2 dA(z) \\
&= \frac{3\delta}{4} - \|\varphi\|_\infty \int_{\mathbb{D}\setminus D(a,\varepsilon)} |k_a(z)|^2 dA(z),
\end{aligned}
$$

where the second inequality comes from the definition of G. Since

$$
\begin{aligned}
\int_{\mathbb{D}\setminus D(a,\varepsilon)} |k_a(z)|^2 dA(z) &= 1 - \int_{D(a,\varepsilon)} |k_a(z)|^2 dA(z) \\
&= 1 - \int_{D(0,\varepsilon)} dA(z) = 1-\varepsilon^2,
\end{aligned}
$$

we can choose $\varepsilon \in (0,1)$ such that

$$
\int_{\mathbb{D}\setminus D(a,\varepsilon)} |k_a(z)|^2 dA(z) < \frac{\delta}{4\|\varphi\|_\infty},
$$

which gives

$$
\frac{4\|\varphi\|_\infty}{(1-|a|)^2} m(G\cap D(a,\varepsilon)) \geqslant \frac{3\delta}{4} - \frac{\delta}{4} = \frac{\delta}{2}.
$$

Thus we obtain

$$m(G \cap D(a,\varepsilon)) \geqslant \frac{\delta(1-|a|)^2}{8\|\varphi\|_\infty} \geqslant \frac{\delta m(D(a,\varepsilon))}{8C_\varepsilon\|\varphi\|_\infty},$$

where the positive constant C_ε follows from the fact that

$$m(D(a,\varepsilon)) = \frac{(1-|z|^2)^2\varepsilon^2}{(1-|z|^2\varepsilon^2)^2}.$$

Here $s = \tanh(\varepsilon) \in (0,1)$. See [28] if necessary. This completes the proof of Theorem 3.5.

Now we turn to the study of the invertibility of Bergman Toeplitz operators with bounded harmonic symbols. Our goal is to establish a Bergman space version of the Chang-Tolokonnikov theorem for Toeplitz operators with bounded harmonic symbols. Before doing this, let us review the analogous question for Hardy Toeplitz operators. On the Hardy space, the invertible Toeplitz operators are completely characterized [5]. But the equivalent characterization for the invertibility of Hardy Toeplitz operators is not so easy to check. In [6], using the homotopy, Douglas showed that for a continuous function φ on the unit circle, T_φ is invertible if the harmonic extension $|\widehat{\varphi}(z)| \geqslant \delta$ for some positive constant δ and for all z in the unit disk, where the harmonic extension $\widehat{\varphi}$ of $\varphi \in L^\infty(\partial\mathbb{D})$ is defined by

$$\widehat{\varphi}(z) = \frac{1}{2\pi}\int_0^{2\pi} \varphi(e^{i\theta})\frac{1-|z|^2}{|1-ze^{-i\theta}|^2}d\theta$$

for z in the unit disk $\mathbb{D}$. In [6] Douglas posed the following question:

Question 4.4 *If φ is in $L^\infty(\partial\mathbb{D})$ and the harmonic extension $|\widehat{\varphi}(z)| \geqslant \delta$ for some positive constant δ and for all z in the unit disk, is $\mathbb{T}_\varphi$ invertible on the Hardy space H^2?*

As mentioned in [7] and [23], Chang and Tolokonnikov obtained a sufficient condition for a Toeplitz operator to be invertible on the Hardy space and showed that if for a constant δ sufficiently close to 1, $\delta \leqslant |\widehat{\varphi}(z)| \leqslant 1$ for all $z \in \mathbb{D}$, then $\mathbb{T}_\varphi$ is invertible. In fact, Tolokonnikov found that $1 > \delta > \frac{45}{46}$ and

$$\|\mathbb{T}_\varphi^{-1}\| \leqslant \sqrt{\frac{1}{46\delta-45}}$$

in [21]. Nikolski [15] proved the invertibility of a Toeplitz operator $\mathbb{T}_\varphi$ (on H^2) with $|\varphi| \leqslant 1$ and estimated

$$\|\mathbb{T}_\varphi^{-1}\| \leqslant \sqrt{\frac{1}{24\delta-23}}$$

under the condition $1 > \delta > \frac{23}{24}$.

Indeed, a slightly better estimation was proved by the reasoning in [15] (see page 374 of [15]):

$$1 > \delta > \sqrt{\frac{4e}{4e+1}} := \Delta$$

is already sufficient for $\mathbb{T}_\varphi$ to be invertible on the Hardy space. It is curious to note that this result of Nikolski follows also from a recent estimate of Hankel operators by Treil (see Theorem 1.1 and its proof in [22]). In a private communication, Nikolski conjectured that the constant Δ defined above is sharp for the invertibility problem of $\mathbb{T}_\varphi$ on the Hardy space H^2 since the methods from [15] and [22] are quite different.

We will show in the following theorem that T_φ is invertible on the Bergman space $L_a^2(\mathbb{D})$ if φ is harmonic on the unit disk $\mathbb{D}$ and

$$\frac{2\sqrt{2}}{3} < \delta \leqslant |\varphi(z)| < 1$$

for all $z \in \mathbb{D}$. It is remarkable to notice that $\frac{2\sqrt{2}}{3} \approx 0.943$ in the above inequality is smaller than the constant $\sqrt{\frac{4e}{4e+1}} \approx 0.957$ obtained by Treil [22]. For a bounded harmonic function φ on the unit disk, we have

$$\tilde{\varphi}(z) = \hat{\varphi^*}(z) = \varphi(z)$$

for z in the unit disk where φ^* is the radial limit of φ on the unit circle.

Theorem 4.5 *Suppose that φ is a bounded harmonic function on $\mathbb{D}$. If there exists a constant $\delta \in (0,1)$ such that*

$$|\tilde{\varphi}(z)| = |\varphi(z)| \geqslant \delta\|\varphi\|_\infty > \frac{2\sqrt{2}}{3}\|\varphi\|_\infty$$

for all $z \in \mathbb{D}$, then T_φ is invertible on $L_a^2(\mathbb{D})$ and we have

$$\|T_\varphi^{-1}\| \leqslant \frac{1}{\sqrt{9\delta^2 - 8}\|\varphi\|_\infty}.$$

The proof of the above theorem is based on the BMO-norm estimation of Hankel operators with bounded harmonic symbols. Recall that BMO is the space of functions φ with

$$\|\varphi\|_{BMO} = \sup_{z\in\mathbb{D}} \left[\widetilde{|\varphi|^2}(z) - |\tilde{\varphi}(z)|^2\right]^{\frac{1}{2}} < +\infty.$$

Using a change of variables one has

$$\|\varphi\|_{BMO} = \sup_{z\in\mathbb{D}} \left[\int_{\mathbb{D}} |\varphi(\varphi_z(w)) - \tilde{\varphi}(z)|^2 dA(w)\right]^{\frac{1}{2}}.$$

Recall that the Hankel operator H_φ with symbol $\varphi \in L^\infty(\mathbb{D})$ is defined by

$$H_\varphi(f) = (I-P)(\varphi f) \quad \left(f \in L_a^2(\mathbb{D})\right),$$

which is a bounded linear operator from $L_a^2(\mathbb{D})$ to $(L_a^2(\mathbb{D}))^{\perp}$. Thus we have the identity

$$\|T_\varphi f\|^2 + \|H_\varphi f\|^2 = \|\varphi f\|^2$$

for all f in $L_a^2(\mathbb{D})$.

Furthermore, there is a norm estimation for Hankel operators with bounded symbols in terms of the BMO-norm; see [28] for its proof.

Theorem 4.6 *Suppose that φ is a bounded function on $\mathbb{D}$. Then we have*

$$\|H_\varphi\| \leqslant C\|\varphi\|_{BMO}$$

for some constant $C > 0$ (independent of φ).

However, if φ is harmonic on $\mathbb{D}$, then we will obtain a better estimate for the positive constant C in the above theorem by some special properties of harmonic functions.

Lemma 4.7 *Let φ be a bounded harmonic function on $\mathbb{D}$. Then we have*

$$\|H_\varphi\| \leqslant 2\sqrt{2}\|\varphi\|_{BMO}.$$

Proof We decompose φ as $\varphi = \varphi_1 + \overline{\varphi_2}$ where φ_1 and φ_2 are both analytic functions since φ is harmonic on $\mathbb{D}$. Then we have by the definition of the Hankel operator that

$$H_\varphi = H_{\varphi_1} + H_{\overline{\varphi_2}} = H_{\overline{\varphi_2}}.$$

To estimate the norm of the Hankel operator $H_{\overline{\varphi_2}}$, we take $g \in L_a^2(\mathbb{D})$ and $h \in C_0^\infty(\mathbb{D})$. Since $(L_a^2)^\perp$ is the closure of

$$\left\{\partial h := \frac{\partial h}{\partial z} : h \in C_0^\infty(\mathbb{D})\right\}$$

in $L^2(\mathbb{D}, dA)$ (see Lemma 8.30 in [28] if necessary), we have

$$\langle H_{\overline{\varphi_2}} g, \partial h\rangle = \langle \overline{\varphi_2} g, \partial h\rangle = \int_{\mathbb{D}} \overline{\varphi_2(z)} g(z) \overline{\partial h(z)} dA(z).$$

Denote $\frac{\partial h}{\partial \bar{z}}$ by $\overline{\partial} h$. Then using the fact that

$$\partial(\overline{\varphi_2} g) = g\overline{\partial}\,\overline{\varphi_2} = g\overline{\partial \varphi_2},$$

we get

$$\begin{aligned}\langle H_{\overline{\varphi_2}} g, \partial h\rangle &= -\int_{\mathbb{D}} \overline{\partial \varphi_2(z)} g(z) \overline{h(z)} dA(z)\\ &= -\int_{\mathbb{D}} \overline{\varphi_2'(z)} g(z) \overline{h(z)} dA(z).\end{aligned}$$

Applying the Cauchy-Schwartz inequality to the above integral, we obtain

$$
\begin{aligned}
\left|\langle H_{\overline{\varphi_2}}g, \partial h\rangle\right| &\leqslant \int_{\mathbb{D}} |\varphi_2'(z)g(z)|\cdot|h(z)|dA(z)\\
&= \int_{\mathbb{D}} \left|(1-|z|^2)\varphi_2'(z)g(z)\right|\cdot\frac{|h(z)|}{1-|z|^2}dA(z)\\
&\leqslant \left(\int_{\mathbb{D}}(1-|z|^2)^2|\varphi_2'(z)|^2\cdot|g(z)|^2dA(z)\right)^{\frac{1}{2}}\cdot\left(\int_{\mathbb{D}}\frac{|h(z)|^2dA(z)}{(1-|z|^2)^2}\right)^{\frac{1}{2}}\\
&\leqslant \sup_{z\in\mathbb{D}}\left[(1-|z|^2)|\varphi_2'(z)|\right]\cdot\|g\|\cdot\left(\int_{\mathbb{D}}\frac{|h(z)|^2dA(z)}{(1-|z|^2)^2}\right)^{\frac{1}{2}}.
\end{aligned}
$$

From the proof of Lemma 8.32 in [28], we get

$$
\begin{aligned}
\int_0^1\frac{|h(re^{i\theta})|^2}{(1-r^2)^2}rdr &\leqslant 2\int_0^1|h(se^{i\theta})|\cdot|\nabla h(se^{i\theta})|\left(\int_0^s\frac{rdr}{(1-r^2)^2}\right)ds\\
&= \int_0^1|h(se^{i\theta})|\cdot|\nabla h(se^{i\theta})|\frac{s^2}{1-s^2}ds\\
&\leqslant \int_0^1|h(se^{i\theta})|\cdot|\nabla h(se^{i\theta})|\frac{s}{1-s^2}ds.
\end{aligned}
$$

Integrating the above inequality with respect to $\theta : 0 \to 2\pi$, we obtain

$$
\int_{\mathbb{D}}\frac{|h(z)|^2dA(z)}{(1-|z|^2)^2} \leqslant \int_{\mathbb{D}}|\nabla h(z)|^2dA(z) = 2\int_{\mathbb{D}}(|\partial h(z)|^2+|\overline{\partial}h(z)|^2)dA(z),
$$

where the last equality follows from the definition of the gradient function. Since ∂h and $\overline{\partial}h$ have the same norm in $L^2(\mathbb{D}, dA)$ (see Lemma 8.31 in [28]), we have

$$
\int_{\mathbb{D}}\frac{|h(z)|^2}{(1-|z|^2)^2}dA(z) \leqslant 4\|\partial h\|^2.
$$

Therefore,

$$
\left|\langle H_{\overline{\varphi_2}}g, \partial h\rangle\right| \leqslant 2\sup_{z\in\mathbb{D}}\left[(1-|z|^2)|\varphi_2'(z)|\right](\|g\|\cdot\|\partial h\|),
$$

to obtain

$$
\|H_{\overline{\varphi}}\| = \|H_{\overline{\varphi_2}}\| \leqslant 2\sup_{z\in\mathbb{D}}\left[(1-|z|^2)|\varphi_2'(z)|\right].
$$

In order to deal with the term

$$
\sup_{z\in\mathbb{D}}\left[(1-|z|^2)|\varphi_2'(z)|\right],
$$

we observe that

$$
\varphi_2'(0) = 2\int_{\mathbb{D}}\left[\varphi_2(w)-\varphi_2(0)\right]\overline{w}dA(w),
$$

which gives us

$$|\varphi_2'(0)| \leqslant 2\Big(\int_{\mathbb{D}} |\varphi_2(w) - \varphi_2(0)|^2 dA(w)\Big)^{\frac{1}{2}} \cdot \Big(\int_{\mathbb{D}} |w|^2 dA(w)\Big)^{\frac{1}{2}}$$
$$= \sqrt{2}\Big(\int_{\mathbb{D}} |\varphi_2(w) - \varphi_2(0)|^2 dA(w)\Big)^{\frac{1}{2}},$$

where the equality above follows from

$$\int_{\mathbb{D}} |w|^2 dA(w) = \frac{1}{\pi}\int_0^{2\pi}\int_0^1 r^2 \cdot r dr d\theta = \frac{1}{2}.$$

Let

$$\varphi_z(w) = \frac{z-w}{1-\bar{z}w}$$

be the Möbius mapping on $\mathbb{D}$. Clearly, $\varphi_z(0) = z$. Now replacing φ_2 by $\varphi_2 \circ \varphi_z$, we have

$$\big|\varphi_2' \circ \varphi_z(0)\big| \cdot \big|\varphi_z'(0)\big| \leqslant \sqrt{2}\Big(\int_{\mathbb{D}} |\varphi_2(\varphi_z(w)) - \varphi_2(z)|^2 dA(w)\Big)^{\frac{1}{2}}.$$

Using

$$\varphi_z'(w) = \frac{1-|z|^2}{(1-\bar{z}w)^2} \qquad \big(z, w \in \mathbb{D}\big),$$

we obtain

$$(1-|z|^2)|\varphi_2'(z)| \leqslant \sqrt{2}\Big(\int_{\mathbb{D}} |\varphi_2(\varphi_z(w)) - \varphi_2(z)|^2 dA(w)\Big)^{\frac{1}{2}}.$$

By the definition of BMO-norm,

$$\sup_{z\in\mathbb{D}} \big[(1-|z|^2)|\varphi_2'(z)|\big] \leqslant \sqrt{2}\|\varphi_2\|_{BMO}.$$

Thus we deduce that

$$\|H_\varphi\| \leqslant 2\sqrt{2}\|\varphi_2\|_{BMO}.$$

To finish the proof of this lemma, we need only to show

$$\|\varphi_2\|_{BMO} \leqslant \|\varphi\|_{BMO}.$$

By (a) of Proposition 2.3, the Berezin transform is a linear mapping, thus we have

$$\begin{aligned}
\widetilde{|\varphi|^2}(z) - |\varphi(z)|^2 &= \widetilde{|\varphi_1|^2}(z) + \widetilde{|\varphi_2|^2}(z) + \widetilde{\varphi_1\varphi_2}(z) + \widetilde{\overline{\varphi_1\varphi_2}}(z) - \big|\varphi_1(z) + \overline{\varphi_2(z)}\big|^2 \\
&= \widetilde{|\varphi_1|^2}(z) + \widetilde{|\varphi_2|^2}(z) + \varphi_1(z)\varphi_2(z) + \overline{\varphi_1(z)\varphi_2(z)} - \big|\varphi_1(z) + \overline{\varphi_2(z)}\big|^2 \\
&= \Big[\widetilde{|\varphi_1|^2}(z) - |\varphi_1(z)|^2\Big] + \Big[\widetilde{|\varphi_2|^2}(z) - |\varphi_2(z)|^2\Big] \\
&\geqslant \widetilde{|\varphi_2|^2}(z) - |\varphi_2(z)|^2
\end{aligned}$$

for all $z \in \mathbb{D}$, where the second equality follows from that $\varphi_1\varphi_2$ is analytic. Taking supremum with respect to $z \in \mathbb{D}$ on both sides of the above inequality, we get the desired inequality by the definition of BMO-norm. This completes the proof of Lemma 4.7.

Based on Lemma 4.7, now we are ready to prove Theorem 4.5.

Proof of Theorem 4.5 Recall that $L^\infty(\mathbb{D}) \subset BMO$ and

$$\|\varphi\|_{BMO}^2 = \sup_{z\in\mathbb{D}} \left[\widetilde{|\varphi|^2}(z) - |\widetilde{\varphi}(z)|^2\right].$$

Thus we have

$$\begin{aligned}
\|T_\varphi f\|^2 &= \|\varphi f\|_2^2 - \|H_\varphi f\|^2 \\
&\geqslant \delta^2\|\varphi\|_\infty^2\|f\|^2 - \|H_\varphi\|^2\|f\|^2 \\
&\geqslant \delta^2\|\varphi\|_\infty^2\|f\|^2 - 8\|\varphi\|_{BMO}^2\|f\|^2 \quad \text{(by Lemma 4.7)} \\
&= \delta^2\|\varphi\|_\infty^2\|f\|^2 - 8\sup_{z\in\mathbb{D}}\left[\widetilde{|\varphi|^2}(z) - |\widetilde{\varphi}(z)|^2\right]\|f\|^2 \\
&\geqslant \delta^2\|\varphi\|_\infty^2\|f\|^2 - 8(\|\varphi\|_\infty^2 - \delta^2\|\varphi\|_\infty^2)\|f\|^2 \quad (\text{since } \varphi \text{ is harmonic}) \\
&= \left[\delta^2 - 8(1-\delta^2)\right]\|\varphi\|_\infty^2\|f\|^2 \\
&= (9\delta^2 - 8)\|\varphi\|_\infty^2\|f\|^2
\end{aligned}$$

for each $f \in L_a^2(\mathbb{D})$, where the last inequality follows from (b) of Proposition 2.3 and (d) of Proposition 2.5.

Since the constant $\delta > \frac{2\sqrt{2}}{3}$, we see that T_φ is bounded below and

$$\|T_\varphi f\| \geqslant \sqrt{9\delta^2 - 8}\,\|\varphi\|_\infty\|f\|$$

for all $f \in L_a^2(\mathbb{D})$. Using (c) of Proposition 2.1, we have $T_{\overline{\varphi}} = T_\varphi^*$ and

$$\|T_\varphi^* f\| \geqslant \sqrt{9\delta^2 - 8}\,\|\varphi\|_\infty\|f\|$$

for all $f \in L_a^2(\mathbb{D})$. Thus T_φ is invertible on the Bergman space $L_a^2(\mathbb{D})$. This completes the proof of Theorem 4.5.

Theorem 4.5 tells us that a Toeplitz operator with harmonic symbol is invertible on the Bergman space if its Berezin transform is bounded below by some positive constant. On the other hand, using the Berezin transform $\widetilde{\varphi}$ and the n-th Berezin transform $B_n(\varphi)$, the authors also gave a sufficient condition for T_φ to be an invertible operator on $L_a^2(\mathbb{D})$ if φ is a bounded function on the disk $\mathbb{D}$; see the theorem below and [26, Theorem 4.5] for the details.

Theorem 4.8 *Let $\varphi \in L^\infty(\mathbb{D})$ and C be the constant in Theorem 4.6. Then there exists some integer $N = N(\varphi)$ (depending only on φ) such that the inequalities*

$$\inf_{z\in\mathbb{D}} |\widetilde{\varphi}(z)| > \delta\|\varphi\|_\infty$$

and

$$|(B_{N_0}\varphi)(z)| \geqslant \varepsilon\|\varphi\|_\infty$$

hold for all $z \in \mathbb{D}$, where $1 > \delta > 0$, $1 > \varepsilon > C\sqrt{1-\delta^2}$ are constants and $N_0 \geqslant N(\varphi)$ imply that the Toeplitz operator T_φ is invertible on the Bergman space $L_a^2(\mathbb{D})$.

The theorem above leads to the following question:

Question 4.9 *Is T_φ invertible on the Bergman space if $|\widetilde{\varphi}(z)| \geqslant \delta_1$ and the n-th Berezin transforms $|(B_n\varphi)(z)| \geqslant \delta_2$ for some positive constants δ_1, δ_2 and for all sufficiently large integers n and all z in the unit disk $\mathbb{D}$?*

Indeed, we will construct a counterexample and show that the answer to Question 4.9 is negative. To do this, we will prove that there exists a function $\varphi \in C(\overline{\mathbb{D}})$ such that $\widetilde{\varphi}$ is invertible and $|B_n(\varphi)|$ are bounded below for all sufficiently large n, but the corresponding Bergman Toeplitz operator is not invertible. Recall that (f) of Proposition 2.5 tells us that

$$\lim_{n\to\infty} \|B_n(\varphi) - \varphi\|_\infty = 0$$

if φ is continuous on the closed disk $\overline{\mathbb{D}}$. Thus, we have if $\varphi \in C(\overline{\mathbb{D}})$ and $|\varphi|$ is bounded below by some positive constant δ, then the n-th Berezin transforms $|B_n(\varphi)|$ are bounded below for all sufficiently large n.

In view of this observation we need only to construct a continuous function φ with the following three properties:

(i) φ is invertible in $L^\infty(\mathbb{D})$;

(ii) the Berezin transform $\widetilde{\varphi}$ is also invertible in $L^\infty(\mathbb{D})$;

(iii) T_φ is not invertible on $L_a^2(\mathbb{D})$.

Note that for a real-valued continuous function φ, if φ is invertible in $L^\infty(\mathbb{D})$, then T_φ is invertible on $L_a^2(\mathbb{D})$, which can be proved easily by the same idea in Proposition 7.18 of [7]. Consequently, we need to find a complex-valued continuous function on $\mathbb{D}$ that satisfies the above three conditions.

Theorem 4.10 *Suppose that $\varphi(z) = |z|^2 + a|z| + b$, where a, b are constants. Then there exist $a, b \in \mathbb{C}\backslash\mathbb{R}$ and constants $\delta_1, \delta_2 > 0$ such that*

$$|\varphi(z)| \geqslant \delta_1 \quad \text{and} \quad |\widetilde{\varphi}(z)| \geqslant \delta_2$$

for all $z \in \mathbb{D}$, but the Toeplitz operator T_φ is not invertible on $L_a^2(\mathbb{D})$.

Proof From the proof of Theorem 3.7, we have that the eigenvalues of T_φ are given by

$$\lambda_n = \frac{2n+2}{2n+4} + a\frac{2n+2}{2n+3} + b \quad (n \geqslant 0).$$

On the other hand, recall that the Berezin transform of φ is given by:

$$\widetilde{\varphi}(z) = \left[2 - \frac{1}{|z|^2} - \frac{(1-|z|^2)^2}{|z|^4}\log(1-|z|^2)\right] + \frac{a}{2}\left[3 - \frac{1}{|z|^2} + \frac{(1-|z|^2)^2}{2|z|^3}\log\frac{1+|z|}{1-|z|}\right] + b \;\; (z \in \mathbb{D}).$$

Now we take $a = 2(1+i)$ and $b = -\frac{34}{15} - \frac{8}{5}i$ to show that $0 = \lambda_1 \in \sigma(T_\varphi)$ and that the continuous functions φ and $\widetilde{\varphi}$ are both invertible in $L^\infty(\mathbb{D})$.

From the expression of λ_n, we have

$$\lambda_1 = \frac{4}{6} + \frac{4}{5}a + b = 0,$$

which implies that T_φ is not invertible on the Bergman space.

To prove that φ is invertible in $L^\infty(\mathbb{D})$, we need to show that φ has no zeros in $\overline{\mathbb{D}}$. Since

$$\varphi(z) = \left(|z|^2 + 2|z| - \frac{34}{15}\right) + 2i\left(|z| - \frac{4}{5}\right),$$

we have that $\varphi(z) = 0$ if and only if $|z| - \frac{4}{5} = 0$ and

$$|z|^2 + 2|z| - \frac{34}{15} = 0.$$

It is clear that the above equations have no solution. Thus φ is invertible in $L^\infty(\mathbb{D})$.

The difficult part is to prove that the Berezin transform $\widetilde{\varphi}$ is invertible. Under the above assumption, we have $\widetilde{\varphi}(z) = P(z) + iQ(z)$, where

$$P(z) = \frac{41}{15} - \frac{2}{|z|^2} + \frac{(1-|z|^2)^2}{2|z|^3}\left[\log\frac{1+|z|}{1-|z|} - \frac{2}{|z|}\log(1-|z|^2)\right]$$

and

$$Q(z) = \frac{7}{5} - \frac{1}{|z|^2} + \frac{(1-|z|^2)^2}{2|z|^3}\log\frac{1+|z|}{1-|z|}.$$

Thus $\widetilde{\varphi}(z) = 0$ if and only if $P(z) = 0$ and $Q(z) = 0$. Letting $t = |z| \in [0,1]$, we consider the following two functions:

$$F(t) = |z|^4 P(z) = \frac{41}{15}t^4 - 2t^2 + \frac{(1-t^2)^2}{2}\left[t\log\frac{1+t}{1-t} - 2\log(1-t^2)\right] \quad (t \in [0,1])$$

and

$$G(t) = |z|^3 Q(z) = \frac{7}{5}t^3 - t + \frac{(1-t^2)^2}{2}\log\frac{1+t}{1-t} \quad (t \in [0,1]).$$

Observe that $F(0) = G(0) = 0$ but

$$\widetilde{\varphi}(0) = \frac{1}{2} + \frac{2}{3}a + b = -\frac{13}{30} - \frac{4}{15}i \neq 0,$$

so we need only to show that the equations $F(t)=0$ and $G(t)=0$ do not have any nonzero solutions.

To do so, our idea is as follows: first we prove that $G(t)=0$ has only one nonzero root t_1 in $(0,1)$; next we show that t_1 is not a root of $F(t)=0$. We consider the monotonicity of the function G. Taking derivative gives

$$\begin{aligned} G'(t) &= 2t(1-t^2)\left[\frac{8t}{5(1-t^2)} - \log\frac{1+t}{1-t}\right] \\ &:= 2t(1-t^2)H(t) \quad \big(t\in(0,1)\big). \end{aligned}$$

On the other hand, we have

$$H'(t) = \frac{2(9t^2-1)}{5(1-t^2)^2} \quad \big(t\in(0,1)\big).$$

From the above computations, we have that $H(t)$ is increasing in $(\frac{1}{3},1)$ and $H(t)$ is decreasing in $(0,\frac{1}{3})$. Note that $H(0)=0$, $H(\frac{1}{3})=\frac{3}{5}-\log 2<0$ and $\lim\limits_{t\to 1^-} H(t)=+\infty$. We obtain that $H(t)=0$ has only one root t_0 in $(0,1)$. This implies that: if $t_0<t<1$, then $H(t)>0$ and so $G'(t)>0$; if $0<t<t_0$, then $H(t)<0$, so we have $G'(t)<0$.

From the arguments above, we get that $G(t)$ is increasing if $t\in(t_0,1)$ and decreasing if $t\in(0,t_0)$. Observe that $G(0)=0$ and $G(1)=\frac{2}{5}$. We have that $G(t)=0$ has exactly one solution t_1 in $(0,1)$. To approximate t_1, we evaluate values of $G(t)$ for some points in $(0,1)$ to get that $G(\frac{17}{25})<0$ and $G(\frac{7}{10})>0$. Thus the intermediate value theorem tells us that $t_1\in\big(\frac{17}{25},\frac{7}{10}\big)$.

Now we are going to show that $F(t_1)\neq 0$. If this is not true, we have that $F(t_1)=0$. We will derive a contradiction. To do so, we need the following function:

$$l(t) = F(t)-tG(t) = t^2\Big(\frac{4}{3}t^2-1\Big) - (1-t^2)^2\log(1-t^2) \quad \big(t\in(0,1)\big).$$

Using $F(t_1)=0$ we have $l(t_1)=F(t_1)-t_1G(t_1)=0$.

Let $1-t^2=x\in(0,1)$. The function l becomes the following

$$L(x) = \frac{(1-x)(1-4x)}{3} - x^2\log x \quad \big(x\in(0,1)\big).$$

One has $L(x_1)=0$, where

$$x_1 = 1-t_1^2 \in \left(0,\ 1-\Big(\frac{17}{25}\Big)^2\right] \subset \left(0,\ \frac{27}{50}\right].$$

Simple calculations give us

$$L'(x) = -2x\left[\frac{5(1-x)}{6x} + \log x\right] := -2xR(x).$$

For the function $R(x)$, we have

$$R'(x) = \frac{1}{x^2}\Big(x-\frac{5}{6}\Big) \quad \big(x\in(0,1)\big).$$

Therefore the function R decreases on $\left(0, \frac{27}{50}\right]$. Hence we have

$$R(x) \geqslant R\left(\frac{27}{50}\right) > 0$$

on this interval, and one gets that L decreases on $\left(0, \frac{27}{50}\right]$.

Since $L(\frac{27}{50}) > 0$, L is positive on $\left(0, \frac{27}{50}\right]$, which contradicts the fact that $x_1 \in \left(0, \frac{27}{50}\right]$ is a root of $L(x) = 0$. The contradiction implies that t_1 is not a zero of F. Thus we have $\widetilde{\varphi}(z) \neq 0$ for any $z \in \overline{\mathbb{D}}$. This implies that the Berezin transform $\widetilde{\varphi}$ is also invertible in $L^\infty(\mathbb{D})$, to complete the proof of the Theorem 4.10.

Acknowledgment

This work was partially supported by NSFC (grant numbers: 11531003, 11701052). The first author was partially supported by the Fundamental Research Funds for the Central Universities (106112016CDJRC000080, 106112017CDJXY100007) and Chongqing Natural Science Foundation (cstc2017jcyjAX0373).

References

[1] S. Axler, Bergman spaces and their operators, *Surveys of Some Recent Results in Operator Theory,* Vol. I, 1-50, Pitman Res. Notes Math. Ser., 171, Longman Sci. Tech., Harlow, 1988.

[2] S. Axler and D. Zheng, Compact operators via the Berezin transform, *Indiana Univ. Math. J.* **47** (1998), 387-400.

[3] L. Coburn, Berezin transform and Weyl-type unitary operators on the Bergman space, *Proc. Amer. Math. Soc.* **140** (2012), 3445-3451.

[4] Ž. Čučković and A. Vasaturo, Carleson measures and Douglas' question on the Bergman space, *Rend. Circ. Mat. Palermo* **67** (2018), 323-336.

[5] A. Devinatz, Toeplitz operators on H^2 spaces, *Tran. Amer. Math. Soc.* **112** (1964), 304-317.

[6] R. Douglas, *Banach Algebra Techniques in the Theory of Toeplitz Operators,* American Mathematical Society, 1980.

[7] R. Douglas, *Banach Algebra Techniques in Operator Theory,* 2nd edn, Graduate Texts in Mathematics, vol. 179, Springer, New York, 1998.

[8] N. S. Faour, Toeplitz operators on Bergman spaces, *Rend. Circ. Mat. Palermo* **35** (1986), 221-232.

[9] K. Guo, X. Zhao and D. Zheng, The spectrum of Toeplitz operators with some harmonic polynomial symbols on the Bergman space, preprint, 2018.

[10] M. Gürdal and F. Söhret, Some results for Toeplitz operators on the Bergman space, *Appl. Math. Comput.* **218** (2011), 789-793.

[11] M. T. Karaev, Berezin symbol and invertibility of operators on the functional Hilbert spaces, *J. Funct. Anal.* **238** (2006), 181-192.

[12] D. Luecking, Inequalities on Bergman spaces, *Illinois J. Math.* **25** (1981), 1-11.

[13] G. McDonald and C. Sundberg, Toeplitz operators on the disc, *Indiana Univ. Math. J.* **28** (1979), 595-611.

[14] F. Nazarov, private communication.

[15] N. K. Nikolskii, *Treatise on the Shift Operator: Spectral Function Theory*, Springer, 1986.

[16] K. Stroethoff, The Berezin Transform and Operators on Spaces of Analytic Functions, *Banach Center Publications* 38, Polish Academy of Sciences, Warsaw, 1997, 361-380.

[17] K. Stroethoff and D. Zheng, Toeplitz and Hankel operators on Bergman spaces, *Tran. Amer. Math. Soc.* **329** (1992), 773-794.

[18] D. Suárez, Approximation and symbolic calculus for Toeplitz algebras on the Bergman space, *Rev. Mat. Iberoam* **20** (2004), 563-610.

[19] D. Suárez, Approximation and the n-Berezin transform of operators on the Bergman space, *J. Reine Angew. Math.* **581** (2005), 175-192.

[20] C. Sundberg and D. Zheng, The spectrum and essential spectrum of Toeplitz operators with harmonic symbols, *Indiana Univ. Math. J.* **59** (2010), 385-394.

[21] V. A. Tolokonnikov, Estimates in the Carleson corona theorem, ideals of the algebra H^∞, a problem of S.-Nagy, *Zapiski Nauchnykh Seminarov POMI* **113** (1981), 178-198.

[22] S. Treil, A remark on the reproducing kernel thesis for Hankel operators, arXiv preprint arXiv: 1201. 0063v2, 2011.

[23] T. H. Wolff, Counterexamples to two variants of the Helson-Szegö theorem, *J. Anal. Math.* **88** (2002), 41-62.

[24] X. Zhao and D. Zheng, Positivity of Toeplitz operators via Berezin transform, *J. Math. Anal. Appl.* **416** (2014), 881-900.

[25] X. Zhao and D. Zheng, The spectrum of Bergman Toeplitz operators with some harmonic symbols, *Sci. China Math.* **59** (2016), 731-740.

[26] X. Zhao and D. Zheng, Invertibility of Toeplitz operators via Berezin transforms, *J. Operator Theory* **75** (2016), 101-121.

[27] X. Zhao, Invertibility of Toeplitz operators with some harmonic polynomial symbols on the Bergman space, preprint, 2018.

[28] K. Zhu, *Operator Theory in Function Spaces*, 2nd edn, Mathematical Surveys and Monographs 138, American Mathematical Society, 2007.

Chapter 11

Towards a Dictionary for the Bargmann Transform

Kehe Zhu
Department of Mathematics and Statistics, SUNY, Albany, NY 12222, USA, and Department of Mathematics, Shantou University, Guangdong 515063, China
kzhu@albany.edu

CONTENTS

11.1 Introduction

The Fock space F^2 is the Hilbert space of all entire functions f such that

$$\|f\|^2 = \int_{\mathbb{C}} |f(z)|^2 \, d\lambda(z) < \infty,$$

where

$$d\lambda(z) = \frac{1}{\pi} e^{-|z|^2} \, dA(z)$$

is the Gaussian measure on the complex plane $\mathbb{C}$. Here dA is the ordinary area measure. The inner product on F^2 is inherited from $L^2(\mathbb{C}, d\lambda)$. The Fock space is a convenient setting for many problems in functional analysis, mathematical physics, and engineering. A sample of early and recent work in these areas includes [3, 6, 7, 8, 24, 27, 28, 30, 31, 32, 33, 34, 35, 36, 37]. See [39] for a recent survey of the mathematical theory of Fock spaces.

Another Hilbert space we consider is $L^2(\mathbb{R}) = L^2(\mathbb{R}, dx)$. We will look at the Fourier transform, the Hilbert transform, and several other operators and concepts on $L^2(\mathbb{R})$. The books [20, 21, 22] are excellent sources of information for these classical subjects.

The Bargmann transform B is the operator $L^2(\mathbb{R}) \to F^2$ defined by

$$Bf(z) = c \int_{\mathbb{R}} f(x) e^{2xz - x^2 - (z^2/2)}\, dx,$$

where $c = (2/\pi)^{1/4}$. It is well known that B is a unitary operator from $L^2(\mathbb{R})$ onto F^2; see [21, 22, 39]. The easiest way to see this is outlined in the next section. Furthermore, the inverse of B is also an integral operator, namely,

$$B^{-1} f(x) = c \int_{\mathbb{C}} f(z) e^{2x\bar{z} - x^2 - (\bar{z}^2/2)}\, d\lambda(z).$$

The Bargmann transform is an old tool in mathematical analysis and mathematical physics. See [1, 2, 8, 14, 21, 22] and references there. In this article we attempt to establish a "dictionary" between $L^2(\mathbb{R})$ and F^2 that is based on the Bargmann transform. Thus we translate several important operators and concepts between these two spaces. It goes without saying that the dictionary is not complete, and it cannot be complete. Nevertheless, it covers some of the most important operators and concepts in Fourier and harmonic analysis, for example, the Fourier transform, the Hilbert transform, Gabor frames, the standard commutation relation, pseudo-differential operators, and the uncertainty principle.

Most results in the paper are classical and should be well known to experts. Thus the paper is expository in nature. However, it is not always easy to identify original references and to find existing proofs for some of the results in the paper. Therefore, for each result I decided to either provide a precise reference or to include a detailed proof. In particular, I chose to omit any proof that can be found in my recent monograph [39]. This makes the paper more or less self-contained and more user-friendly for newcomers to the field.

I thank Hans Feichtinger and Bruno Torresani for their invitation to visit CIRM/Luminy in the fall semester of 2014. This paper was motivated by discussions with several visitors during my stay at CIRM.

11.2 Hermite polynomials

The standard monomial orthonormal basis for F^2 is given by

$$e_n(z) = \sqrt{\frac{1}{n!}}\, z^n, \qquad n \geq 0.$$

Thus the reproducing kernel of F^2 is

$$K(z,w) = \sum_{n=0}^{\infty} e_n(z)\overline{e_n(w)} = \sum_{n=0}^{\infty} \frac{(z\overline{w})^n}{n!} = e^{z\overline{w}}.$$

The normalized reproducing kernel of F^2 at the point a is given by

$$k_a(z) = e^{-\frac{|a|^2}{2} + z\overline{a}}.$$

Each k_a is a unit vector in F^2.

To exhibit an orthonormal basis for $L^2(\mathbb{R})$, recall that for any $n \geq 0$ the function

$$H_n(x) = (-1)^n e^{x^2} \frac{d^n}{dx^n} e^{-x^2}$$

is called the nth Hermite polynomial. The history of Hermite polynomials is rich and long. For example, an online look at Wikipedia will reveal tons of interesting information and references about Hermite polynomials. In particular, it is well known that the functions

$$h_n(x) = \frac{c}{\sqrt{2^n n!}} e^{-x^2} H_n(\sqrt{2}x), \qquad n \geq 0,$$

form an orthonormal basis for $L^2(\mathbb{R})$, where $c = (2/\pi)^{1/4}$ again.

Theorem 2.1 *For every $n \geq 0$ we have $Bh_n = e_n$.*

Proof See Theorem 6.8 of [39].

Corollary 2.2 *The Bargmann transform is a unitary operator from $L^2(\mathbb{R})$ onto F^2, and it maps the normalized Gauss function*

$$g(z) = (2/\pi)^{1/4} e^{-x^2},$$

which is a unit vector in $L^2(\mathbb{R})$, to the constant function 1 *in F^2.*

That B maps the Gauss function to a constant is the key ingredient when we later translate Gabor frames with the Gauss window to analytic atoms in the Fock space.

11.3 The Fourier transform

There are several normalizations for the Fourier transform. We define the Fourier transform by

$$F(f)(x) = \frac{1}{\sqrt{\pi}} \int_{\mathbb{R}} f(t) e^{2ixt}\, dt.$$

It is well known that the Fourier transform acts as a bounded linear operator on $L^2(\mathbb{R})$. In fact, Plancherel's formula tells us that F is a unitary operator on $L^2(\mathbb{R})$, and its inverse is given by

$$F^{-1}(f)(x) = \frac{1}{\sqrt{\pi}} \int_{\mathbb{R}} f(t) e^{-2ixt}\, dt.$$

It is not at all clear from the definition that F is bounded and invertible on $L^2(\mathbb{R})$. There are also issues concerning convergence: it is not clear that the integral defining $F(f)$ converges in $L^2(\mathbb{R})$ for $f \in L^2(\mathbb{R})$. The situation will change dramatically once we translate F to an operator on the Fock space. In other words, we will show that, under the Bargmann transform, the operator $F : L^2(\mathbb{R}) \to L^2(\mathbb{R})$ is unitarily equivalent to an extremely simple operator on the Fock space F^2.

Lemma 3.1 *We have*

$$\int_{\mathbb{R}} e^{-(x-z)^2}\, dx = \sqrt{\pi}$$

for every complex number z.

Proof Write

$$I(z) = \int_{\mathbb{R}} e^{-(x+z)^2}\, dx, \qquad z \in \mathbb{C}.$$

It is clear that $I(z)$ is an entire function and

$$I'(z) = -2 \int_{\mathbb{R}} (x+z) e^{-(x+z)^2}\, dx = e^{-(x+z)^2} \Big|_{-\infty}^{+\infty} = 0$$

for $z \in \mathbb{C}$. It follows that

$$I(z) = I(0) = \int_{\mathbb{R}} e^{-x^2}\, dx = \sqrt{\pi}$$

for all $z \in \mathbb{C}$.

This elementary result along with some of its close relatives will be used many times later in the paper without being mentioned. The following result is well known to experts in the field of Fourier analysis and its applications. In particular, this result can be found in the introduction of Bargmann's paper [2].

Theorem 3.2 *The operator*

$$T = BFB^{-1} : F^2 \to F^2$$

is given by $Tf(z) = f(iz)$ *for all* $f \in F^2$. *Consequently, the operator*

$$T^{-1} = BF^{-1}B^{-1} : F^2 \to F^2,$$

where F^{-1} *is the inverse Fourier transform, is given by* $T^{-1}f(z) = f(-iz)$ *for all* $f \in F^2$.

Proof We consider a family of Fourier transforms as follows:

$$F_\sigma(f)(x) = \sqrt{\frac{\sigma}{\pi}} \int_{\mathbb{R}} f(t) e^{2i\sigma xt}\, dt,$$

where σ is a real parameter with the convention that $\sqrt{-1} = i$.

For the purpose of applying Fubini's theorem in the calculations below, we assume that f is any polynomial (recall that the polynomials are dense in F^2, and under the inverse Bargmann transform, they become the Hermite polynomials times the Gauss function, which have very good integrability properties on the real line). For $c = (2/\pi)^{1/4}$ again, we have

$$\begin{aligned}
F_\sigma(B^{-1}f)(x) &= c\sqrt{\frac{\sigma}{\pi}} \int_{\mathbb{R}} e^{2i\sigma xt}\, dt \int_{\mathbb{C}} f(z) e^{2t\bar{z} - t^2 - \frac{\bar{z}^2}{2}}\, d\lambda(z) \\
&= c\sqrt{\frac{\sigma}{\pi}} \int_{\mathbb{C}} f(z) e^{-\frac{\bar{z}^2}{2}}\, d\lambda(z) \int_{\mathbb{R}} e^{2t(i\sigma x + \bar{z}) - t^2}\, dt \\
&= c\sqrt{\frac{\sigma}{\pi}} \int_{\mathbb{C}} f(z) e^{-\frac{\bar{z}^2}{2} + (i\sigma x + \bar{z})^2}\, d\lambda(z) \int_{\mathbb{R}} e^{-(t - i\sigma x - \bar{z})^2}\, dt \\
&= c\sqrt{\sigma} e^{-\sigma^2 x} \int_{\mathbb{C}} f(w) e^{\frac{\bar{w}^2}{2} + 2i\sigma x\bar{w}}\, d\lambda(w).
\end{aligned}$$

To shorten the displayed equations below, we write $g(z) = BF_\sigma B^{-1} f(z)$. Then

$$\begin{aligned}
g(z) &= c^2\sqrt{\sigma} \int_{\mathbb{R}} e^{2xz - x^2 - \frac{z^2}{2} - \sigma^2 x^2}\, dx \int_{\mathbb{C}} f(w) e^{\frac{\bar{w}^2}{2} + 2i\sigma x\bar{w}}\, d\lambda(w) \\
&= c^2\sqrt{\sigma} e^{-\frac{z^2}{2}} \int_{\mathbb{R}} e^{2xz - (1+\sigma^2)x^2}\, dx \int_{\mathbb{C}} f(w) e^{\frac{\bar{w}^2}{2} + 2i\sigma x\bar{w}}\, d\lambda(w) \\
&= c^2\sqrt{\sigma} e^{-\frac{z^2}{2}} \int_{\mathbb{C}} f(w) e^{\frac{\bar{w}^2}{2}}\, d\lambda(w) \int_{\mathbb{R}} e^{2x(z + i\sigma\bar{w}) - (1+\sigma^2)x^2}\, dx \\
&= c^2\sqrt{\frac{\sigma}{\sigma^2+1}} e^{-\frac{z^2}{2}} \int_{\mathbb{C}} f(w) e^{\frac{\bar{w}^2}{2}}\, d\lambda(w) \int_{\mathbb{R}} e^{2\cdot\frac{z + i\sigma\bar{w}}{\sqrt{\sigma^2+1}}\cdot t - t^2}\, dt \\
&= c^2\sqrt{\frac{\sigma\pi}{\sigma^2+1}} e^{-\frac{z^2}{2}} \int_{\mathbb{C}} f(w) e^{\frac{\bar{w}^2}{2} + \frac{(z + i\sigma\bar{w})^2}{\sigma^2+1}}\, d\lambda(w) \\
&= \sqrt{\frac{2\sigma}{\sigma^2+1}} e^{\left(\frac{1}{\sigma^2+1} - \frac{1}{2}\right)z^2} \int_{\mathbb{C}} f(w) e^{\left(\frac{1}{2} - \frac{\sigma^2}{\sigma^2+1}\right)\bar{w}^2 + \frac{2i\sigma}{\sigma^2+1} z\bar{w}}\, d\lambda(w).
\end{aligned}$$

In the case $\sigma = 1$, we have $F_\sigma = F$, so that

$$(BFB^{-1}f)(z) = \int_{\mathbb{C}} f(w)e^{iz\overline{w}}\,d\lambda(w) = f(iz).$$

It is then clear that $BF^{-1}B^{-1}f(z) = f(-iz)$. This completes the proof of the theorem.

As a consequence of Theorem 3.2, we obtain an alternative proof of the Fourier inversion formula and Plancherel's formula.

Corollary 3.3 *The Fourier transform is a unitary operator on $L^2(\mathbb{R})$: it is one-to-one, onto, and isometric in the sense that*

$$\int_{\mathbb{R}} |F(f)|^2\,dx = \int_{\mathbb{R}} |f|^2\,dx$$

for all $f \in L^2(\mathbb{R})$. Furthermore, the inverse Fourier transform is given by

$$F^{-1}(f)(x) = \frac{1}{\sqrt{\pi}}\int_{\mathbb{R}} f(t)e^{-2ixt}\,dt.$$

Proof It is obvious that $f(z) \mapsto f(iz)$ is a unitary operator on F^2. Thus the Fourier transform is a unitary operator on $L^2(\mathbb{R})$, hence Plancherel's formula.

By the proof of Theorem 3.2, we have

$$BF_{-1}B^{-1}f(z) = i\int_{\mathbb{C}} f(w)e^{-iz\overline{w}}\,d\lambda(w) = if(-iz) = iBF^{-1}B^{-1}f(z).$$

Thus $F^{-1} = -iF_{-1}$, namely,

$$F^{-1}f(x) = -i\sqrt{\frac{-1}{\pi}}\int_{\mathbb{R}} f(t)e^{-2ixt}\,dt = \frac{1}{\sqrt{\pi}}\int_{\mathbb{R}} f(t)e^{-2ixt}\,dt,$$

which is the Fourier inversion formula.

Note that it is a little vague how the Fourier transform is defined for a function in $L^2(\mathbb{R})$. But there is absolutely no ambiguity for the unitary operator $f(z) \mapsto f(iz)$ on F^2. This operator is clearly well defined for every $f \in F^2$ and it is clearly a unitary operator.

The following result is clear from our new representation of the Fourier transform on the Fock space, because an entire function uniquely determines its Taylor coefficients.

Corollary 3.4 *For each $n \geq 0$ the Hermite function h_n is an eigenvector of the Fourier transform and the corresponding eigenvalue is i^n. In particular, the fixed points of the Fourier transform are exactly functions of the form*

$$f(x) = \sum_{n=0}^{\infty} c_n h_{4n}(x), \qquad \{c_n\} \in l^2.$$

To go one step further, we can completely determine the spectral properties of the Fourier transform as a unitary operator on $L^2(\mathbb{R})$.

Corollary 3.5 *For each* $0 \leq k \leq 3$ *let* X_k *denote the closed subspace of* $L^2(\mathbb{R})$ *spanned by the Hermite functions* h_{k+4m}, $m \geq 0$, *and let* $P_k : L^2(\mathbb{R}) \to X_k$ *be the orthogonal projection. Then*

$$L^2(\mathbb{R}) = \bigoplus_{k=0}^{3} X_k,$$

and the corresponding spectral decomposition for the unitary operator $F : L^2(\mathbb{R}) \to L^2(\mathbb{R})$ *is given by* $F = P_0 + iP_1 - P_2 - iP_3$.

In particular, each X_k is nothing but the eigenspace of the Fourier transform corresponding to the eigenvalue i^k.

Here we first proved that the Fourier transform becomes an operator of rotation on the Fock space after the Bargmann transform and then deduced that each i^k is an eigenvalue of the Fourier transform and the corresponding eigenfunctions are the Hermite functions. Alternatively, we could have first proved that the Hermite functions of the Fourier transform are eigenfunctions corresponding to the eigenvalues i^k (this could be done using integration by parts for example) and then obtain Theorem 3.2 as a corollary.

Finally in this section we mention that for any real θ the operator U_θ defined by $U_\theta f(z) = f(e^{i\theta}z)$ is clearly a unitary operator on F^2. For $\theta = \pm\pi/2$ the resulting operators are unitarily equivalent to the Fourier and inverse Fourier transforms on $L^2(\mathbb{R})$. When θ is a rational multiple of π, the structure of U_θ is relatively simple. However, if θ is an irrational multiple of π, the structure of such an "irrational rotation operator" U_θ is highly non-trivial.

It is interesting to note that the operators on $L^2(\mathbb{R})$ that are unitarily equivalent to U_θ via the Bargmann transform are nothing but the fractional Fourier transforms. See [18].

11.4 Dilation, translation, and modulation operators

For any positive r we consider the dilation operator $D_r : L^2(\mathbb{R}) \to L^2(\mathbb{R})$ defined by $D_r f(x) = \sqrt{r} f(rx)$. It is obvious that D_r is a unitary operator on $L^2(\mathbb{R})$. Furthermore, it follows from the definitions of F and F_r in the previous section that

$$F_r f(x) = \frac{\sqrt{r}}{\sqrt{\pi}} \int_{\mathbb{R}} f(t) e^{2irxt}\, dt = \sqrt{r} F f(rx) = D_r F f(x).$$

Therefore, each $F_r = D_r F$ is a unitary operator on $L^2(\mathbb{R})$ with

$$F_r^{-1} = F^{-1} D_r^{-1} = F^{-1} D_{1/r},$$

that is,

$$F_r^{-1}f(x) = \frac{1}{\sqrt{\pi r}}\int_{\mathbb{R}} f(t/r)e^{-2ixt}\,dt = \sqrt{\frac{r}{\pi}}\int_{\mathbb{R}} f(t)e^{-2irxt}\,dt,$$

which is a slightly generalized version of the Fourier inversion formula. The following result gives the equivalent form of the dilation operator D_r on the Fock space.

Theorem 4.1 *For any $r > 0$ let $T_r = BD_rB^{-1}$ on F^2. Then*

$$T_rf(z) = \sqrt{\frac{2r}{1+r^2}}e^{\left(\frac{1}{1+r^2}-\frac{1}{2}\right)z^2}\int_{\mathbb{C}} f(w)e^{\left(\frac{r^2}{1+r^2}-\frac{1}{2}\right)\overline{w}^2}e^{\frac{2rz}{1+r^2}\overline{w}}\,d\lambda(w)$$

for all $f \in F^2$.

Proof Since $D_r = F_rF^{-1}$, we have

$$T_r = BF_rB^{-1}BF^{-1}B^{-1}.$$

By Theorem 3.2, $BF^{-1}B^{-1}f(z) = f(-iz)$. Combining this with the integral formula for BF_rB^{-1} that was obtained in the proof of Theorem 3.2, we obtain

$$\begin{aligned} T_rf(z) &= \sqrt{\frac{2r}{1+r^2}}e^{\left(\frac{1}{1+r^2}-\frac{1}{2}\right)z^2}\int_{\mathbb{C}} f(-iw)e^{\left(\frac{1}{2}-\frac{r^2}{1+r^2}\right)\overline{w}^2+\frac{2irz}{1+r^2}\overline{w}}\,d\lambda(w) \\ &= \sqrt{\frac{2r}{1+r^2}}e^{\left(\frac{1}{1+r^2}-\frac{1}{2}\right)z^2}\int_{\mathbb{C}} f(w)e^{\left(\frac{r^2}{1+r^2}-\frac{1}{2}\right)\overline{w}^2}e^{\frac{2rz}{1+r^2}\overline{w}}\,d\lambda(w). \end{aligned}$$

This proves the desired formula for T_r on F^2.

Thus in the case of dilation, the operator is much simpler on the space $L^2(\mathbb{R})$ than on F^2. Therefore, it is unlikely that the Fock space will be helpful in the study of dilation operators on $L^2(\mathbb{R})$. Since the theory of wavelets depends on dilation in a critical way, we expect the Fock space (and its associated complex analysis) to be of limited use for wavelet analysis. See [14, 19, 29] for an application of the Bargmann transform to the study of the wavelet localization operators.

The situation is completely different for Gabor analysis (see next section), where critical roles are played by the so-called translation and modulation operators, whose representations on F^2 via the Bargmann transform are much more useful. More specifically, for any real numbers a and b we define two unitary operators T_a and M_b on $L^2(\mathbb{R})$ as follows:

$$T_af(x) = f(x-a), \qquad M_bf(x) = e^{2\pi bix}f(x).$$

It is traditional to call T_a a translation operator and M_b a modulation operator. See [22] for more information on such operators.

To identify the equivalent form of T_a and M_b on the Fock space, we need the classical Weyl operators on F^2. Recall that for any complex number a the Weyl operator W_a on F^2 is defined by

$$W_a f(z) = f(z-a)k_a(z) = f(z-a)e^{z\overline{a}-\frac{|a|^2}{2}},$$

where k_a is the normalized reproducing kernel of F^2 at a. It is well known, and it follows easily from a change of variables, that each W_a is a unitary operator on F^2. See [39] for more information about the Weyl operators.

Theorem 4.2 *For any $a \in \mathbb{R}$ we have $BT_aB^{-1} = W_a$.*

Proof For any polynomial f in F^2 we have

$$T_aB^{-1}f(x) = \left(\frac{2}{\pi}\right)^{\frac{1}{4}} \int_{\mathbb{C}} f(z)e^{-(x-a-\overline{z})^2+(\overline{z}^2/2)}\, d\lambda(z),$$

and

$$\begin{aligned} BT_aB^{-1}f(z) &= \sqrt{\frac{2}{\pi}}e^{\frac{z^2}{2}} \int_{\mathbb{R}} e^{-(x-z)^2}\, dx \int_{\mathbb{C}} f(u)e^{-(x-a-\overline{u})^2+\frac{\overline{u}^2}{2}}\, d\lambda(u) \\ &= \sqrt{\frac{2}{\pi}}e^{\frac{z^2}{2}} \int_{\mathbb{C}} f(u)e^{\frac{\overline{u}^2}{2}}\, d\lambda(u) \int_{\mathbb{R}} e^{-(x-z)^2-(x-a-\overline{u})^2}\, dx \\ &= \sqrt{\frac{2}{\pi}}e^{-\frac{z^2}{2}} \int_{\mathbb{C}} f(u)e^{-\frac{1}{2}\overline{u}^2-a^2-2a\overline{u}} I(u,z)\, d\lambda(u), \end{aligned}$$

where

$$\begin{aligned} I(u,z) &= \int_{\mathbb{R}} e^{-2x^2+2xz+2ax+2x\overline{u}}\, dx \\ &= e^{(z+a+\overline{u})^2/2} \int_{\mathbb{R}} e^{-2[x-(z+a+\overline{u})/2]^2}\, dx \\ &= \sqrt{\frac{\pi}{2}}e^{(z+a+\overline{u})^2/2}. \end{aligned}$$

It follows that

$$\begin{aligned} BT_aB^{-1}f(z) &= e^{-\frac{z^2}{2}} \int_{\mathbb{C}} f(u)e^{-\frac{1}{2}\overline{u}^2-a^2-2a\overline{u}+\frac{1}{2}(z+a+\overline{u})^2}\, d\lambda(u) \\ &= e^{-\frac{1}{2}a^2+az} \int_{\mathbb{C}} f(u)e^{(z-a)\overline{u}}\, d\lambda(u) \\ &= e^{-\frac{1}{2}a^2+az} f(z-a) = W_a f(z). \end{aligned}$$

This completes the proof of the theorem.

Theorem 4.3 *For any real b we have $BM_bB^{-1} = W_{-\pi bi}$.*

Proof For any polynomial f in F^2 we have

$$M_bB^{-1}f(x) = \left(\frac{2}{\pi}\right)^{\frac{1}{4}} e^{2\pi bix} \int_{\mathbb{C}} f(z)e^{-(x-\bar{z})^2+\frac{\bar{z}^2}{2}}\, d\lambda(z).$$

Let $F(z) = BM_bB^{-1}f(z)$ to shorten the displayed equations below. Then

$$\begin{aligned}
F(z) &= \sqrt{\frac{2}{\pi}} e^{\frac{z^2}{2}} \int_{\mathbb{R}} e^{-(x-z)^2+2\pi bix}\, dx \int_{\mathbb{C}} f(u)e^{-(x-\bar{u})^2+\frac{\bar{u}^2}{2}}\, d\lambda(u) \\
&= \sqrt{\frac{2}{\pi}} e^{\frac{z^2}{2}} \int_{\mathbb{C}} f(u)e^{\frac{1}{2}\bar{u}^2}\, d\lambda(u) \int_{\mathbb{R}} e^{-(x-z)^2-(x-\bar{u})^2+2\pi bix}\, dx \\
&= \sqrt{\frac{2}{\pi}} e^{-\frac{z^2}{2}} \int_{\mathbb{C}} f(u)e^{-\frac{\bar{u}^2}{2}}\, d\lambda(u) \int_{\mathbb{R}} e^{-2x^2+2x(z+\bar{u}+\pi bi)}\, dx \\
&= \sqrt{\frac{2}{\pi}} e^{-\frac{z^2}{2}} \int_{\mathbb{C}} f(u)e^{-\frac{\bar{u}^2}{2}+\frac{1}{2}(z+\bar{u}+\pi bi)^2}\, d\lambda(u) \int_{\mathbb{R}} e^{-2[x-(z+\bar{u}+\pi bi)/2]^2}\, dx \\
&= e^{-\frac{z^2}{2}} \int_{\mathbb{C}} f(u)e^{-\frac{\bar{u}^2}{2}+\frac{1}{2}(z+\bar{u}+\pi bi)^2}\, d\lambda(u) \\
&= e^{-\frac{1}{2}\pi^2b^2+\pi biz} \int_{\mathbb{C}} f(u)e^{(z+\pi bi)\bar{u}}\, d\lambda(u) \\
&= e^{-\frac{1}{2}\pi^2b^2+\pi biz} f(z+\pi bi) = W_{-\pi bi}f(z).
\end{aligned}$$

This proves the desired result.

Corollary 4.4 *Let a and b be any pair of real numbers. Then we have*

$$B(M_bT_a)B^{-1} = e^{-\pi abi}W_{a-\pi bi}.$$

Consequently, if $g \in L^2(\mathbb{R}, dx)$ and $f = Bg$. Then the Bargmann transform maps the function M_bT_ag to the function $e^{-\pi abi}W_{a-\pi bi}f$.

Proof This follows from the two theorems above and the identity (2.22) on page 61 of [39].

11.5 Gabor frames

A sequence $\{f_n\}$ in a Hilbert space H is called a frame if there exists a positive constant C such that

$$C^{-1}\|f\|^2 \le \sum_{n=1}^{\infty} |\langle f, f_n\rangle|^2 \le C\|f\|^2$$

for all $f \in H$. The following result contains the most important properties of frames in a Hilbert space.

Theorem 5.1 *Every frame $\{f_n\}$ in a Hilbert space H has the following properties.*

(i) For any $\{c_n\} \in l^2$ the series $\sum c_n f_n$ converges in the norm topology of H.

(ii) For any $f \in H$, there exists a sequence $\{c_n\} \in l^2$ such that $f = \sum c_n f_n$.

The result above is well known and is the foundation for the theory of frames. See [13] for this and other properties of general frames in Hilbert spaces.

Let $\{a_n\}$ and $\{b_n\}$ be two sequences of real numbers, each consisting of distinct values, and let $g \in L^2(\mathbb{R})$. Write $g_n = M_{b_n} T_{a_n} g$ for $n \geq 1$. We say that $\{g_n\}$ is a Gabor frame for $L^2(\mathbb{R})$ if there exists a positive constant C such that

$$C^{-1}\|f\|^2 \leq \sum_{n=1}^{\infty} |\langle f, g_n\rangle|^2 \leq C\|f\|^2$$

for all $f \in L^2(\mathbb{R})$. In this case, g is called the window function of the Gabor frame $\{g_n\}$. The book [22] is an excellent source of information for Gabor frames and other topics of time-frequency analysis.

By Theorem 5.1 above, if $\{g_n\}$ is a Gabor frame, then every function $f \in L^2(\mathbb{R})$ can be represented in the form

$$f = \sum_{n=1}^{\infty} c_n g_n,$$

where $\{c_n\} \in l^2$ and the series converges in the norm topology of $L^2(\mathbb{R})$. Conversely, if $\{g_n\}$ is a Gabor frame and $\{c_n\} \in l^2$, then the series above converges to some $f \in L^2(\mathbb{R})$ in the norm topology of $L^2(\mathbb{R})$. The representation is generally not unique though.

There are two classical examples of window functions in Gabor analysis.

Let $g(x)$ be the characteristic function of the unit interval $[0,1)$ and let $\{z_n = a_n + ib_n\}$ denote any fixed arrangement of the square lattice $\mathbb{Z}^2$ into a sequence. Then $\{g_n\}$ is not only a Gabor frame, but it is also an orthonormal basis. To see this, it is more transparent to use two indices:

$$g_{nm}(x) = \chi_{[n,n+1)}(x) e^{2m\pi i x}, \qquad (n,m) \in \mathbb{Z}^2.$$

Now it is clear that

$$f(x) = \sum_{n\in\mathbb{Z}} f(x)\chi_{[n,n+1)}(x) = \sum_{n\in\mathbb{Z}} f_n(x),$$

where the functions $f_n(x) = f(x)\chi_{[n,n+1)}(x)$ are mutually orthogonal in $L^2(\mathbb{R})$. On the other hand, for each $n \in \mathbb{Z}$, the function $f_n(x)$ is compactly supported on $[n, n+1)$, so it can be expanded into a Fourier series:

$$f_n(x) = \sum_{m\in\mathbb{Z}} c_{nm} e^{2m\pi ix}, \qquad n \le x < n+1,$$

whose terms are mutually orthogonal over $[n, n+1)$. Therefore, we have

$$f(x) = \sum_{(n,m)\in\mathbb{Z}^2} c_{nm}\chi_{[n,n+1)}(x)e^{2m\pi ix} = \sum_{(n,m)\in\mathbb{Z}^2} c_{nm} g_{nm}(x),$$

where $g_{nm} = M_m T_n g$ are mutually orthogonal in $L^2(\mathbb{R})$.

Thus the characteristic function of the unit interval $[0,1)$ is a Gabor window, which is usually called a box window.

Another important Gabor window is the Gauss function $g(x) = e^{-x^2}$, which can be thought of as a smooth analog of the box window. However, it is not a trivial matter to see that the Gauss function is a Gabor window. It will become clear once we establish the connection with the Fock space. Furthermore, using the theory of Fock spaces, we will be able to know exactly which translation and modulation sequences give rise to Gabor frames for the Gauss window.

Recall from Corollary 2.2 that the Bargmann transform maps the Gauss window function to a constant in the Fock space. Thus the following result is a consequence of Corollary 4.4.

Corollary 5.2 *Let $\{a_n\}$ and $\{b_n\}$ be two sequences of real numbers, each consisting of distinct points in $\mathbb{R}$. Let $g(x) = e^{-x^2}$ be the Gauss window function. Then the system $\{g_n = M_{b_n} T_{a_n} g\}$ is a Gabor frame for $L^2(\mathbb{R})$ if and only if the sequence $\{k_{z_n}\}$ is a frame for F^2, where $z_n = a_n - \pi b i_n$ and k_{z_n} is the normalized reproducing kernel of F^2 at z_n.*

There are many fundamental questions that Gabor analysis tries to address. Among them we mention the following two.

(1) Characterize all window functions. Recall that we say a function $g \in L^2(\mathbb{R})$ is a Gabor window if there exist two sequences $\{a_n\}$ and $\{b_n\}$ such that $\{g_n = M_{b_n} T_{a_n} g\}$ is a Gabor frame.

(2) Given a Gabor window g, characterize all sequences $\{a_n\}$ and $\{b_n\}$ such that $\{M_{b_n} T_{a_n} g\}$ is a Gabor frame.

The most popular Gabor frames are constructed using rectangular lattices $a\mathbb{Z} \times b\mathbb{Z}$. Such Gabor frames will be called regular. Slightly more general are lattices based on congruent parallelograms: $\omega_1\mathbb{Z} + \omega_2\mathbb{Z}$, where ω_1 and ω_2 are two nonzero complex numbers which are linearly independent in $\mathbb{R}^2$. Gabor frames based on such lattices will also be called regular.

Semi-regular lattices are of the form $\{a_n\}\omega_1 + \{b_m\}\omega_2$, where $\{a_n\}$ and $\{b_m\}$ are two sequences of distinct real numbers, and ω_1 and ω_2 are two complex numbers that are linearly independent in $\mathbb{R}^2$. These are lattices based on parallelograms of different sizes. The resulting Gabor frames will be called semi-regular.

Irregular Gabor frames are constructed using an arbitrary sequence $z_n = (a_n, b_n)$ in $\mathbb{R}^2$.

To understand the two questions raised earlier about Gabor frames, we need the notions of atomic decomposition and sampling sequences for F^2.

Let $\{z_n\}$ denote a sequence of (distinct) points in $\mathbb{C}$. We say that atomic decomposition for F^2 holds on the sequence $\{z_n\}$ if

(1) For any sequence $\{c_n\} \in l^2$ the series

$$\sum_{n=1}^{\infty} c_n k_{z_n}(z)$$

converges in the norm topology of F^2.

(2) For any function $f \in F^2$ there exists a sequence $\{c_n\} \in l^2$ such that

$$f(z) = \sum_{n=1}^{\infty} c_n k_{z_n}.$$

See [27, 39] for more information about atomic decomposition for Fock spaces.

We say that the sequence $\{z_n\}$ is sampling for F^2 if there exists a positive constant C such that

$$C^{-1}\|f\|^2 \le \sum_{n=1}^{\infty} |f(z_n)|^2 e^{-|z_n|^2} \le C\|f\|^2$$

for all $f \in F^2$. See [36, 37, 39] for more information about sampling sequences in F^2.

Theorem 5.3 *Given a sequence $\{z_n = a_n - \pi b_n i\}$ of distinct points in $\mathbb{C}$, the following three conditions are equivalent.*

(1) The system $\{g_n = M_{b_n} T_{a_n} g\}$ is a Gabor frame for $L^2(\mathbb{R})$, where g is the Gauss window.

(2) The sequence $\{z_n\}$ is sampling for F^2.

(3) Atomic decomposition for F^2 holds on F^2.

Proof Recall from Corollary 5.2 that $\{g_n\}$ is a Gabor frame for $L^2(\mathbb{R})$ iff $\{k_{z_n}\}$ is a frame for F^2. Since $\langle f, k_z\rangle = f(z)e^{-\frac{1}{2}|z|^2}$, we see that $\{k_{z_n}\}$ is a frame for F^2 if and only if $\{z_n\}$ is a sampling sequence for F^2. It is well known that $\{z_n\}$ is a sampling sequence if and only if atomic decomposition holds on $\{z_n\}$; see [39] for example.

More generally, we have the following.

Theorem 5.4 *Suppose $\{a_n\}$ and $\{b_n\}$ are sequences of real numbers and $g \in L^2(\mathbb{R})$. Then $\{M_{b_n}T_{a_n}g\}$ is a Gabor frame for $L^2(\mathbb{R})$ if and only if $\{W_{z_n}f\}$ is a frame in F^2, where $z_n = a_n - \pi bi$, W_{z_n} are the Weyl operators, and $f = Bg$.*

Proof This follows from Corollaries 2.2 and 4.4 again.

Thus the study of Gabor frames for $L^2(\mathbb{R})$ is equivalent to the study of frames of the form $\{W_{z_n}f\}$ in F^2. In particular, Gabor frames with the Gauss window correspond to frames $\{k_{z_n}\}$ in F^2 which in turn correspond to sampling sequences $\{z_n\}$ for F^2. The idea of studying Gabor frames with the help of the Bargmann transform is not new. See [9, 23, 30, 31, 32, 33] for example.

Sampling sequences have been completely characterized by Seip and Wallstén. To describe their results, we need to introduce a certain notion of density for sequences in the complex plane. Thus for a sequence $Z = \{z_n\}$ of distinct points in the complex plane, we define

$$D^+(Z) = \limsup_{R\to\infty} \sup_{z\in\mathbb{C}} \frac{|Z\cap B(z,R)|}{\pi R^2},$$

and

$$D^-(Z) = \liminf_{R\to\infty} \inf_{z\in\mathbb{C}} \frac{|Z\cap B(z,R)|}{\pi R^2},$$

where

$$B(z,R) = \{w \in \mathbb{C} : |w-z| < R\}$$

and $|Z\cap B(z,R)|$ denotes the cardinality of $Z\cap B(z,R)$. These are called the Beurling upper and lower densities of Z, respectively. The following characterization of sampling sequences for F^2 can be found in [36, 37].

Theorem 5.5 *A sequence Z of distinct points in $\mathbb{C}$ is sampling for F^2 if and only if the following two conditions are satisfied.*

(a) Z is the union of finitely many subsequences each of which is separated in the Euclidean metric.

(b) Z contains a subsequence Z' that is separated in the Euclidean metric such that $D^-(Z') > 1/\pi$.

As a consequence of Theorems 5.3 and 5.5 we obtain the following complete characterization of Gabor frames in $L^2(\mathbb{R})$ associated with the Gauss window.

Theorem 5.6 *Let $Z = \{z_n = a_n - \pi i b_n : n \geq 0\}$ be a sequence of points in $\mathbb{C} = \mathbb{R}^2$ and let $g(x)$ be the Gauss window. Then $\{M_{b_n}T_{a_n}g\}$ is a Gabor frame in $L^2(\mathbb{R})$ if and only if the following two conditions are satisfied:*

(a) Z is the union of finitely many subsequences each of which is separated in the Euclidean metric.

(b) Z contains a subsequence Z' that is separated in the Euclidean metric such that $D^-(Z') > 1/\pi$.

Corollary 5.7 *Let g be the Gauss window and $Z = \{z_n = a_n - \pi i b_n : n \geq 0\}$ be a sequence in $\mathbb{C}$ that is separated in the Euclidean metric. Then $\{M_{b_n} T_{a_n} g\}$ is a Gabor frame for $L^2(\mathbb{R})$ if and only if $D^-(Z) > 1/\pi$.*

The most classical case is when the sequence Z is a rectangular lattice in $\mathbb{C}$, which is clearly separated in the Euclidean metric. Thus we have the following.

Corollary 5.8 *Let g be the Gauss window and a, b be positive constants. Then $\{M_{mb} T_{na} g : m, n \in \mathbb{Z}\}$ is a Gabor frame if and only if $ab < 1$.*

Proof Let $Z = \{na - \pi mbi : (m,n) \in \mathbb{Z}^2\}$ be the rectangular lattice spanned by a and b. When R is very large, it is easy to see that $|Z \cap B(z,R)|$ is roughly $\pi R^2/(\pi ab)$. It follows that

$$D^-(Z) = D^+(Z) = 1/(\pi ab).$$

Therefore, according to Corollary 5.7, $\{M_{mb} T_{na} g : m, n \in \mathbb{Z}\}$ is a Gabor frame if and only if $1/(\pi ab) > 1/\pi$, or $ab < 1$.

It is interesting to observe that, for the Gauss window, whether or not a rectangular lattice $Z = \{na + mib : n, m \in \mathbb{Z}\}$ generates a Gabor frame only depends on the product ab. This was already a well-known fact (and a conjecture for a long time) before Seip and Wallstén gave a complete characterization of sampling sequences for the Fock space. The Bargmann transform allows us to obtain a complete characterization of (regular AND irregular) Gabor frames corresponding to the Gauss window.

In addition to the Gauss window, we also mentioned the box window $g(x) = \chi_{[0,1)}(x)$ earlier. In this case, the Gabor frame

$$g_{mn}(x) = \chi_{[n,n+1)}(x) e^{2m\pi i x}, \qquad (m,n) \in \mathbb{Z}^2,$$

is actually an orthogonal basis for $L^2(\mathbb{R})$. Via the Bargmann transform, we have

$$f(z) = Bg(z) = c \int_0^1 e^{2xz - x^2 - (z^2/2)}\, dx,$$

and the corresponding orthogonal basis for F^2 is given by

$$f_{mn}(z) = W_{z_{mn}} f(z) = f(z - z_{mn}) k_{z_{mn}}(z) = f(z - z_{mn}) e^{z\bar{z}_{mn} - \frac{|z_{mn}|^2}{2}},$$

where $z_{mn} = n - m\pi i$ for $(m,n) \in \mathbb{Z}^2$. Although the box window and the associated orthogonal basis $\{g_{mn}\}$ are very natural in Gabor analysis, their counterparts in the Fock space have not been fully studied.

11.6 The canonical commutation relation

There are two unbounded operators that are very important in the study of $L^2(\mathbb{R})$, namely, the operator of multiplication by x and the operator of differentiation. In particular, these two operators play essential roles in classical Fourier analysis and in the more modern time-frequency analysis, which forms the basis for signal analysis and many other engineering applications.

Thus for this section we write

$$Tf(x) = xf(x), \qquad Sf(x) = f'(x), \qquad f \in L^2(\mathbb{R}).$$

It is clear that both of them are unbounded but densely defined. In the real-variable theory of $L^2(\mathbb{R})$, the operator T is clearly self-adjoint, and it is easy to check that iS is self-adjoint as well. We will identify the operators on F^2 that correspond to these two operators under the Bargmann transform.

Theorem 6.1 *We have*

$$BTB^{-1}f(z) = \frac{1}{2}\left[zf(z) + f'(z)\right]$$

for all $f \in F^2$.

Proof Let $c = (2/\pi)^{1/4}$ and $f \in F^2$ (a polynomial for the sake of using Fubini's theorem). We have

$$TB^{-1}f(x) = cx\int_{\mathbb{C}} f(w)e^{2x\overline{w}-x^2-\frac{\overline{w}^2}{2}}\,d\lambda(w),$$

and so

$$\begin{aligned}
BTB^{-1}f(z) &= c^2\int_{\mathbb{R}} xe^{2xz-x^2-(z^2/2)}\,dx\int_{\mathbb{C}} f(w)e^{2x\overline{w}-x^2-(\overline{w}^2/2)}\,d\lambda(w)\\
&= c^2e^{-z^2/2}\int_{\mathbb{C}} f(w)e^{-\overline{w}^2/2}d\lambda(w)\int_{\mathbb{R}} xe^{-2x^2+2x(z+\overline{w})}\,dx\\
&= \frac{c^2}{2}e^{-\frac{z^2}{2}}\int_{\mathbb{C}} f(w)e^{-\frac{\overline{w}^2}{2}}\,d\lambda(w)\int_{\mathbb{R}} xe^{-x^2+2x(z+\overline{w})/\sqrt{2}}dx\\
&= \frac{c^2}{2}\int_{\mathbb{C}} f(w)e^{z\overline{w}}\,d\lambda(w)\int_{\mathbb{R}} xe^{-[x-(z+\overline{w})/\sqrt{2}]^2}\,dx\\
&= \frac{c^2}{2}\int_{\mathbb{C}} f(w)e^{z\overline{w}}\,d\lambda(w)\int_{\mathbb{R}}\left(x+\frac{z+\overline{w}}{\sqrt{2}}\right)e^{-x^2}\,dx\\
&= \frac{1}{2}\int_{\mathbb{C}} f(w)(z+\overline{w})e^{z\overline{w}}\,d\lambda(w)\\
&= \frac{1}{2}\left[zf(z) + \frac{d}{dz}\int_{\mathbb{C}} f(w)e^{z\overline{w}}\,d\lambda(w)\right]\\
&= \frac{1}{2}[zf(z) + f'(z)].
\end{aligned}$$

This proves the desired result.

Theorem 6.2 *We have*

$$BSB^{-1}f(z) = f'(z) - zf(z)$$

for all $f \in F^2$.

Proof Let $f \in F^2$ and $c = (2/\pi)^{1/4}$ again. We have

$$\begin{aligned} SB^{-1}f(x) &= c\int_{\mathbb{C}} f(w)(2\overline{w} - 2x)e^{2x\overline{w} - x^2 - (\overline{w}^2/2)}\, d\lambda(w) \\ &= 2c\int_{\mathbb{C}} \overline{w} f(w)e^{2x\overline{w} - x^2 - (\overline{w}^2/2)}\, d\lambda(w) - 2xB^{-1}f(x). \end{aligned}$$

Let $g(x)$ denote the first term above; the second term will be handled with the help of Theorem 6.1. We have

$$\begin{aligned} Bg(z) &= 2c^2 \int_{\mathbb{R}} e^{2xz - x^2 - (z^2/2)}\, dx \int_{\mathbb{C}} \overline{w} f(w)e^{2x\overline{w} - x^2 - (\overline{w}^2/2)}\, d\lambda(w) \\ &= 2c^2 e^{-\frac{z^2}{2}} \int_{\mathbb{C}} \overline{w} f(w)e^{-\frac{\overline{w}^2}{2}}\, d\lambda(w) \int_{\mathbb{R}} e^{-2x^2 + 2x(z+\overline{w})}\, dx \\ &= \sqrt{2}c^2 e^{-\frac{z^2}{2}} \int_{\mathbb{C}} \overline{w} f(w)e^{-\frac{\overline{w}^2}{2}}\, d\lambda(w) \int_{\mathbb{R}} e^{-x^2 + 2x(z+\overline{w})/\sqrt{2}}\, dx \\ &= \sqrt{2}c^2 \int_{\mathbb{C}} \overline{w} f(w)e^{z\overline{w}}\, d\lambda(w) \int_{\mathbb{R}} e^{-[x - (z+\overline{w})/\sqrt{2}]^2}\, dx \\ &= \sqrt{2\pi}c^2 \int_{\mathbb{C}} \overline{w} f(w)e^{z\overline{w}}\, d\lambda(w) \\ &= 2f'(z). \end{aligned}$$

It follows that

$$BSB^{-1}f(z) = 2f'(z) - (zf(z) + f'(z)) = f'(z) - zf(z).$$

This proves the desired result.

Let $A_1 = BTB^{-1}$ and $A_2 = BSB^{-1}$. Thus

$$A_1 f(z) = \frac{1}{2}[f'(z) + zf(z)], \qquad A_2 f(z) = f'(z) - zf(z).$$

It is easy to verify that

$$(A_2A_1 - A_1A_2)f = f, \qquad f \in F^2.$$

This gives the classical commutation relation as follows.

Corollary 6.3 *On the space* $L^2(\mathbb{R})$ *we have* $[S,T] = I$*, and on the space* F^2 *we have* $[A_2, A_1] = I$*. Here* I *denotes the identity operator on the respective spaces.*

Note that if we simply define

$$Mf(z) = zf(z), \qquad Df(z) = f'(z), \qquad f \in F^2,$$

then we also have

$$DM - MD = [D, M] = I.$$

So the following problem becomes interesting in functional analysis: characterize all operator pairs (potentially unbounded) on a Hilbert space such that their commutator is equal to the identity operator.

It is also interesting to observe here that $D^* = M$ on F^2. In the quantum theory of harmonic oscillators, if $[D, D^*] = I$, then D is called an annihilation operator, and the associated $M = D^*$ is called a creation operator. To a certain extent the Fock space was invented because these operators satisfy the Heisenberg commutation relation and are adjoints of each other.

11.7 Uncertainty principles

With our normalization of the Fourier transform, namely,

$$\widehat{f}(x) = \frac{1}{\sqrt{\pi}} \int_{\mathbb{R}} f(t) e^{2ixt}\, dt,$$

the classical uncertainty principle in Fourier analysis takes the following form:

$$\frac{1}{4}\|f\|^2 \le \|(x-a)f\|\|(x-b)\widehat{f}\|,$$

where the norm is taken in $L^2(\mathbb{R})$, and a and b are arbitrary real numbers. See [21, 22] for this and for exactly when equality holds.

Thus the corresponding result for the Fock space is

$$\frac{1}{4}\|f\|^2 \le \|(T-a)f\|\|(T-b)\widehat{f}\|,$$

where the norm is taken in the Fock space, the operator T (unitarily equivalent to the operator of multiplication by x on $L^2(\mathbb{R})$) is given by

$$Tf(z) = \frac{1}{2}(f'(z) + zf(z)),$$

and a and b are arbitrary real numbers. Recall from Theorem 3.2 that in the context of the Fock space we have $\widehat{f}(z) = f(iz)$. Therefore, the corresponding uncertainty principle for the Fock space is

$$\|f'(z) + zf(z) - af(z)\|\|if'(iz) + zf(iz) - bf(iz)\| \ge \|f\|^2,$$

where a and b are arbitrary real numbers. Rewrite this as

$$\|f'(z)+zf(z)-af(z)\|\|f'(iz)-izf(iz)+bif(iz)\| \geq \|f\|^2.$$

Since the norm in F^2 is rotation invariant, we can rewrite the above as

$$\|f'+zf-af\|\|f'-zf+bif\| \geq \|f\|^2.$$

Changing b to $-b$, we obtain the following version of the uncertainty principle for the Fock space.

Theorem 7.1 *Suppose a and b are real constants. Then*

$$\|f'+zf-af\|\|f'-zf-bif\| \geq \|f\|^2$$

for all $f \in F^2$ (with the understanding that the left-hand side may be infinite). Moreover, equality holds if and only if

$$f(z) = C\exp\left(\frac{c-1}{2(c+1)}z^2 + \frac{a+ibc}{c+1}z\right),$$

where C is any complex constant and c is any positive constant.

Proof We only need to figure out exactly when equality occurs. This can be done with the help of the Bargmann transform and the known condition about when equality occurs in the classical version of the uncertainty principle in Fourier analysis. However, it is actually easier to do it directly in the context of the Fock space.

More specifically, we consider the following two self-adjoint operators on F^2:

$$S_1 f = f' + zf, \qquad S_2 f = i(f' - zf), \qquad f \in F^2.$$

It is easy to check that S_1 and S_2 satisfy the commutation relation

$$[S_1, S_2] = S_1S_2 - S_2S_1 = -2iI.$$

By the well-known functional analysis result on which the uncertainty principle is usually based (see [21, 22] for example), we have

$$\|(S_1-a)f\|\|(S_2-b)f\| \geq \frac{1}{2}|\langle [S_1,S_2]f, f\rangle| = \|f\|^2$$

for all $f \in F^2$ and all real constants a and b. Moreover, equality holds if and only if $(S_1-a)f$ and $(S_2-b)f$ are purely imaginary scalar multiples of one another. It follows that

$$\|f'+zf-af\|\|f'-zf-ibf\| \geq \|f\|^2,$$

with equality if and only if

$$f'+zf-af = ic[i(f'-zf)+bf], \tag{11.1}$$

or

$$i(f'-zf)+bf=ic[f'+zf-af], \tag{11.2}$$

where c is a real constant. In the first case, we can rewrite the equality condition as

$$(1+c)f'+[(1-c)z-(a+ibc)]f=0.$$

If $c=-1$, the only solution is $f=0$, which can be written in the form (11.3) below with $C=0$ and arbitrary positive c. If $c\neq -1$, then it is elementary to solve the first order linear ODE to get

$$f(z)=C\exp\left(\frac{c-1}{2(c+1)}z^2+\frac{a+ibc}{c+1}z\right), \tag{11.3}$$

where C is any complex constant. It is well known that every function $f\in F^2$ must satisfy the growth condition

$$\lim_{z\to\infty} f(z)e^{-|z|^2/2}=0. \tag{11.4}$$

See page 38 of [39] for example. Therefore, a necessary condition for the function in (11.3) to be in F^2 is $C=0$ or $|c-1|\leq|c+1|$. Since c is real, we must have either $C=0$ or $c\geq 0$. When $c=0$, the function in (11.3) becomes

$$f(z)=C\exp\left(-\frac{1}{2}z^2+az\right),$$

which together with (11.4) forces $C=0$. The case of (11.2) is dealt with in a similar manner. This completes the proof of the theorem.

The result above was published in Chinese in [11], where several other versions of the uncertainty principle were also obtained. We included some details here for the convenience of those readers who are not familiar with Chinese.

11.8 The Hilbert transform

The Hilbert transform is the operator on $L^2(\mathbb{R})$ defined by

$$Hf(x)=\frac{1}{\pi}\int_{\mathbb{R}}\frac{f(t)\,dt}{t-x},$$

where the improper integral is taken in the sense of "principal value". This is a typical "singular integral operator".

The Hilbert transform is one of the most studied objects in harmonic analysis. It is well known that H is a bounded linear operator on $L^p(\mathbb{R})$ for every $1<p<\infty$, and it is actually a unitary operator on $L^2(\mathbb{R})$. See [21] for example.

In order to identify the corresponding operator on the Fock space, we need the entire function

$$A(z)=\int_0^z e^{u^2}\,du, \qquad z\in\mathbb{C},$$

which is the antiderivative of e^{z^2} satisfying $A(0)=0$.

For $c=(2/\pi)^{1/4}$ again and $f\in F^2$ (we may start out with a polynomial in order to justify the use of Fubini's theorem), we have

$$\begin{aligned}
HB^{-1}f(x) &= \frac{1}{\pi}\int_{\mathbb{R}}\frac{B^{-1}f(t)\,dt}{t-x}\\
&= \frac{c}{\pi}\int_{\mathbb{R}}\frac{dt}{t-x}\int_{\mathbb{C}} f(z)e^{2t\bar{z}-t^2-(\bar{z}^2/2)}\,d\lambda(z)\\
&= \frac{c}{\pi}\int_{\mathbb{C}} f(z)e^{-\frac{\bar{z}^2}{2}}\,d\lambda(z)\int_{\mathbb{R}}\frac{e^{2t\bar{z}-t^2}}{t-x}\,dt\\
&= \frac{c}{\pi}\int_{\mathbb{C}} f(z)e^{2x\bar{z}-x^2-\frac{\bar{z}^2}{2}}\,d\lambda(z)\int_{\mathbb{R}}\frac{e^{-t^2+2t(\bar{z}-x)}}{t}\,dt,
\end{aligned}$$

where each integral over $\mathbb{R}$ above is a "principle value" integral. Let us consider the entire function

$$h(u)=\int_{\mathbb{R}}\frac{e^{-t^2+2tu}}{t}\,dt.$$

We can rewrite this PV-integral in the form of an ordinary integral as follows:

$$h(u)=\int_{\mathbb{R}}\frac{e^{-t^2}(e^{2tu}-1)}{t}\,dt.$$

The singularity at $t=0$ and the singularity at infinity are both gone. Thus we can differentiate inside the integral sign to get

$$h'(u)=2\int_{\mathbb{R}}e^{-t^2+2tu}\,dt=2e^{u^2}\int_{\mathbb{R}}e^{-(t-u)^2}\,dt=2\sqrt{\pi}e^{u^2}.$$

Since $h(0)=0$, we must have $h(u)=2\sqrt{\pi}A(u)$. Thus

$$HB^{-1}f(x)=\frac{2c}{\sqrt{\pi}}\int_{\mathbb{C}} f(w)e^{2x\bar{w}-x^2-\frac{\bar{w}^2}{2}}A(\bar{w}-x)\,d\lambda(w).$$

Therefore,

$$\begin{aligned}
&BHB^{-1}f(z)\\
&= \frac{2\sqrt{2}}{\pi}\int_{\mathbb{R}}e^{2xz-x^2-\frac{z^2}{2}}\,dx\int_{\mathbb{C}} f(w)e^{2x\bar{w}-x^2-\frac{\bar{w}^2}{2}}A(\bar{w}-x)\,d\lambda(w)\\
&= \frac{2\sqrt{2}}{\pi}e^{-\frac{z^2}{2}}\int_{\mathbb{C}} f(w)e^{-\frac{\bar{w}^2}{2}}\,d\lambda(w)\int_{\mathbb{R}}e^{-2x^2+2x(z+\bar{w})}A(\bar{w}-x)\,dx.
\end{aligned}$$

Fix w and consider the entire function

$$J(z) = \int_{\mathbb{R}} e^{-2x^2+2x(\overline{w}+z)} A(\overline{w}-x)\,dx.$$

We have

$$\begin{aligned} J'(z) &= 2\int_{\mathbb{R}} xe^{-2x^2+2x(\overline{w}+z)} A(\overline{w}-x)\,dx \\ &= \int_{\mathbb{R}} [2x-(\overline{w}+z)+(\overline{w}+z)]e^{-2x^2+2x(\overline{w}+z)} A(\overline{w}-x)\,dx \\ &= -\frac{1}{2}\int_{\mathbb{R}} A(\overline{w}-x)de^{-2x^2+2x(\overline{w}+z)} + \\ &\quad +(\overline{w}+z)\int_{\mathbb{R}} e^{-2x^2+2x(\overline{w}+z)} A(\overline{w}-x)\,dx \\ &= -\frac{1}{2}\int_{\mathbb{R}} e^{-2x^2+2x(\overline{w}+z)} e^{(\overline{w}-x)^2}\,dx + (\overline{w}+z)J(z) \\ &= -\frac{1}{2}e^{\overline{w}^2+z^2}\int_{\mathbb{R}} e^{-(x-z)^2}\,dx + (\overline{w}+z)J(z) \\ &= -\frac{\sqrt{\pi}}{2}e^{\overline{w}^2+z^2} + (\overline{w}+z)J(z). \end{aligned}$$

We can rewrite this in the following form:

$$\frac{d}{dz}\left[J(z)e^{-\frac{1}{2}(\overline{w}+z)^2}\right] = -\frac{\sqrt{\pi}}{2}e^{\frac{1}{2}(\overline{w}-z)^2}.$$

It follows that

$$J(z)e^{-\frac{1}{2}(\overline{w}+z)^2} = -\sqrt{\frac{\pi}{2}}A\left(\frac{z-\overline{w}}{\sqrt{2}}\right) + C(w).$$

We are going to show that $C(w) = 0$. To this end, let $z = -\overline{w}$ in the identity above. We obtain

$$C(w) = J(-\overline{w}) + \sqrt{\frac{\pi}{2}}A(-\sqrt{2}\,\overline{w}).$$

Let

$$F(\overline{w}) = J(-\overline{w}) = \int_{\mathbb{R}} e^{-2x^2} A(\overline{w}-x)\,dx,$$

or

$$F(u) = \int_{\mathbb{R}} e^{-2x^2} A(u-x)\,dx, \qquad u \in \mathbb{C}.$$

We have

$$F'(u) = \int_{\mathbb{R}} e^{-2x^2} e^{(u-x)^2}\,dx = e^{2u^2}\int_{\mathbb{R}} e^{-(x+u)^2}\,dx = \sqrt{\pi}e^{(\sqrt{2}u)^2}.$$

It follows that

$$F(u) = \sqrt{\frac{\pi}{2}}A(\sqrt{2}u) + C.$$

Since $A(0) = 0$ and $A(u)$ is odd (because e^{u^2} is even), we have

$$F(0) = \int_{\mathbb{R}} e^{-2x^2} A(-x)\,dx = 0.$$

This shows that $C = 0$, so that

$$F(u) = \sqrt{\frac{\pi}{2}} A(\sqrt{2}u).$$

Going back to the formula for $C(w)$, we obtain

$$C(w) = \sqrt{\frac{\pi}{2}} A(\sqrt{2}\,\overline{w}) + \sqrt{\frac{\pi}{2}} A(-\sqrt{2}\,\overline{w}) = 0,$$

because $A(u)$ is odd again. Therefore,

$$J(z) = -\sqrt{\frac{\pi}{2}} A\left(\frac{z-\overline{w}}{\sqrt{2}}\right) e^{\frac{1}{2}(\overline{w}+z)^2},$$

and

$$\begin{aligned} BHB^{-1}f(z) &= -\frac{2}{\sqrt{\pi}} e^{-\frac{z^2}{2}} \int_{\mathbb{C}} f(w) e^{-\frac{1}{2}\overline{w}^2 + \frac{1}{2}(\overline{w}+z)^2} A\left(\frac{z-\overline{w}}{\sqrt{2}}\right) d\lambda(w) \\ &= -\frac{2}{\sqrt{\pi}} \int_{\mathbb{C}} f(w) e^{z\overline{w}} A\left(\frac{z-\overline{w}}{\sqrt{2}}\right) d\lambda(w). \end{aligned}$$

We summarize the result of this analysis as the following theorem.

Theorem 8.1 *Suppose $A(z)$ is the anti-derivative of e^{z^2} with $A(0) = 0$ and $T = BHB^{-1}$. Then*

$$Tf(z) = -\frac{2}{\sqrt{\pi}} \int_{\mathbb{C}} f(w) e^{z\overline{w}} A\left(\frac{z-\overline{w}}{\sqrt{2}}\right) d\lambda(w)$$

for $f \in F^2$ and $z \in \mathbb{C}$.

This appears to be a very interesting integral operator on the Fock space. Note that we clearly have

$$T(1)(z) = -\frac{2}{\sqrt{\pi}} A\left(\frac{z}{\sqrt{2}}\right),$$

which must be a function in F^2. The following calculation gives an alternative proof that this function is indeed in F^2.

Lemma 8.2 *The function*

$$f(z) = A\left(\frac{z}{\sqrt{2}}\right)$$

belongs to the Fock space F^2.

Proof Since

$$e^{z^2} = \sum_{n=0}^{\infty} \frac{1}{n!} z^{2n},$$

we have

$$A(z) = \sum_{n=0}^{\infty} \frac{1}{(2n+1)n!} z^{2n+1},$$

and so

$$\begin{aligned} f(z) &= \frac{1}{\sqrt{2}} \sum_{n=0}^{\infty} \frac{1}{(2n+1)2^n n!} z^{2n+1} \\ &= \frac{1}{\sqrt{2}} \sum_{n=0}^{\infty} \frac{\sqrt{(2n+1)!}}{(2n+1)2^n n!} e_{n+1}(z), \end{aligned}$$

where $\{e_n\}$ is the standard monomial orthonormal basis for F^2. It follows that

$$\|f\|^2 = \frac{1}{2} \sum_{n=0}^{\infty} \frac{(2n+1)!}{(2n+1)^2 4^n (n!)^2}.$$

By Stirling's formula, it is easy to check that

$$\|f\|^2 \sim \sum_{n=1}^{\infty} \frac{1}{n^{3/2}} < \infty.$$

This proves the desired result.

A natural problem here is to study the spectral properties of the integral operator T above (or equivalently, the Hilbert transform as an operator on $L^2(\mathbb{R})$): fixed-points, eigenvalues, spectrum, invariant subspaces, etc. Not much appears to be known, which is in sharp contrast to the case of the Fourier transform. Recall that Corollary 3.5 gives a complete spectral picture for the Fourier transform F as an operator on $L^2(\mathbb{R})$.

Motivated by Theorem 8.1, we consider more general "singular integral operators" on F^2 of the form

$$S_\varphi f(z) = \int_{\mathbb{C}} f(w) e^{z\overline{w}} \varphi(z - \overline{w})\, d\lambda(w),$$

where φ is any function in F^2. The most fundamental problem here is to characterize those $\varphi \in F^2$ such that S_φ is bounded on F^2.

To show that the problem is interesting and non-trivial, we present several examples in the rest of this section.

If $\varphi = 1$, it follows from the reproducing property of the kernel function $e^{z\overline{w}}$ that S_φ is the identity operator.

If $\varphi(z) = z$, it is easy to verify that

$$S_\varphi f(z) = zf(z) - f'(z),$$

which shows that S_φ is unbounded on F^2; see [12]. This may seem discouraging, as the function $\varphi(z) = z$ appears to be as nice as it can be (except for constant functions) in F^2. But we will see that there are many other nice functions φ that induce bounded operators S_φ.

First consider functions of the form $\varphi(z) = e^{z\overline{a}}$. An easy calculation in [40] shows that

$$S_\varphi f(z) = e^{(\overline{a}-a)z+\frac{|a|^2}{2}} W_{\overline{a}} f(z),$$

where W_z are the Weyl operators (which are unitary on F^2). Thus S_φ is bounded if and only if a is real, because the only point-wise multipliers of the Fock space are constants.

Next, it was shown in [40] again that the operator S_φ induced by $\varphi(z) = e^{az^2}$, where $0 < a < 1/2$, is bounded on F^2. Furthermore, the range for a above is best possible.

Finally, a necessary condition was obtained in [40] for the boundedness of S_φ. To state the condition, we need a notion that has been popular in the study of operators on reproducing kernel Hilbert spaces. More specifically, if H is a reproducing kernel Hilbert space of analytic functions on a domain Ω and k_z are the normalized reproducing kernels, then every bounded operator T on H gives rise to a bounded function $\widetilde{T}$ on Ω:

$$\widetilde{T}(z) = \langle Tk_z, k_z \rangle, \qquad z \in \Omega.$$

This notion was first introduce in [4, 5] and was made popular in [6, 7, 8]. It was observed in [40] that the Berezin transform of S_φ is given by

$$\langle S_\varphi k_z, k_z \rangle = \varphi(z - \overline{z}), \qquad z \in \mathbb{C}.$$

Therefore, a necessary condition for S_φ to be bounded on F^2 is that the function φ be bounded on the imaginary axis. It would be nice to know how far away is the condition from being sufficient as well.

The problem of characterizing bounded singular integeral operators S_φ on the Fock space has attracted the attention of several analysts. In particular, substantial progress on the problem was made in the recent paper [26].

11.9 Pseudo-differential operators

In this section we explain that, under the Bargmann transform and with mild assumptions on the symbol functions, Toeplitz operators on the Fock space are unitarily equivalent to pseudo-differential operators on $L^2(\mathbb{R})$.

Recall that if $\varphi = \varphi(z)$ is a symbol function on the complex plane, the Toeplitz operator T_φ on F^2 is defined by $T_\varphi f = P(\varphi f)$, where

$$P : L^2(\mathbb{C}, d\lambda) \to F^2$$

is the orthogonal projection. We assume that φ is good enough so that the operator T_φ is at least densely defined on F^2. See [3, 6, 7, 8, 16, 24, 38, 39] for more information about Toeplitz operators.

To be consistent with the theory of pseudo-differential operators, in this section we use X and D to denote the following operators on $L^2(\mathbb{R})$:

$$Xf(x) = xf(x), \qquad Df(x) = \frac{f'(x)}{2i}.$$

Both X and D are self-adjoint and densely defined on $L^2(\mathbb{R})$. We also consider the following operators on F^2:

$$Z = X + iD, \qquad Z^* = X - iD.$$

There are several notions of pseudo-differential operators. We mention two of them here. First, if

$$\sigma = \sigma(z, \overline{z}) = \sum a_{mn} z^n \overline{z}^m$$

is a real-analytic polynomial on the complex plane, we define

$$\sigma(Z, Z^*) = \sum a_{mn} Z^n Z^{*m}$$

and call it the anti-Wick pseudo-differential operator with symbol σ. See [21] for basic information about pseudo-differential operators.

Theorem 9.1 *Let $\sigma = \sigma(z, \overline{z})$ be a real-analytic polynomial and $\varphi(z) = \sigma(\overline{z}, z)$. Then $B\sigma(Z, Z^*)B^{-1} = T_\varphi$.*

This was proved in [39], so we omit the details here.

A more widely used notion of pseudo-differential operators is defined in terms of the Fourier transform. More specifically, if $\sigma = \sigma(\zeta, x)$ is a symbol function on $\mathbb{R}^2 = \mathbb{C}$ and if $f \in L^2(\mathbb{R})$, we define

$$\sigma(D, X)f(x) = \frac{1}{\pi} \int_{\mathbb{R}} \int_{\mathbb{R}} \sigma\left(\zeta, \frac{x+y}{2}\right) e^{2i(x-y)\zeta} f(y)\, dy\, d\zeta.$$

We assume that σ is good enough so that the operator $\sigma(D, X)$ is at least densely defined on $L^2(\mathbb{R})$. We call $\sigma(D, X)$ the Weyl pseudo-differential operator with symbol σ.

Theorem 9.2 *Suppose $\varphi = \varphi(z)$ is a (reasonably good) function on the complex plane. For $z = x + i\zeta$ we define*

$$\sigma(z) = \sigma(\zeta, x) = \frac{2}{\pi} \int_{\mathbb{C}} \varphi(\overline{w}) e^{-2|z-w|^2}\, dA(w).$$

Then $B\sigma(D, X)B^{-1} = T_\varphi$.

Again, this was proved in [39]. Furthermore, the result was used in [40] to study Toeplitz operators on F^2 with the help of the more mature theory of pseudo-differential operators on $L^2(\mathbb{R})$.

11.10 Further results and remarks

For any $0 < p \leq \infty$ let F^p denote the space of entire functions f such that the function $f(z)e^{-\frac{1}{2}|z|^2}$ belongs to $L^p(\mathbb{C}, dA)$. For $p < \infty$ and $f \in F^p$ we write

$$\|f\|_{F^p}^p = \frac{p}{2\pi} \int_{\mathbb{C}} \left| f(z)e^{-\frac{1}{2}|z|^2} \right|^p dA(z).$$

For $f \in F^\infty$ we define

$$\|f\|_{F^\infty} = \sup_{z\in\mathbb{C}} |f(z)|e^{-\frac{1}{2}|z|^2}.$$

These spaces F^p are also called Fock spaces.

Let $f \in L^\infty(\mathbb{R})$ and let $c = (2/\pi)^{1/4}$. We have

$$\begin{aligned} |Bf(z)| &\leq c\|f\|_\infty |e^{-z^2/2}| \int_{\mathbb{R}} \left| e^{2xz-x^2} \right| dx \\ &= c\|f\|_\infty e^{-\frac{1}{2}\text{Re } (z^2)+(\text{Re } z)^2} \int_{\mathbb{R}} e^{-(x-\text{Re } z)^2} dx \\ &= c\sqrt{\pi}\|f\|_\infty e^{\frac{1}{2}|z|^2}. \end{aligned}$$

Here we used the elementary identity

$$-\frac{1}{2}\text{Re } (z^2) + (\text{Re } z)^2 = \frac{1}{2}|z|^2,$$

which can be verified easily by writing $z = u + iv$. Therefore, we have shown that

$$\|Bf\|_{F^\infty} \leq c\sqrt{\pi}\|f\|_\infty$$

for all $f \in L^\infty(\mathbb{R})$. In other words, the Bargmann transform B is a bounded linear operator from $L^\infty(\mathbb{R})$ into the Fock space F^∞. By complex interpolation (see [39]), we have proved the following.

Theorem 10.1 *For any $2 \leq p \leq \infty$, the Bargmann transform maps $L^p(\mathbb{R})$ boundedly into F^p.*

It turns out that the range $2 \leq p \leq \infty$ is best possible. In fact, it was recently shown in [10] that for any $0 < p < 2$, the Bargmann transform does not map $L^p(\mathbb{R})$ boundedly into F^p. But for $1 \leq p < 2$, the Bargmann transform maps $L^p(\mathbb{R})$ boundedly into F^q, where $1/p + 1/q = 1$.

Among other mapping properties of the Bargmann transform we mention the following two: if S_0 is the Schwarz class, what is the image of S_0 in F^2 under the Bargmann transform? And what properties does Bf have if f is a function in $L^2(\mathbb{R})$ with compact support?

One of the fundamental questions in Gabor analysis is to characterize all possible window functions. Via the Bargmann transform we know that this is equivalent to the following problem: characterize all functions $g \in F^2$ such that $\{W_{z_n}g\}$ is a Gabor frame for some $\{z_n\} \subset \mathbb{C}$. In terms of "atomic decomposition", this is also equivalent to the following problem: characterize functions $g \in F^2$ such that for some $\{z_n\}$ we have "atomic decomposition" for F^2: functions of the form

$$f(z) = \sum_{n=1}^{\infty} c_n W_{z_n} g(z) = \sum_{n=1}^{\infty} c_n e^{-\frac{1}{2}|z_n|^2 + z\bar{z}_n} g(z - z_n), \qquad \{c_n\} \in l^2,$$

represent exactly the space F^2.

Another well-known problem in time-frequency analysis is the so-called "linear independence" problem; see [25]. More specifically, given any nonzero vector $g \in L^2(\mathbb{R})$ and distinct points $z_k = a_k + ib_k$, $1 \leq k \leq N$, are the vectors

$$g_k(x) = M_{b_k} T_{a_k} g(x) = e^{2\pi b_k ix} g(x - a_k), \qquad 1 \leq k \leq N,$$

linearly independent in $L^2(\mathbb{R})$? Via the Bargmann transform, this is equivalent to the following problem for the Fock space: given any nonzero function $f \in F^2$ and distinct points $z_k \in \mathbb{C}$, $1 \leq k \leq N$, are the functions

$$W_{z_k} f(z) = e^{z\bar{z}_k - \frac{|z_k|^2}{2}} f(z - z_k), \qquad 1 \leq k \leq N,$$

always linearly independent in F^2? Equivalently, are the functions

$$f_k(z) = e^{z\bar{z}_k} f(z - z_k), \qquad 1 \leq k \leq N,$$

always linearly independent in F^2?

Acknowledgment

The author's research is supported by the National Natural Science Foundation of China (Grant Number 11720101003) and STU Scientific Research Foundation for Talents (Grant Number NTF17009).

References

[1] V. Bargmann, On a Hilbert space of analytic functions and an associated integral transform I, *Comm. Pure Appl. Math.* **14** (1961), 187-214.

[2] V. Bargmann, On a Hilbert space of analytic functions and an associated integral transform II, *Comm. Pure Appl. Math.* **20** (1967), 1-101.

[3] W. Bauer, L. Coburn, and J. Isralowitz, Heat flow, BMO, and compactness of Toeplitz operators, *J. Funct. Anal.* **259** (2010), 57-78.

[4] F. Berezin, Covariant and contra-variant symbols of operators, *Math. USSR-Izv.* **6** (1972), 1117-1151.

[5] F. Berezin, Quantization, *Math. USSR-Izv.* **8** (1974), 1109-1163.

[6] C. Berger and L. Coburn, Toeplitz operators and quantum mechanics, *J. Funct. Anal.* **68** (1986), 273-299.

[7] C. Berger and L. Coburn, Toeplitz operators on the Segal-Bargmann space, *Trans. Amer. Math. Soc.* **301** (1987), 813-829.

[8] C. Berger and L. Coburn, Heat flow and Berezin-Toeplitz estimates, *Amer. J. Math.* **116** (1994), 563-590.

[9] A. Borichev, K. Gröchenig, Y. Lyubarskii, Frame constants of Gabor frames near the critical density, *Math. Pures Appl.* **94** (2010), 170-182.

[10] G. Cao, L. He, and S. Hou, The Bargmann transform on $L^p(\mathbb{R})$, to appear in *J. Math. Anal. Appl.*.

[11] Y. Chen and K. Zhu, Uncertainty principles for the Fock space, *Sci. China. Math.* **45** (2015), 1847-1854

[12] H. Cho and K. Zhu, Fock-Sobolev spaces and their Carleson measures, *J. Funct. Anal.* **263** (2012), 2483-2506.

[13] O. Christensen, *An Introduction to Frames and Riesz Basis*, Birkhauser, 2003.

[14] L. Coburn, The Bargmann isometry and Gabor-Daubechies wavelet localization operators, in *Systems, Approximation, Singular Integral Operators, and Related Topics*, (Bordeaux 2000), 169-178; *Oper. Theory Adv. Appl.* **129**, Birkhauser, Basel, 2001.

[15] L. Coburn, Berezin-Toeplitz quantization, in *Algebraic Methods in Operator Theory*, 101-108, Birkhauser, Boston, 1994.

[16] L. Coburn, J. Isralowitz, and B. Li, Toeplitz operators with BMO symbols on the Segal-Bargmann space, *Trans. Amer. Math. Soc.* **363** (2011), 3015-3030.

[17] I. Daubechies and A. Grossmann, Frames in the Bargmann space of entire functions, *Comm. Pure. Appl. Math.* **41** (1988), 151-164.

[18] X. Dong and K. Zhu, The Fourier and Hilbert transforms under the Bargmann transform, *Complex Var. Elliptic Equ.* **63** (2018), 517-531.

[19] M. Englis, Toeplitz operators and localization operators, *Trans. Amer. Math. Soc.* **361** (2009), 1039-1052.

[20] G. Folland, *Fourier Analysis and Its Applications*, Brooks/Cole Publishing Company, 1992.

[21] G. Folland, *Harmonic Analysis in Phase Space*, *Ann. Math. Studies* **122**, Princeton University Press, 1989.

[22] K. Gröchenig, *Foundations of Time-Frequency Analysis*, Birkhäuser, Boston, 2001.

[23] K. Gröchenig and D. Walnut, A Riesz basis for the Bargmann-Fock space related to sampling and interpolation, *Ark. Math.* **30** (1992), 283-295.

[24] V. Guillemin, Toeplitz operators in n-dimensions, *Integr. Equat. Oper. Theory* **7** (1984), 145-205.

[25] C. Heil, History and evolution of the density theorem for Gabor frames, *J. Fourier Anal. Appl.* **13** (2007), 113-166.

[26] S. Hou, S. Wei, L. Zhao, and X. Zhu, Bounded singular integral operators on the Fock space, preprint, 2018.

[27] S. Janson, J. Peetre, and R. Rochberg, Hankel forms and the Fock space, *Revista Mat. Ibero-Amer.* **3** (1987), 58-80.

[28] H. Landau, Necessary density conditions for sampling and interpolation of certain entire functions, *Acta Math.* **117** (1967), 37-52.

[29] M. Lo, The Bargmann transform and windowed Fourier localization, *Integr. Equat. Oper. Theory* **57** (2007), 397-412.

[30] Y. Lyubarskii, Frames in the Bargmann space of entire functions, *Adv. Soviet Math.* **429** (1992), 107-113.

[31] Y. Lyubarskii and K. Seip, Sampling and interpolation of entire functions and exponential systems in convex domains, *Ark. Mat.* **32** (1994), 157-194.

[32] Y. Lyubarskii and K. Seip, Complete interpolating sequences for Paley- Wiener spaces and Munckenhoupt's A_p condition, *Rev. Mat. Iberoamer.* **13** (1997), 361-376.

[33] J. Ortega-Cerdá and K. Seip, Fourier frames, *Ann. Math.* **155** (2002), 789-806.

[34] A. Perelomov, *Generalized Coherent States and Their Applications*, Springer, Berlin, 1986.

[35] K. Seip, Reproducing formulas and double orthogonality in Bargmann and Bergman spaces, *SIAM J. Math. Anal.* **22** (1991), 856-876.

[36] K. Seip, Density theorems for sampling and interpolation in the Bargmann-Fock space I, *J. Reine Angew. Math.* **429** (1992), 91-106.

[37] K. Seip and R. Wallstén, Density theorems for sampling and interpolation in the Bargmann-Fock space II, *J. Reine Angew. Math.* **429** (1992), 107-113.

[38] K. Zhu, *Operator Theory in Function Spaces*, American Mathematical Society, 2007.

[39] K. Zhu, *Analysis on Fock Spaces*, Springer, New York, 2012.

[40] K. Zhu, Singular integral operators on the Fock space, *Integr. Equat. Oper. Theory* **81** (2015), 451-454.

Index